About Island Press

Since 1984, the nonprofit Island Press has been stimulating, shaping, and communicating the ideas that are essential for solving environmental problems worldwide. With more than 800 titles in print and some 40 new releases each year, we are the nation's leading publisher on environmental issues. We identify innovative thinkers and emerging trends in the environmental field. We work with world-renowned experts and authors to develop cross-disciplinary solutions to environmental challenges.

Island Press designs and implements coordinated book publication campaigns in order to communicate our critical messages in print, in person, and online using the latest technologies, programs, and the media. Our goal: to reach targeted audiences—scientists, policymakers, environmental advocates, the media, and concerned citizens—who can and will take action to protect the plants and animals that enrich our world, the ecosystems we need to survive, the water we drink, and the air we breathe.

Island Press gratefully acknowledges the support of its work by the Agua Fund, Inc., The Margaret A. Cargill Foundation, Betsy and Jesse Fink Foundation, The William and Flora Hewlett Foundation, The Kresge Foundation, The Forrest and Frances Lattner Foundation, The Andrew W. Mellon Foundation, The Curtis and Edith Munson Foundation, The Overbrook Foundation, The David and Lucile Packard Foundation, The Summit Foundation, Trust for Architectural Easements, The Winslow Foundation, and other generous donors.

The opinions expressed in this book are those of the author(s) and do not necessarily reflect the views of our donors.

ABOUT THE CENTER FOR PLANT CONSERVATION

The Center for Plant Conservation (CPC), founded in 1984 as a national non-profit organization, is dedicated to establishing and supporting a science-based network of community-based institutions (botanic gardens, arboreta, museums) devoted to preventing extinction and achieving recovery for imperiled plants native to the United States of America. The thirty-six participating institutions secure seed and living collections off site for conservation, conduct scientific research, work with land managers to monitor and restore rare species populations in the wild, and engender public support for conservation through educational outreach. Collectively, CPC conservation officers have worked with nearly 200 reintroductions throughout the United States in the last 20 years. CPC hosts the *CPC International Reintroduction Registry* that is an ongoing resource for conservation practitioners.

Previous works by CPC include *The Genetics and Conservation of Rare Plants*, edited by D. A. Falk and K. E. Holsinger (1991), which reviewed genetics and population biology critical for the management and conservation of endangered plant species; *Restoring Diversity*, edited by D. A. Falk, C. I. Millar, and M. Olwell (1996), which encapsulated the contemporary understanding of reintroductions from both political and biological contexts, and *Ex Situ Plant Conservation*, edited by E. O. Guerrant Jr., K. Havens, and M. Maunder (2004), which explored the important aspects of seed banking and botanical garden collections as a link to conserving species in nature. These contributions from the CPC have been influential resources in conservation science and practice.

Plant Reintroduction in a Changing Climate: Promises and Perils is the fourth book in the CPC series and updates and extends all previous work. As is true of all CPC publications, this book and its appendices are intended as a useful resource for conservation scientists and practitioners.

PLANT REINTRODUCTION IN A CHANGING CLIMATE

Plant Reintroduction in a Changing Climate

Promises and Perils

Edited by Joyce Maschinski and Kristin E. Haskins
Foreword by Peter H. Raven

CENTER FOR PLANT CONSERVATION

Washington | Covelo | London

Library of Congress Cataloging-in-Publication Data

Plant reintroduction in a changing climate : promises and perils / edited by Joyce Maschinski and Kristin E. Haskins ; foreword by Peter H. Raven.
    p. cm.
"Center for Plant Conservation".
Based on a symposium held in fall 2009 in Saint Louis, Missouri.
Includes bibliographical references and index.
ISBN-13: 978-1-59726-830-1 (cloth : alk. paper)
ISBN-10: 1-59726-830-5 (cloth : alk. paper)
ISBN-13: 978-1-59726-831-8 (pbk. : alk. paper)
ISBN-10: 1-59726-831-3 (pbk. : alk. paper) 1. Plant reintroduction. 2. Plant conservation. 3. Endangered plants—Climatic factors. I. Maschinski, Joyce. II. Haskins, Kristin E., 1969– III. Center for Plant Conservation (Saint Louis, Mo.)
QK86.4.P53 2011
639.9'9—dc23

2011026668

Printed on recycled, acid-free paper

Manufactured in the United States of America
10 9 8 7 6 5 4 3 2 1

Keywords: Island Press, rare plants, plant reintroduction, plant conservation, botanical gardens, ecological restoration, restoration ecology, conservation biology, managed relocation, assisted migration, global warming, climate change, extinction risk, endangered species, biodiversity

What matters is not to know the world but to change it.
—Frantz Fanon

To the next generation of plant conservationists,
especially our children, Mira, Ray, and AJ,
and the new CPC babies, Liam, Jack, and Carys,
born while this book was being written, that they may inherit
a world of diverse rare flora.

CONTENTS

For scores of species with small and declining populations, human intervention is usually needed to ensure their survival. We are responsible for the whole planet, gardeners in effect, and can find sustainability only through continuing, effective efforts. The Center for Plant Conservation (CPC) and its partners in science and land management have aggressively promoted conservation of the plants of the United States for the past quarter century. These efforts have international support and global appeal. The present volume has been made possible because of CPC leadership; its appearance is both timely and welcome.

The successful reintroduction of plant species into natural habitats has never been a simple matter. In their conservation efforts, botanical gardens have long used this strategy, because it is the most cost effective and long lasting that we have devised. Even at a basic level, however, reintroduction is chancy because of stochastic events such as fire, flood, and drought; the appearance of invasive species of plants and animals; the spread of diseases and pests; competition with other plants; and now climate change.

Global climate change, limit its extent as we may, makes the prospects for species survival and successful reintroduction bleaker than we have thought. For example, it is projected that more than half of the roughly 2,400 endemic plant species of California may be on the way to extinction by the end of the century as their habitats are wiped out (Loarie et al. 2008). Similar calculations have been made for other areas, such as habitats at higher, cooler elevations, and coastal species subject to sea-level rise (Walther et al. 2002). We need precise models to identify areas suitable for reintroduction, and we need our best efforts to get it right. Seedbanks will also be of increasing importance in the future.

A critical situation arises when suitable habitats are completely eliminated in the areas to which the plants of concern are restricted. For example, by any of the accepted models of climate change, all above-timberline habitats in the lower forty-eight states will be eliminated in the twenty-first century (Grace et al. 2002; IPCC 2007; Williams et al. 2007). Should we then attempt to establish these plants further north? What about plants from the cooler parts of the Southern Hemisphere, or the Arctic? There appears to be nowhere to establish them in the world of the future. If we introduce plants to areas where they have not grown before, what will be the consequences? The science and practice of conservation ecology will need to be refined and developed greatly if we are to make sound decisions in these areas over the decades to come.

No matter what problems we are facing now or will encounter in the future, we clearly must continue our efforts to understand the factors involved in successful reintroduction. Although the process remains challenging, often yielding uneven results, we are making progress by using our best knowledge of ecology, population biology, genetics, horticulture, and other relevant fields and by working collaboratively. Overall, the present volume makes a particularly valuable contribution, and one that is sure to be widely appreciated.

Peter H. Raven
Missouri Botanical Garden, St. Louis, Missouri

Many helped to bring this book together. We are grateful to our dedicated colleagues at the National Office of the Center for Plant Conservation (CPC), especially Kathryn Kennedy and Rick Luhman, who along with Matthew Albrecht, Ed Guerrant, and Joyce Maschinski helped create the *CPC International Reintroduction Registry*, from which two of the meta-analyses were made possible. We thank Don Falk, Connie Millar, and Peggy Olwell, the editors of CPC's first volume examining reintroduction, *Restoring Diversity*, for their encouragement to undertake the effort to revisit and update our experiences. We also thank the conservation officers working in the thirty-six organizations within the CPC network who contributed time, expertise, and enthusiasm. Lively discussions at CPC national meetings, including those of CPC Science Advisory Council members Chris Walters and Eric Menges, were invaluable. There are countless professional colleagues outside the network who participated in the symposium as speakers, attendees, and friends, gave valued advice, or shared their work.

A book becomes reality only through the generosity of its authors. For their valuable insight, their hard work to synthesize and share their professional expertise and experience, and their patience through the proposal, writing, and editing process, we especially thank those who contributed chapters to this volume.

For their contributions to the *CPC Best Reintroduction Practice Guidelines*, we thank Matthew Albrecht, Eric Menges, Leonie Monks, Dieter Wilken, and symposium participants T. Abeli, P. Ackerman, H. Barnes, D. Bender, S. Birnbaum, T. Bittner, M. Bowles, M. Bruegmann, B. Brumback, J. Brunson, J. Busco, G. Call, S. Calloway, R. Curtis, S. Dalrymple, R. Dodge, J. Dollard, E. Douglas, J. Durkin, A. Dziergowski, J. Fish, H. Forbes, W. Gibble, M. Gisler, E. O. Guerrant Jr., K. Havens, J. Jacobi, K. Kawelo, T. Kaye, B. Keel, K. Kennedy, T. Knight,

A. Kramer, C. Krause, L. Krueger, M. Kunz, N. Li, K. Lindsey, R. Luhman, E. Mayo, K. McCue, J. Midgley, L. Millar, S. Murray, J. Neale, S. Overstreet, V. Pence, C. Pollack, J. Poole, J. Possley, D. Powell, J. Randall, S. Reichard, M. Rose, D. Roth, J. Rynear, P. Schleuning, K. Schutz, E. Seymour, A. Strong, N. Sugii, D. Suiter, W. Sun, A. Thorpe, A. Tiller, P. Vitt, M. Underwood, C. Wells, and P. Williamson.

Attendees at the 2010 Florida Rare Plant Task Force also offered lively discussions about reintroduction and managed relocation. We especially thank Kevin Rice for his insights about genetic issues and our regulating agency and academic panelists: Annie Dziergowski, Dave Bender, Dennis Hardin, Greg Kaufmann, Vivian Negron-Ortiz, and Jack Stout.

We thank the following scientists who served as external reviewers of the chapters: G. Allan, M. Albrecht, P. Quintana-Ascencio, T. Bell, D. Bender, C. Brigham, R. Edelstein, J. Fisher, M. Freedman, D. Gordon, S. Hyndman, H. Liu, K. McCue, L. Mehrhoff, P. Moon, J. Mukherjee, C. Peterson, J. Possley, R. Preszler, P. W. Price, J. Randall, P. Rundel, S. Soto, J. Springer, S. Squires, K. Wendelberger, E. Von Wettberg, and L. Woolstenhulme.

Joyce Maschinski is ever grateful for her wonderful staff, Jennifer Possley, Devon Powell, Sam Wright, and former postdoc Lisa Krueger, who helped with many aspects of this book project and whose work inspires her every day. Tony de Luz, Andi Kleinman, Kristie Wendelberger, and Jennifer Possley helped perfect the figures. Tiffany Lum helped organize the literature cited. Adriana Cantillo helped type numerous studies into the *CPC International Reintroduction Registry*. Libing Zhang assisted with communication with our Chinese colleagues. Kristin Haskins is very appreciative of Sheila Murray, her dedicated botanist, who obligingly picked up the slack while this book was gestating.

Financial support for the *CPC International Reintroduction Registry* and this book came from the CPC, the U.S. Bureau of Land Management, the U.S. Fish and Wildlife Service, and Fairchild Tropical Botanic Garden.

These organizations' and individuals' dedication to progress in the science of plant conservation are a manifestation of what works best about CPC's model of collaborative work.

Chapter 1

Introduction

JOYCE MASCHINSKI AND KRISTIN E. HASKINS

In the face of mounting numbers of plant species at risk of extinction (Gilbert 2010), increasing rates of habitat destruction, spreading invasive species, and effects of climate change (Tilman and Lehman 2001; Walther et al. 2002; Karl and Trenberth 2003), there is a great need for urgent action to preserve species before they are extirpated. This book tells a story of the good fight to save and restore some of the rarest plant species in the world. The story is a complicated but hopeful one.

The practice of plant reintroduction is preceded by many other activities. They include documenting rare species' current distributions, gathering seeds and propagules from living plants for ex situ collections, and researching species' biology and threats. In the best circumstances conservation practitioners work in tandem with land managers and the public to restore healthy, wild habitats and populations of rare species. But has plant reintroduction worked as a conservation tool? And how have plant reintroductions contributed to reintroduction science and practice?

For two decades, responsible agencies have promoted using reintroduction as a recovery strategy for endangered plant species (e.g., US Fish and Wildlife Service 1999). It is considered an essential worldwide conservation tool, and the efficacy of this conservation strategy for animals has been reviewed recently (Fischer and Lindenmayer 2000; Seddon et al. 2007), but such a critical review of plant reintroductions had not been done at the time we began writing this volume (but see Godefroid et al. 2011). A review of plant reintroductions is paramount because plant management techniques are fundamentally different from those used for animals. Herein we take stock of our progress with reintroduction in an effort to facilitate the wise decisions needed to preserve future biodiversity.

As one of the national leaders in plant conservation, the Center for Plant Conservation (CPC) and its thirty-six participating institutions are devoted to preventing extinction and achieving recovery for imperiled plants native to the United States of America. Working with land managers and the public, CPC conservation officers actively secure seed and living ex situ collections and monitor, research, and restore rare species populations in the wild. Collectively they have conducted nearly 200 reintroductions throughout the United States in the last 20 years.

Fifteen years ago, CPC published *Restoring Diversity* with Island Press (Falk et al. 1996), which encapsulated the contemporary understanding of plant reintroductions from both political and biological perspectives. It provided a set of guidelines intended to help practitioners implement plant reintroductions. Since its publication, knowledge of the practice of restoring endangered plants has increased exponentially. This volume presents a comprehensive review of reintroduction projects and practices, the circumstances of their successes or failures, the lessons learned, and the potential role for reintroductions in preserving species threatened by climate change. These findings culminate in a revised set of guidelines for best reintroduction practice.

To assess the current status of plant reintroductions worldwide and to gain a broad sample of reintroduction circumstances, CPC initiated a multipronged approach: a web-based registry, a symposium, and this volume. The extensive *CPC International Reintroduction Registry*, launched in the spring of 2009, aimed to document published and unpublished reintroductions. In fall 2009, the international symposium "Evaluating Plant Reintroductions as a Plant Conservation Strategy: Two Decades of Evidence" convened in Saint Louis, Missouri. Oral presentations by plant reintroduction experts formed the basis of this volume. Attended by government personnel responsible for implementing the Endangered Species Act, environmental consultants, academicians, and botanical garden scientists, the symposium served as a focal event to review plant reintroduction practice and science. Participants provided suggestions for *CPC Best Reintroduction Practice Guidelines* and made recommendations for future directions pertaining to managed relocation (MR). This volume, organized in four parts, represents the ideas discussed at the symposium, often with passion and commitment but with good stewardship of biodiversity foremost in mind. It is noteworthy that the contributors to this volume represent professional conservationists with first-hand reintroduction experience working in government agencies and botanical gardens around the world.

Part I comprises two reviews of plant reintroductions conducted independently. Guerrant (chap. 2) presents a general overview of the database projects reported in the *CPC International Reintroduction Registry*. He typifies the nature and attributes of reintroduction projects, which is an essential first step for identifying information gaps. In an independent and rigorous meta-analysis of a different set of reintroductions, Dalrymple and colleagues (chap. 3) assess the efficacy of reintroductions for establishing sustainable populations. Together these provide a general context for our current knowledge about plant reintroduction science and future research needs. Because the results of any analysis will be influenced by the studies sampled, we offer a list of the specific studies used in each analysis and citations (appendix 2). These studies represent a resource for techniques and practitioners with whom readers are encouraged to correspond for informal discussions and peer review of reintroduction proposals.

Part II presents new insights related to the science and practice of reintroduction. We purposefully begin this section by outlining the critical role the public plays in plant conservation. Maschinski and colleagues (chap. 4) document the mutual benefits of volunteerism and rare plant conservation and give examples of great programs underway throughout the world. Connecting human populations to nature is vital for plant conservation.

Reintroduction requires careful planning before execution. Among the factors that may determine whether a reintroduction will be successful and acceptable to land managers is careful attention to genetics. Neale (chap. 5) reviews the tenets of restoration genetics and the limitations and benefits of various genetic techniques, and she describes protocols and future avenues for research. She provides guidelines for selection of source populations for reintroducing rare plant populations. Haskins and Pence (chap. 6) review several underused and perhaps underappreciated aspects of plant horticulture related to reintroductions, specifically focusing on tissue culture, roots, and soil microbial mutualists.

One of the most critical factors for reintroduction success emerging from these reviews is the selection of optimal habitat. Maschinski and colleagues (chap. 7) review theory that helps explain species' distributions and give examples of experimental reintroductions that demonstrate how fine- and broad-scale variation influences population persistence. They recommend approaches for assessing appropriate sites and microhabitat for reintroductions and provide three approaches for evaluating and prioritizing potential reintroduction sites for any species.

Conserving biodiversity will face new challenges in our changing world. Refining modeling techniques to project accurately where suitable habitat for endemic plants will be under future climates, Krause and Pennington (chap. 8) use geographic information system data linked with distribution modeling algorithms. They address how these tools can be used in decision-making strategies for conserving species on public lands.

Ecological theory predicts that the greater the number of founding individuals, the more likely new populations will colonize and establish successfully. But how many is enough, and what is the best stage class to use? Addressing these questions from different perspectives, Knight (chap. 9) examines population viability analysis of natural populations, and Albrecht and Maschinski (chap. 10) analyze reintroduced populations from published literature and the *CPC International Reintroduction Registry*.

To declare that reintroductions are a success and have truly created viable populations, it is often necessary to compare long-term demography of the reintroduced population with that of reference natural populations. For long-lived species this requires significant ongoing resources for extended timeframes. Based on their extensive personal experience with plant reintroductions, Monks and colleagues (chap. 11) present an overview of alternative approaches for assessing reintroduction success in long-lived plants.

The world's rarest plant populations require special consideration. Kawelo and colleagues (chap. 12) provide examples of reintroductions of taxa that have fewer than fifty individuals and very limited ranges in Hawaii. Their exemplary work on military land demonstrates the challenges of on-the-ground plant conservation and the need for ongoing human assistance to remove threats, conserve ex situ collections, maintain genetic diversity, pollinate, and disperse seeds of these rare gems.

Conscientious land managers are faced with making critical decisions in a timely manner about the conservation of species destined for extirpation within the next century. With the degraded state of some habitats and overdevelopment of others, the value of introducing stocks of endangered species, which lack viable sites within their historic range, to novel sites outside of range is now being considered (Maunder 1992), but there are many objections to the idea. In Part III, Haskins and Keel (chap. 13) review the history and growing debate surrounding the use of managed relocation (MR). As an illustration of the arguments and counterarguments for MR, they present the lively discussions that resulted from the 2009 CPC symposium. They offer criteria for ranking good and bad MR proposals and suggest that any MR proposal for rare plants include documentation of and adherence to *CPC Best Reintroduction Practice Guidelines* (appendix 1). Reichard and colleagues (chap. 14) address one of the major criticisms of MR: risk of invasion. Using weed risk assessments, they test the invasiveness of plants long established within botanical gardens in South Florida and make recommendations related to MR proposals.

In Part IV, Kennedy and colleagues synthesize findings from this volume and suggest future directions, including research needs and clarification of the possibilities and cautions for MR (chap. 15). The meta-analyses, symposium discussions, and chapter contributions culminated in the *CPC Best Reintroduction Practice Guidelines* (appendix 1). It is our sincere hope that these guidelines will improve reintroduction success and improve species recovery while reducing frustration, costs, and labor. We invite our readers to give us feedback.

How to Use This Book

This book is intended to be a resource for students, practitioners, and conservation scientists. Each of the chapters in the first two parts has a review component, a body that relates to reintroduction practice, suggestions for future research needs, and the implications for MR. Each chapter has a conclusions section. The summary of the whole is presented in the *CPC Best Reintroduction Practice Guidelines*. Our goal is to provide a quick reference for practitioners to use when planning and executing rare plant reintroductions. These guidelines include not only recommended actions but also checklists of questions to consider that were influenced by Australian researchers Vallee and colleagues (2004). We have made an effort to make the presented information accessible regardless of whether English is the readers' most familiar language. To this end, we've provided a glossary of terms.

PART I

Review of Plant Reintroductions

"The only source of knowledge is experience."

—Albert Einstein

Worldwide reintroductions are being used to combat the ongoing and massive loss of biodiversity (IUCN/SSC Re-introduction Specialist Group 2007). The intention of reintroduction practice is to establish viable wild populations in their natural habitats. This section of the book is devoted to taking a look backward so that forward progress can be made. We can learn lessons in plant reintroduction only if we take an honest look at what we have already attempted, both the successes and the failures. Critically reviewing previous reintroductions is a crucial first step in assessing the role reintroduction plays as a conservation strategy. Part I presents two such reviews and the insight that has been gained.

Guerrant (chap. 2) summarizes and reviews the kinds of plant reintroductions that have been conducted primarily by practitioners from the United States of America and Australia. Guerrant used the web-based *CPC International Reintroduction Registry*, which includes projects with diverse plant taxa representing many life forms and life histories, conducted in many native habitats. He reviews reintroductions that were conducted in an experimental context, the hypotheses, and factors that have been tested.

Dalrymple and colleagues (chap. 3) expanded the geographic range of the plant reintroductions reviewed to include more studies from Europe and South Africa. Using peer-reviewed and gray literature, this chapter presents an overview of reintroductions done for approximately 700 taxa, a subset of which was included in a metadata analysis. The metadata analysis of 301 attempted reintroductions of 128 plant taxa generated relative measures of reintroduction success based on propagule survival, population persistence, and next-generation recruitment. They address current use of reintroductions, short-term and long-term establishment, whether ex situ–derived propagule performance differs from that of wild-collected propagules, whether it is necessary to remove threats before reintroduction attempts, whether species with broad ranges have different establishment success from species with narrow ranges, whether the number of donor populations used as source material makes a difference for population persistence, whether introductions within historic range have different success from those conducted outside the historic range, and what reasons for failures are reported.

The chapters indicate that plant reintroduction, still a young science, is increasing and perhaps improving, yet there is much to learn. These reviews indicate that practitioners have been addressing primarily population-level questions related to establishment and persistence of reintroduced populations, as has been suggested by Armstrong and Seddon (2007). Few metapopulation and ecosystem studies have been done, limiting our ability to do rigorous statistical evaluation in some cases. As is true for animal reintroductions, improved documentation is expected to increase the power of future meta-analyses (Seddon et al. 2007). Importantly, these reviews identified gaps in some areas of plant reintroduction practice and paved the way for insights presented in the rest of this volume.

Chapter 2

Characterizing Two Decades of Rare Plant Reintroductions

EDWARD O. GUERRANT JR.

There is enormous potential conservation value in the ability to establish new populations of rare plants. When combined with ex situ source material, reintroduction offers valuable and otherwise unavailable conservation options. Indeed, reintroduction of material stored ex situ has made the difference between extinction in the wild and continued survival.

It is almost axiomatic to argue that reintroduction projects are best structured as scientific experiments designed to test specific and explicit hypotheses (e.g., Falk et al. 1996; Kaye 2008). The reason for this is perhaps best viewed in the context of Pavlik's (1996) distinction between *biological* and *project* success in reintroduction projects. Ultimately, biological success, the establishment of self-sustaining populations, is the primary goal. But not all projects achieve biological success. Regardless of the biological outcome, however, reintroduction attempts that are structured as designed experiments have a much greater likelihood of generating reliable information about what worked, what didn't, and why, rather than projects that do only one treatment or do multiple treatments in a haphazard or undocumented manner.

To be truly successful in the long run, reintroduced populations must necessarily and repeatedly complete their entire life cycle without assistance. Guerrant postulates that anything that minimizes the extinction risk or that maximizes population growth rate of individuals in the founding population should increase long-term survival prospects (Guerrant 1996a). It is in this context that much experimentation by manipulation of environmental factors experienced by the founding populations can be viewed.

The purpose of this chapter is to present an overview of a substantial sample of reintroduction projects that were documented in the *CPC International*

Reintroduction Registry (CPCIRR; Center for Plant Conservation 2009). The goal of the registry and database is to assist the conservation community in its ongoing efforts to advance the science and improve the practice of conservation. The effort is ongoing, and readers are invited to provide information about additional projects by accessing the registry on the Center for Plant Conservation website. Any reported projects can provide additional insights to the practice of reintroduction, its successes and failures.

Materials and Methods

The CPCIRR is a tremendous resource describing work conducted by many practitioners, mostly in the United States and Australia. The earliest project about which we have information involves *Pediocactus knowltonii* (Cactaceae), an endangered cactus that was outplanted as seeds and cuttings in 1985 (Olwell et al. 1987; Cully 1996; Sivinski 2008). Plants first reached sexual maturity in 1997, with the first successful seedling establishment documented in 2006 (Sivinski 2008). The most recent projects in the registry were planted in 2008.

Albrecht and Maschinski (this volume) used CPCIRR and incorporated additional studies from the literature in their analyses. Dalrymple and colleagues (this volume) provide an overview of a similar but independent effort to record and learn from reintroduction events and include many projects from Europe and South Africa. (Appendix 2 provides a summary overview of the taxa used in these chapters, along with references used by each.) We suspect that, taken together, they constitute a modest sampling of the reintroduction work that has been conducted but has not been reported.

Data described here reflect projects entered into in the CPCIRR, along with published and gray literature. At the time of writing, approximately two-thirds of the reintroduction projects had been planted in the past decade, and the other third had been planted between 10 and 20 years ago, with only three projects more than 20 years old. (See appendix 2 for details.) The treatment offered here is largely descriptive and provides an account of the types and attributes of reintroductions that have been done.

Results and Discussion

As of fall 2009, the registry contained information on 145 projects that included at least basic taxonomic, location, and data source information and a narrative project description. The amount of information available about these projects varies dramatically. Eighty-nine projects included specific information about what treatments were applied and how they were implemented and reported at least some monitoring results in addition to a written description. An additional fifty-six projects had descriptions but generally lacked specific information on individual treatments and therefore were less useful for analysis.

Eighty projects initiated between 1985 and 2008 have shown varying levels of population establishment (fig. 2.1). The fate in 2009 is known for forty-nine projects, of which forty-five (almost 92%) were known to be alive in 2009. Only four are known to have no surviving plants. Note that three of four that failed did so not long after planting, although the other that failed biologically survived for more than 15 years. There are an additional thirty-one projects about which some monitoring data are available, but the status in 2009 is not known.

Of the forty-nine projects of known status in 2009, 76% had attained reproductive adulthood, 33% produced a second generation, and 16% had reproductive adults in the next generation. With a survival rate of more than 90% (among projects of known status), it is tempting to declare victory. However, caution is warranted because most projects, even the oldest reported, have only been in the ground for a short period of time. Furthermore, there may well be an underlying bias in the data toward successful projects and against failed projects (Pavlik 1996). Valuable information can be gleaned from biologically unsuccessful projects, which are probably underrepresented in this data set, and we encourage those with information about such projects to contribute to our growing base of knowledge about the science and practice of reintroduction.

Phylogeny, Geography, Life Form, and History

Taxonomically, the database comprises information about 107 terminal taxa (i.e., species, subspecies, or varieties) distributed among eighty-seven genera, in forty-nine families, and twenty-five orders (appendix 2). From a phylogenetic perspective, tracheophytes, or vascular plants, are the most exclusive clade represented, with *Isoetes* in the Lycopsida being the most basal taxon. Within tracheophytes some groups are well represented, whereas others are either poorly represented or not represented at all. There are no ferns (or so-called fern allies), gymnosperms, or any of the four most basal groups of angiosperms: Amborellales, Nympheales, Austrobaileyales, or Magnoliids. The next most exclusive group includes the monocots and eudicots. Four orders of monocots are represented, though not the more basal clades.

Some major clades (e.g., ferns, gymnosperms), a basal order of monocots (Alismatales), and two orders of eudicots (Vitales, Dipsacales) were not represented in the CPCIRR or in the analysis by Dalrymple and colleagues (this volume;

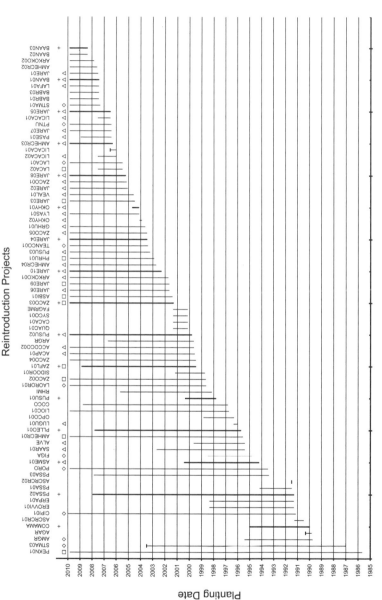

Figure 2.1. Graphic representation of selected projects in order of planting date from 1985 to 2010. Duration of each project is indicated by the origin of the vertical line for each taxon. Reintroduction projects are indicated by a four- to six-letter taxon code corresponding to the first two letters of the genus, species, and subspecific epithet. Biological failure is indicated by a short perpendicular dash (e.g., the second project, *Stephanomeria malheurensis*, STMA03). Project lines without a perpendicular dash extend to the latest date at which some individuals in the reintroduced population were observed as being alive. If project lines do not extend to 2010, more recent data were not available. Symbols along the top edge indicate the greatest demographic benchmark reached by a project: reproductive adult (triangle), next generation (square), and next generation with reproductive adults (diamond). As visual aids, every fifth line is wider than the others and has a corresponding plus sign between the reproductive status symbols and taxon key corresponding to the darkened lines.

appendix 2). However, note that Sun and colleagues (box 4.2) report a *Magnolia sinica* reintroduction in this volume, which was not available for inclusion in this study.

Of the sixty-five taxa for which information about Raunkiaer plant life forms was provided, a substantial majority were either phanerophytes (plants with perennating buds or shoot apices borne well above ground, 46%) or chamaephytes (plants with perennating buds or shoot apices remaining close to but above ground level, 28%). Most of the remaining taxa were cryptophytes (plants with buds or shoot apices below ground surface, 14%) or hemicryptophytes (plants with perennating buds or shoot apices remaining at ground surface, 8%). The remaining 5% were therophytes, or annuals, which may survive the unfavorable season as dormant seeds.

The data set included life history information for eighty-nine taxa, of which 84% are polycarpic (iteroparous) and 16% monocarpic (semelparous). Forty-four percent of the total were long-lived polycarpic taxa (more than 10 years), another 26% short-lived (less than 10 years), and an additional 15% were of uncertain longevity.

The central message of this section is that the empirical evidence we have about reintroductions, though substantial and growing, is limited and not necessarily representative of species, life forms, life histories, or geographic locations. Nonetheless, there is much to be learned in this initial review of the projects about which we do have information.

Types of Reintroductions

The types of reintroductions recognized in the registry include traditional widely accepted methods, such as reintroduction in the narrow sense, augmentation, and introduction within the historic range, and also methods that are considered to be more controversial: introduction outside a species' range and translocation (see Glossary). Approximately half the projects in the registry involved restoring a species to a previously occupied site (reintroduction, 20%) or increasing the size and genetic diversity of an extant population (augmentation, 30%). The other half of the projects involved attempts to establish a taxon in new, previously unoccupied sites (introduction within, 32%, or outside, 8%, the historic range), with 8% involving translocation of naturally occurring plants from one site to another, either within or outside the range. The remaining 2% could not be placed unambiguously into any one category. As necessary and useful as categorization can be, it may also create the mistaken impression that the categories are sharply distinct, without any overlap or gradation between them.

Different types of reintroduction address a variety of problems and purposes and evoke a range of views on their potential conservation value and ethical propriety. Reintroduction and augmentation have the most straightforward goals and general acceptance by the conservation community, but even these have their strategic and ethical dilemmas. The reintroduction of *Stephanomeria malheurensis* (Asteraceae) represents the archetypical example of a clearly necessary and appropriate project: the return of genetic material collected from the single known naturally occurring population after that species had become extinct in the wild (Brauner 1988; Guerrant 1996b). Reintroduction also includes the establishment of a species in a site from which it has become extirpated, using genetic material from another site (e.g., *Pseudophoenix sargentii* [Arecaceae]; Maschinski and Duquesnel 2007). The strategic and ethical appropriateness of such a reintroduction depends on the availability, proximity, ecology, and genetics of source material in relation to the recipient site (see appendix 1). Using source material from populations with similar geography, ecology, and genetics is generally preferred over choosing distant and potentially distinctive stock (box 2.1). The latter situation is exemplified by the reintroduction of *Castilleja levisecta* (Scrophulariaceae), which was native to lowlands of the Willamette Valley in Oregon, north through the Puget Trough of Washington, with a few island populations in adjacent British Columbia, Canada. It had become extirpated in the southern half of its historic range. The attempt to reestablish populations in formerly occupied territory using source stock from the northern part of the range illustrates the not altogether sharp boundaries between reintroduction and introduction (Lawrence and Kaye 2009).

An example of a strategically and ethically appropriate use of ex situ material is an augmentation of a naturally occurring population using source material taken from the same population, such as *Arabis koehleri* var. *koehleri* (Brassicaceae) (Guerrant and Kaye 2007). However, given that a small number of maternal plants contributed multiple individuals to the founding population, at least for the first planting in 2001, it is conceivable that the augmentation may have had the perverse effect of reducing the effective population size while increasing the census population size (Robichaux et al. 1997). Augmentation of naturally occurring populations using genetic material from other source populations can be necessary in some situations. For example, De Mauro (1994) had to use material from another state to restore fertility to a population of *Hymenoxys acaulis* (Asteraceae) that had lost all but one incompatibility allele, in the only Indiana population.

Introduction to previously unoccupied sites within the historic range of a species was the most common project type in our sample, representing almost a third of the total. Strategic and ethical issues revolve around the relation of source material to the reintroduction site and the possibility of exchanging genes with any extant populations in the vicinity (see Neale, this volume).

Box 2.1. Reintroduction of *Ptilimnium nodosum* to the Deep River, North Carolina

Contributed by: Johnny Randall and Mike Kunz
Species Name: *Ptilimnium nodosum*
Common Name: Harperella
Family: Apiaceae
Reintroduction Initiated: 2006
Location: North Carolina, USA
Length of Monitoring: 2006–ongoing
Factors Tested: Streambed stabilization materials

Ptilimnium nodosum (Harperella) is a federally endangered emergent aquatic plant with thirteen remaining populations in the eastern United States—down from twenty-six in 1988—with only one natural occurrence in North Carolina, on the Tar River. Historically, two other populations were located along the Deep River in North Carolina, but because of severe population decline, in 1997 we rescued the eight remaining individuals from one of these populations for ex situ propagation at the North Carolina Botanical Garden. The second population went extinct.

The purpose of our project was to create a self-sustaining Harperella population near the historical site on the Deep River, using ex situ–propagated plants. We designed our reintroduction to test the efficacy of using streambed stabilization treatments (TerraCell® and coconut fabric) in comparison to planting in natural cobble. We chose these treatments because it is particularly challenging to establish an emergent aquatic plant that occupies the uncertain habitat of riverine gravel bars. TerraCell is a plastic-celled stabilization material in which Harperella can be planted, and coconut fabric is a combination of coconut fibers and fine plastic netting through which vegetation can be planted. The natural cobble represented a quasi-control treatment, where Harperella was planted in an unprotected but natural substrate.

In June 2006, we planted seventy individuals in each of nine experimental plots, representing three replicates of three treatments, for a total of 630 plants (fig. 2.2). The reintroduction site was approximately 100 meters downstream of the historical Harperella station within the Triangle Land Conservancy's Second Island Preserve in a semipermanent cobble bar.

Of the 630 original plants, 314 survived the first growing season: 139 in the coconut fabric, 106 in the TerraCell, and 69 in the natural cobble. Note

15

BOX 2.1. CONTINUED

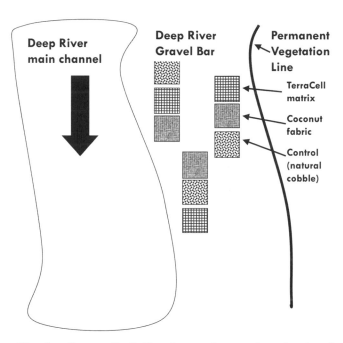

FIGURE 2.2. Planting diagram for *Ptilimnium nodosum* reintroduction that shows the positioning of each plot and within-plot treatments with seventy plants in each plot per treatment (TerraCell®, coconut fabric, and natural cobble).

that two flood events occurred 1 and 2 days after the reintroduction, both covering the plants with approximately 1 meter of water. Additional plants were undoubtedly lost to repeated floods over the first growing season and unexpected herbivory by deer, Canada geese, and muskrats. However, the results are clear that the coconut fabric stabilization treatment was most effective over the establishment phase and afterward (fig. 2.3).

In the first growing season, considerable flowering and vegetative reproduction occurred. Approximately 800 inflorescences were produced that contained an estimated 320,000 flowers. Although we did not collect seed production data, we estimate that more than 600,000 seeds were produced over the growing season.

Because of Harperella's vegetative reproduction habit, following individuals in subsequent years was not possible. We therefore established 1-square-meter permanent plots, subdivided into 100 10-square-centimeter cells, in

BOX 2.1. CONTINUED

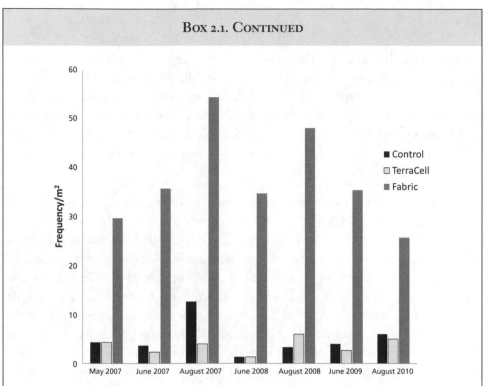

FIGURE 2.3. Frequency of *Ptilimnium nodosum* per plot and treatment type.

each treatment and replicate, and scored for Harperella presence or absence. We noted flowering and fruiting but did not make counts. Our monitoring data (2007–2010) show that there has been a general decline in the permanent plots except for one coconut fabric plot (plot 1), one TerraCell plot (plot 2), and one control plot (plot 1), where there is a slight increase (fig. 2.3). It is important to note that we have seen numerous individuals outside the permanent plots and in adjacent permanent plots, which shows that recruitment is occurring and that the appearance of general decline within the reintroduction site may be in part an artifact of constraining our sampling within plots. Tracking population expansion will entail modifying the spatial extent of our sampling to determine whether the reintroduction established a self-sustaining Harperella population.

The predicted medium- to long-term success of our reintroduction is based on several factors. Because our Harperella reintroduction occurs in a semistable cobble bar within the primary Deep River channel, flood scouring and debris deposition will present ongoing threats and opportunities for

---

BOX 2.1. CONTINUED

expansion. And because Harperella is an emergent plant, herbivory from waterfowl, mammals, fish, and turtles is possible. But because Harperella is an herbaceous perennial with adventitious root growth, is a profligate seed producer, and can reproduce vegetatively from fragmentation and plantlet formation, on-site survival and dispersal to other locations is likely.

Perhaps the most difficult types of reintroductions to evaluate strategically and ethically are translocation and introduction outside a species' historic range. Although both methods may have appropriate uses, they are controversial because they can be used in ways that may not advance or could even detract from the conservation of the species under consideration. The strategic and ethical challenges with translocation, which is used here to mean the movement of naturally occurring individual plants from their native site to another location, depend on the reason for doing the translocation. The basic question to ask concerns whether the translocation is being attempted to save plants from a population that, for unrelated and unavoidable reasons, is destined for destruction, or whether the population is being moved to facilitate an alternative land use. The former is much easier to justify than is the latter. An example of the former type is a translocation done by Brumback and Fyler (1996) in response to imminent habitat destruction involving the state and federally endangered *Isotria medioloides* (Orchidaceae), in which individual plants and a generous amount of accompanying soil were translocated in 1986. Even though initial emergence the next year was encouraging, long-term monitoring results led the authors to conclude that translocation of this taxon is not a viable alternative to in situ land management.

The final type of reintroduction considered here is introduction of new populations outside a species' historic range (Falk et al. 1996). This has been considered by many to be outside the range of generally accepted practice, but there are exceptions where it may be the best if not the only option. Managed relocation (MR) in response to anticipated global climate change is controversial (Haskins and Keel, this volume), but climate change is not the only reason for attempting to establish new populations outside a species' historic range. For example, Monks and colleagues (this volume) have introduced at least three species of Proteaceae (CPCIRR; *Banksia anatona*, *B. brownii*, and *Grevillea humifusa*) outside their historic ranges, in order to find suitable habitat free of the pathogenic agent of dieback disease, *Phytophthora cinnamomi*, which has decimated these and many other taxa in Australia.

Source Populations: Number and Relative Location

Of the sixty-six projects for which the number of source populations could be determined, more than three quarters (77%) used only a single source population for propagules, a practice that has been traditionally supported by reintroduction guidelines (e.g., Falk et al. 1996; Vallee et al. 2004). The proportions of projects using two through five source populations were 9%, 6%, 5%, and 2%, respectively. One other project, involving *Arenaria grandiflora* (Caryophyllaceae), used seventeen source populations (Bottin et al. 2007).

There is not necessarily any single right answer about how many source populations to use in a reintroduction (see Guerrant 1996a and citations therein). Arguments supporting the use of a single source population revolve around maintaining the genetic integrity of lineages with coadapted gene complexes or of specific adaptations to local biotic or abiotic conditions. Conversely, arguments favoring the use of multiple source stocks are more diverse and may have to do with either the lack of a suitable single source or the lack of a suitable match between a single source and potential reintroduction site (see Neale, this volume).

Propagule Types

Given the indeterminate or open growth habit of plants, their modular construction, and their widespread ability to regenerate whole organisms from various parts and even tissues, the range of propagule types available for use in reintroduction is great. The overwhelming majority of projects used whole plants (shoot plus roots, 69%) or seeds (embryonic whole plants, 26%), or both whole plants and seeds (9%). The proportion of projects that used seeds to grow whole plants for founders plus the number of projects that used seeds as founders directly represent more than three quarters (78%). Two projects used shoot tissue only in the form of "cactus pads" (CPCIRR; *Consolea corallicola*, syn. *Opuntia corallicola*, Cactaceae) and one with *Rhus michauxii* (Anacardiaceae) used wild dug root tissue only (Braham et al. 2006). Given the ability of many vascular plants to produce either root-borne shoots or shoot-borne roots (Groff and Kaplan 1988), the range of propagule types useful for reintroduction is great. Except for the use of whole plants grown from cuttings and in vitro methods used on leaf or other tissue samples, practitioners may be missing options created by the ability of some plants to make root-borne shoots or shoot-borne roots.

Founder Stages

Five demographic stages are recognized in the database: seeds, seedlings, juveniles, vegetative adults, and reproductive adults. Founding population stage distributions are known for sixty-four projects. Of these, 78% used only a single stage, 17% used individuals of two different stages, 3% used three stages, and only 2% used individuals of four different demographic stages. For projects using only one stage, the modal demographic stage used was vegetative adults, followed by seeds, seedlings, juveniles, and reproductive adults. Overall, for projects using one or more stages, vegetative adults (36%) were again the most common stage used, followed in order by seedlings (33%), seeds (25%), juveniles (22%), and reproductive adults (13%).

Founder Population Sizes and Numbers of Attempts

Founding population size is a key variable in determining success of a reintroduction attempt. All else being equal, larger founding populations stand a greater chance of survival and growth than do smaller ones (see Albrecht and Maschinski, this volume). That said, there are practical constraints on how large a founding population can be, not the least of which is simply the availability of suitable propagules. There is also the potential impact of collection on the sampled populations to consider (see Menges et al. 2004). The range of founding population sizes among the eighty-nine projects for which we have information varies greatly, from a low of three individuals in *Pseudophoenix sargentii* to a high of more than 4 million in *Abronia umbellata* ssp. *breviflora* (Nyctaginaceae), with a median value of 190 individuals (fig. 2.4a).

Again, all else being equal, the greater number of outplanting attempts made per project should increase the probability of at least one being successful. Of the seventy-two projects for which data are available, 67% consisted of a single outplanting event, with another 28% comprising either two (15%) or three (13%) outplantings (fig. 2.4b). Only one project had more than ten attempts.

Hypotheses and Experimental Results

The database allows an initial assessment to be made of the degree to which reintroductions are being done as explicit experimental tests of particular hypotheses. Of the eighty-nine most data-rich projects in the database, 70% describe one or more explicit hypotheses tested, for a total of 111 tested hypotheses. In addition, 84% of the finalized projects list one or more factors that were tested, without describing any hypotheses.

Each of the 111 hypotheses was initially categorized as having to do with genetic (15%), demographic (14%), or environmental factors (70%). Environmental factors were further subdivided into observational factors (46%), as opposed to

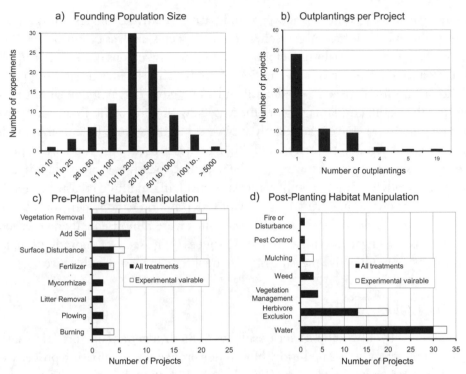

FIGURE 2.4. Four summary bar graphs of projects and experimental treatments. (a) Founding population size distribution among eighty-eight experiments; note that the size class categories are unequal and increase as founding population size increases. (b) Distribution of number of outplantings per project for seventy-two projects. (c) Numbers of projects that involved the use of preplanting habitat manipulation for forty-eight projects. (d) Numbers of projects that involved the use of post-planting habitat manipulation for sixty-five projects. Black portions of bars indicate the number of times a particular manipulation was used that involved all treatments, and therefore was not an experimental factor, and hollow portions of bars indicate the number of projects in which a factor was an experimental variable among treatments in a project. For example, twenty-one projects engaged in vegetation removal before planting, but in only two of these projects was vegetation removal an experimental variable; the remaining nineteen had vegetation removal for all founders.

the environmental factors that were purposefully manipulated by the researcher (24%).

Response variables can be divided into two large categories: survival versus various measures of growth, size, reproductive condition, or health of survivors (see Monks et al., this volume). By far the most common metric used to evaluate success was survival, which was used by 69% of 111 hypotheses tested. Germination of seeds and initial establishment after planting were also reported as a metric of project success. Because the aboveground parts of many geophytes normally die back at the end of the growing season, and some species, such as *Lilium occidentale* (Liliaceae), can remain dormant for up to four growing seasons (Guerrant, unpublished data), emergence was used in lieu of survival. This is because it is not possible to distinguish between a dead plant and a living but dormant one without disrupting the habitat or damaging or otherwise affecting the plant. Plants that have emerged in any given year are clearly survivors, but the status of plants that fail to emerge is not certain. About a fifth of the hypotheses used some measure of plant size as a response variable, involving either a simple measure, such as plant height or crown width, or a more complex measure involving multiple measures such as an index of volume involving height, width, depth, and canopy density. The remaining 10% of the response variable measures were almost equally divided between those having to do with reproductive success (e.g., seed set, reproduction, recruitment) or a subjective measure of plant health or a measure of herbivory.

Genetic Factors

Seventeen sets of hypotheses and results pertained to genetic factors. Six examined the effects of breeding history of founder propagules, comparing progeny of self-pollinated versus outbred individuals or of intentional crosses within or between sites, or controlled versus open pollination. The potential effects of different maternal lines were examined in seven hypotheses, source population effects were the subject of three, and one hypothesis compared the effects of provenance, wild collected versus plants propagated off site. One study asked whether greater genetic diversity improved chances of success in a degraded habitat.

Of the six hypotheses that dealt with breeding history of propagules, four related to augmentation of *Silene douglasii* var. *oraria* (Caryophyllaceae), in which Kephart (2004) asked whether outbred progeny exhibit higher survival and reproduction than inbred or open-pollinated progeny in experimental reintroductions. She found that progeny of intentionally cross-pollinated flowers had significantly greater survival, growth in size, and reproductive output than did offspring of selfed or open-pollinated flowers. In two different attempts with *Jacquemontia reclinata* (Convolvulaceae), Maschinski and colleagues (2005, 2007) found that survival was greater in progeny of plants crossed with more distant neighbors, and also in outcrossed versus selfed propagules.

In none of the six experiments that examined the effects of maternal line on the survival or growth rate of founders was a clear impact documented. Two projects examined the effects of source population, and although differences were found, they were significant in only one year. In one study of *Silene douglasii* var.

oraria, Kephart (2004) asked whether greater genetic diversity increased the likelihood of successful reintroduction in degraded habitats. She did not detect differences in survival. Wendelberger and colleagues (2008) asked whether wild or ex situ source propagules would have greater survival, growth, and reproduction after translocation of *Amorpha herbacea* var. *crenulata* (Fabaceae) to a restored pine rockland. Wild sources survived better than ex situ stock, but the differences were not significant.

Potential effects of source population were the subject of three hypotheses involving two taxa. In their introduction of *Astragalus bibullatus* (Fabaceae), Albrecht and McCue (2010) asked whether seedlings derived from different population sources differentially affect the demographic vital rates of individuals in the introduced population. They found no effect of population source, of which there were three, on transplant survival or transition into reproductive adulthood. To examine the extent to which source population might affect emergence and growth rates, Guerrant (2001) used four of the closest known populations for source material in an introduction of the endangered *Lilium occidentale* (Liliaceae). Source population had an early but transient effect on emergence. No source population effect on size was found until 3 years after planting. It persisted for 3 years and has been sporadic thereafter (Guerrant, unpublished data).

Demographic Factors: Propagule Type, Age, and Horticultural History

Thirteen sets of hypotheses concerned founder propagule type, age, or horticultural history. Four of these explicitly compared plants and seeds, with three having survival as the response variable and one having growth rate of survivors. In all four cases plants were superior to seeds as founders (see Albrecht and Maschinski, Dalrymple et al., this volume). Two others compared survival rates for different sizes or stages of founders, and in both cases larger plants outperformed smaller plants. In an introduction of *Lilium occidentale*, Guerrant (2001) compared seed that had been stored off site for 1 or 2 years (old seed) with seed sowed the same season it was produced (new seed). From the second year after planting, and for more than a decade, new seeds resulted in more plants than old seed, but the plants derived from old seed were larger than those from new. The differences have declined over time, and neither is clearly superior to the other after a decade of growth. In a reintroduction involving *Acacia cochlocarpa* ssp. *cochlocarpa* (Mimosaceae), Monks (CPCIRR) found that 1-year-old seedlings survived and grew better than 2-year-old seedlings. In two other instances, there has been insufficient time for results to become manifest. In another study, involving *Lambertia orbifolia* ssp. *orbifolia* (Proteaceae), Monks (CPCIRR) found that when plants grown from seed were compared with plants grown from cuttings, no significant differences were found in either survival or growth rates.

Environmental Factors

Diverse environmental hypotheses focused largely on how various aspects of the environment might affect success. Naturally occurring factors encompassed a range from narrowly circumscribed and measurable characteristics, such as light levels, to complex multidimensional differences. The latter comparisons ranged from subtle differences in microhabitats to clearly different geographic locations. Differences in timing, such as year or season of outplanting or propagule collecting, are also included. In addition to hypotheses designed to test naturally occurring environmental factors, many projects investigated the effects of environmental factors that were intentionally manipulated. Hypotheses involving environmental manipulation generally used a naturally occurring counterpart for comparison. In contrast to genetic or demographic factors, environmental factors may not always have a predictable theoretical expectation.

Hypotheses comparing different habitats or microhabitats were the most common among naturally occurring environmental factors. They ranged from very specific, measurable differences to more complex and subjective suites of conditions. As an example of the former, Possley and colleagues (2009; box 7.1) examined the effects of the percentage of photosynthetically active radiation (PAR) at time of planting of *Lantana canescens* (Verbenaceae). A more complex factor is represented by Pipoly and colleagues' (2006) work with *Okenia hypogaea* (Nyctaginaceae), in which they compared germination and survival on the lee and windward sides of its foredune habitat along the Florida coast. Seedling establishment, survival, and reproduction were all greater on the lee side of the foredune, but the particular aspects of those habitats most responsible for the differences were not determined.

The first reintroduction attempt of the annual plant *Stephanomeria malheurensis* survived for 16 years but ultimately failed (Parenti and Guerrant 1990; Currin et al. 2007; Currin and Meinke 2008). In a second attempt, Currin and colleagues reintroduced it again to the sole known wild site, called the Narrows, and also introduced it to another nearby location called the Dunes with apparently suitable habitat. Both of these efforts have been successful for the first 2 years. Initial results suggest that both location and year had an effect; the reintroduction site had greater success than the introduction site, and the 2008 planting had initially greater survival and reproduction than the 2007 planting. Recall that most projects consist of a single outplanting.

In addition to planting in different years, other studies compared planting in different seasons. For example, Albrecht and McCue (2010) compared *Astragalus bibulatus* (Fabaceae) plants introduced in the spring with those planted in the fall. Plants introduced in the fall exhibited greater survival and probability of growing into larger stage classes than did those planted in the spring.

Projects involving environmental factors that were intentionally manipulated to examine specific hypotheses were half as abundant as those examining naturally occurring environmental factors and about as numerous as those testing genetic and demographic factors combined. Environmental manipulation projects tested propagation procedures before planting, after planting, or both (fig. 2.4c, 2.4d).

In arid climates, a major factor influencing survival and growth of outplants is water availability. Monks (CPCIRR) hypothesized that watering plants over the first summer (November to April) after planting would increase survival and growth of introduced *Grevillea humifusa* seedlings. Plant survival was significantly greater in watered plants, and although surviving plants were not significantly taller, they did have significantly broader canopy widths. In another study, Monks (CPCIRR) found that watering seedlings of *Acacia cochlocarpa* ssp. *cochlocarpa* over the first summer after planting did not result in significantly greater survivorship than in plants that did not receive supplemental water.

Similarly, Monks (CPCIRR) hypothesized that adding mulch around *Acacia aprica* (Mimosaceae) outplants would increase survival and growth of introduced seedlings. She found no significant difference between mulched and control plants for survival or height, but surviving mulched plants had wider crowns than controls. The implications of such findings are not clear. Mulching is a common horticultural practice and can have significant benefits with respect to moisture retention in the soil, which can be very important in increasing establishment rates. These results are probably best seen as preliminary and not definitive.

Aquatic habitats represent a different extreme. Disturbance frequency and intensity can affect the spatial and temporal suitability of the reintroduction site. Randall and Kunz present a creative experimental approach designed to tackle a situation subject to sporadic and very intense disturbance events (box 2.1).

Other studies manipulated environmental factors to investigate the effects of interspecific interactions on success, including competition, herbivory, and symbiosis. The effects of competition have been examined by experimentally removing potential competitors. Brauner (1988) investigated whether presence or density of the invasive grass *Bromus tectorum* affected survival and reproduction of *Stephanomeria malheurensis*. He found that *Stephanomeria* plants in plots with 50% to 100% *B. tectorum* cover were significantly smaller, slower to bolt and flower, and less fecund than the plants established in plots in which B. *tectorum* had been manually removed (see also Parenti and Guerrant 1990).

Several studies examined the effects of potential herbivores by erecting cages or fences around plants, comparing them with controls. In separate projects, Monks (CPCIRR) compared survival and growth of newly planted *Acacia colocarpha* ssp. *colocarpha* and *A. aprica* seedlings that had been either caged or not. For both taxa, cages increased plant survival and growth.

In an experiment that examined the effects of artificial fertilizer and arbuscular mycorrhizal fungal inoculum obtained from wild plants on survival of seedlings outplanted to a novel habitat, Fisher and Jayachandran (2002) found that although survivorship depended on plant size, arbuscular mycorrhizal fungal inoculum improved both plant size and survival, and fertilization of seedlings yielded the highest survival of outplanted material.

The distinction between what constitutes a natural and manipulated site is not always clear, and several studies illustrate a dilemma that reintroduction practitioners may increasingly have to face. What options are available to us when there is no available suitable habitat, or suitable habitat needs ongoing maintenance?

Habitat destruction has long been recognized as a serious threat to biodiversity, and in an attempt to find a suitable home for *Lantana canescens* (Verbenaceae), Possley and colleagues (2009; box 7.1) asked whether it could survive and establish better in more "natural" historic ecotones maintained by mechanical thinning or in discrete restoration areas? Survival of outplants was roughly comparable, but seedling establishment was greater at the restored site.

Roncal and colleagues (in press) examined whether *Amorpha herbacea* var. *crenulata* (Fabaceae) could establish and grow equally well in four microhabitats along a pineland–glade gradient outside its historic range with suitable habitat. Overall survival was 77% after 2 years; plants established well in three of four microhabitats, but growth was greatest in pine rockland, a habitat in which it was not known to occur historically.

Not Everything Must Be Tested Experimentally

Even though an experimental framework is the most effective way to generate useful information about how best to conduct reintroductions, not everything practitioners did or should have done was conducted in an experimental format. Many projects applied pre-planting or post-planting manipulations to all founders in an effort to increase success because prior knowledge, experimental data, or professional opinion indicated a particular action might be effective (fig. 2.4c, 2.4d). For example, existing vegetation was removed before planting in twenty-one projects, but in only two of those was it an experimental variable. Another nineteen projects had vegetation removed for all treatments, presumably to reduce interspecific competition, because of either earlier experimentation or simply a professional judgment that it would increase the probability of biological success. Only five of forty-four projects (11%) that used pre-planting manipulation and twelve of sixty-five projects (18%) that applied post-planting care tested the factors experimentally. Adding water (thirty-three projects) and excluding herbivores (twenty) were the most common post-planting habitat manipulation techniques, yet watering and caging or fencing were experimental variables in only three and seven projects, respectively.

Reintroduction and Climate Change

Reintroduction can make a substantial and significant difference. But it is expensive and therefore limited largely to the more economically wealthy regions of the planet, which have the resources necessary to use ex situ plant conservation (Guerrant et al. 2004b). As plant endangerment increases, so does the need for more reintroduction projects. But the ability of conservation practitioners to respond to the global needs for reintroduction is a complex function of geography, demography, politics, and natural and economic wealth (Guerrant et al. 2004b). The information in this chapter and elsewhere in this volume provides strong evidence in support of the notion that reintroduction, especially in combination with ex situ conservation, is a tool that can go a long way toward meeting the needs it was intended to address.

Reintroduction attempts offer an opportunity to incorporate into experimental designs questions that address climate change issues. Willis and colleagues (2008) combined flowering time and other phenological data collected originally by Henry David Thoreau at Walden Pond with more modern records to show that climate change has apparently led to major changes in phenology of some but not all taxa. They also compared local survival or extinction of the taxa Thoreau recorded and found that there are distinct phylogenetic patterns to species loss that appear to have been driven by climate change. Are there predictable phylogenetic patterns related to changes in temperature or rainfall? Are predicted future conditions conducive to latitudinal migration? The questions are endless, and now is the time to begin using reintroductions to answer broader questions than we have done so far.

Something fundamental has changed over the last couple of decades in our understanding of the nature of the threats to biodiversity. Habitat destruction and the resulting permanent fragmentation of suitable habitat, along with the effects of invasive species, are still with us and will be for the foreseeable future. What is new is the realization that human beings have changed the composition of the global atmosphere. Global climate change is here. The website of the Arbor Day Foundation (2010) provides a particularly clear graphic that compares the 1990 USDA Plant Hardiness Zone map next to the 2006 revision. Based on minimum temperatures, approximately half the United States has changed hardiness zone designation. Most regions have become warmer, but some small areas cooled, which shows the complexity of predicting the impact of global climate change for any particular area. Rainfall patterns are likely to change as well. The nature, magnitude, and sheer scale of global change pose a serious challenge to those working to preserve biodiversity. I suggest that reintroduction, as it is currently being practiced, is most likely to constitute a focused tool of significant but limited value to a limited number of taxa. Although reintroduction in some form may be necessary, it is not likely to be a sufficient or perhaps even a major factor in what are yet to be clearly articulated and agreed-upon strategies for dealing with the large-scale community- or ecosystem-level disruptions we can expect to result from global climate change.

Nevertheless, and given the apparent phylogenetic patterns in species loss documented by Willis and colleagues (2008), very large-scale reintroduction efforts in the form of massive reciprocal common garden experiments, across wide latitudinal and elevation gradients, perhaps patterned after the classic series by Clausen and colleagues (1940), might be extremely valuable for developing an effective positive response to the impacts of global climate change on biodiversity.

Summary

Reintroduction is a logical culmination of a coherent integrated plant conservation strategy (e.g., Falk 1990). Reintroduction is a deliberate course of action conceived as a pragmatic response to population declines and high extinction risks caused by extensive habitat loss, fragmentation, and ecological impacts of invasive species. And, as this volume shows, reintroduction is a solution that works, at least in some circumstances and in some cases. The apparent high success rate reported in the CPCIRR notwithstanding, it may well take many decades to determine whether reintroduced populations show they can survive and successfully reproduce themselves while maintaining sufficient genetic diversity to enable them to adapt to changing conditions (Monks and colleagues, this volume). Long-term monitoring is therefore an essential component of any reintroduction attempt (appendix 1, #37, 38). It took more than a decade for the first reintroduction attempt of the annual plant *Stephanomeria malheurensis* to fail biologically. Plants with other life histories might reasonably fail after even longer periods of time. Despite over two decades of experience with reintroduction, this represents a relatively short span of ecological time. Monitoring intensity and frequency can perhaps be reduced over time, but for the foreseeable future, all reintroduction attempts are best monitored into the indefinite future. Because our cumulative experience with reintroductions has covered neither plant groups nor circumstances in totality, there is still much to be learned from future plant reintroductions.

I thank Kathryn Kennedy, Rick Luhman, Anna Strong, and Maria Bradford for making this volume possible. Joyce Maschinski and Kris Haskins provided valuable discussion and feedback. Matthew Albrecht deserves special recognition for his assistance in helping to make sense of the CPCIRR. Sarah Dalrymple provided a valuable perspective and many interesting comments. Three anonymous reviewers provided particularly helpful feedback on the manuscript.

Chapter 3

A Meta-Analysis of Threatened Plant Reintroductions from across the Globe

SARAH E. DALRYMPLE, ESTHER BANKS, GAVIN B. STEWART, AND ANDREW S. PULLIN

Reintroductions and associated methods have been recommended as techniques for mitigating or redressing threatened plant species declines for several decades. Their use continues to increase as an option for overcoming problems associated with habitat loss, habitat fragmentation, and reproductive isolation (Quinn et al. 1994). However, these approaches have been criticized for the lack of monitoring and central recording, inappropriateness of the action due to genetic considerations, a lack of demographic knowledge of the donor populations, and inadequate information on the species' habitat needs (Pearman and Walker 2004).

In the United Kingdom, the debate over the use of reintroductions has been influenced by the long history of moving plants from gardens and wild populations to apparently suitable habitat with mixed success. We are able to identify past reintroductions primarily through the efforts of the Botanical Society of the British Isles (BSBI), who have used their extensive cataloging system to identify these events and distinguish between naturally occurring populations and deliberate introductions. Some of these events undoubtedly would be classified as reintroductions in the modern definition of the term (see Glossary), but many more would be classified as conservation introductions, although they have been undertaken outside (or before) prescribed conservation programs (e.g., *Aster linosyris*, *Helianthemum apennium*, and *Lobelia urens*; BSBI unpublished database). It is certainly the case that the unregulated use of seed and plant translocations has confused the debate and provides evidence that reintroductions for conservation of threatened plants are prone to failure and a waste of resources (Pearman and Walker 2004). Subsequently, the debate surrounding reintroductions in the United Kingdom has been informed predominantly by anecdotal evidence from introductions that lacked any rigorous framework at the point of initiation and were often monitored only sporadically due to resource restrictions. Indeed, practitioners are still often unaware of guidelines at national and international levels, and although good networks exist that practically and strategically support work on threatened plants, there is no requirement to record movements of native species except when evaluated against legislative targets for halting species declines. This information is not readily available even within the conservation community; subsequently, many reintroductions are attempted by practitioners working in isolation without benefiting from the experience of colleagues.

Because of the practical similarities between reintroductions and out-of-range translocations (i.e., selection of suitable propagules, identifying recipient sites, and postintervention management and monitoring), the use of reintroductions has again come under scrutiny, albeit through the lens of managed relocation (MR) and climate change. With current climatic shifts and nonclimate stressors showing no signs of abating, we must assume that conservationists will have to deal with an ever-increasing list of threatened species. For practitioners charged with halting the decline of rare and threatened species, reintroductions are an option for meeting management objectives where other options are not available. However, climate change is expected to reduce the extent of historic ranges that remain suitable, and increasingly, the options for reintroductions will decrease while MR will become a more plausible management option for species that are imminently threatened by climatic shifts.

Here we use *reintroduction* as an umbrella term for attempts to create populations at the target taxon's site of extirpation, at sites identified as being within the historic range, and sites of perceived suitable habitat that might be outside the historic range. The authors are aware that according to accepted definitions of the term, the latter effort is normally called *conservation introduction* or the like. However, for the purposes of brevity, *reintroduction* refers to all these interventions unless the authors explicitly describe them otherwise. We searched the libraries of Natural England, Scottish Natural Heritage, the Countryside Council for Wales, and the Joint Nature Conservancy Council, and the following electronic databases: ISI Web of Knowledge, including ISI Web of Science (Science Citation Index expanded 1945–present) and ISI Proceedings (Science and Technology Proceedings 1990–present), JSTOR, Index to Theses Online (1970–present), Digital Dissertations Online, Dogpile Meta-search (Internet search), Google Scholar (Internet search), Copac, Scirus, Scopus, and ConservationEvidence.com. We used the following search terms: plant* AND re-introduc*, reintroduc*, introduc*, translocation*, establish*, re-establish*, restor*, reinstat*, regenerat*, or assisted migration. In addition, we directly contacted practitioners using the IUCN Species Survival Commission Reintroduction Specialist Group Members Database (IUCN 1998b). We included only studies of vascular plants that had undergone a deliberate reintroduction of individual plants or seeds to sites that were unoccupied at the time of translocation. We required that studies include data on propagule number and type (seed, juvenile, or adult plant), number of propagules translocated, and number surviving over a monitoring period described in the study. We also included whether next-generation recruitment had occurred.

Because of the increased risk of mortality when using seeds, we analyzed reintroduction attempts of seeds, juveniles, and adults separately. We used risk ratios to provide meaningful effect measures, where the larger the risk ratio, the higher the probability of death of individuals and hence failure of the reintroduction program, similar to the methods of Wacholder (1986). To generate pooled estimates of effect, we used DerSimonian and Laird (1986) random effects meta-analysis.

Formal meta-analysis of time to event outcomes could not be undertaken using standard techniques because survival curves could not be derived for individual studies, but time was included as an important covariate using meta-regression (Higgins and Thompson 2004). We performed all analyses in Stata version 11.0 (StataCorp USA 2011).

The need for an evidence-based approach to the use of reintroductions is timely. Through consultation with UK practitioners, we recognized that a global review had the potential to describe the extent to which reintroduction had been used and would facilitate greater practitioner communication, evaluate the success of attempted reintroductions, and identify areas of improvement for future implementation. According to guidelines recommended by the Collaboration for Environmental Evidence (2009), we systematically reviewed literature associated with reintroductions between August 2006 and October 2009. Peer-reviewed literature, gray literature, and information directly from plant conservation practitioners provided evidence for the intended or attempted reintroduction of 708 taxa from across the world. We define *intended* to mean the reintroductions for which there is evidence that a reintroduction program has been initiated but not yet implemented. For example, where documentation describes the preparatory stages of a reintroduction program (e.g., reported ex situ cultivation intended for outplanting in the wild but did not report on the outplanting itself) we have included the example in our data set. This is useful because it provides an overview of reintroduction activity. This chapter summarizes the current use of reintroductions, their geographic and taxonomic scope, the main threats that prompt the use of reintroductions, and the target taxa's conservation status. This summary is followed by a review of results from a meta-analysis of a subset of studies identified from the search process.

Current Use of Reintroductions: What, Why, Where, and How?

Reintroductions have been planned or attempted in thirty-two countries, with the earliest record from 1955 (BSBI unpublished). However, botanists have been outplanting individuals of rare plants in suitable habitat (but not necessarily within the historic range) since 1783, according to BSBI records. The 708 taxa that form the foci of proposed or implemented reintroductions are representatives of 124 families (appendix 2). Asteraceae is the most numerous (ninety-one taxa), followed by Orchidaceae (fifty-four taxa) and Rosaceae (thirty-two taxa). The target taxa were further categorized by level of endemism, type of threat, and level of conservation protection in order to characterize the impetus and justification for using reintroduction techniques.

Reintroductions are often regarded as being a technique for rare species management, but the data set used in this analysis (677 taxa) indicates that reintroductions may be undertaken even when the species is widespread (table 3.1). Unsurprisingly, many target taxa do have narrow ranges (271 taxa when "local" and "site" endemics are combined) and therefore qualify for conservation protection because of their limited distribution. Perhaps more surprising, 118 taxa have (or at least had) distributions that spread across continental scales and another 119 have been recorded as occurring in more than one continent. It is assumed that the taxa with formerly widespread distributions are subject to reintroduction projects because they are in decline in part of their range and may have been lost entirely from a significant number of sites at a regional scale. This assumption is supported by the fact that the majority of taxa are identified as needing conservation protection at national levels (440 taxa, although evidence for the presence or absence of national conservation protection could be found for only 498 taxa). Reintroductions are being used as a way of meeting national objectives for species protection and are not limited to preventing species-level extinctions. In fact, 618 taxa have not been evaluated for inclusion in the IUCN Red List, and although there is no legislative obligation to complete threat assessments for the IUCN, this can be taken to indicate that international priorities rarely provide the impetus for initiating reintroduction projects.

The majority of taxa included in reintroduction projects are associated with habitats that fall within temperate broadleaf forest biomes according to the United Nations Educational, Scientific and Cultural Organization (UNESCO; Udvardy 1975) major biome type classification. We suggest that this is more indicative of where reintroductions are considered to be a conservation option for well-funded practitioners but does not indicate where extinctions of threatened plants are most imminent. Certainly, reintroductions are most commonly carried out by practitioners in Europe (twenty-one countries; 354 reintroductions), followed by the United States (228) and Australia (61). Three biomes account for 77.7% of taxa subject to reintroduction projects: temperate broadleaf forest, mixed island systems, and evergreen sclerophyllous forest, scrub, or woodland (table 3.1). In our data set, only thirty-nine taxa were foci of reintroduction programs in South

TABLE 3.1

Summary of reviewed material generated by literature search, including total number of examples and breakdown by category.

Descriptor Categories	Variables Included in Analysis	Number of Examples for Which Data Were Obtained	Data Categories and Frequency of Attempts in Each Category (n)
Species descriptors	Taxa.	708 taxa	123 taxa included in meta-analysis, many associated with multiple attempts or sites. Data on remainder of taxa were insufficient for quantitative meta-analysis. Taxonomic designations of authors were used.
	Life cycle.	301 attempts*	Annual (31), biennial (3), perennial (267); some species show mixed strategies, most typical strategy recorded here.
	Biome (data taken from WWF Atlas, site of reintroduction or region of concern used to determine biome where species occurs across several regions).	698 taxa	Tundra (0), temperate needle leaf forest (10), temperate broadleaf forest (313), temperate grasslands (27), cold winter deserts (7), evergreen sclerophyllous forest, scrub or woodland (108), tropical grasslands and savanna (1), warm deserts and semideserts (8), tropical dry or deciduous forest or woodland including monsoon forests (4), subtropical and temperate rainforests or woodlands (49), tropical humid forests (3), mixed mountain and highland systems with complex zonation (40), mixed island systems (121), river and lake systems (7).
	Endemism.	677 taxa (296 attempts*)	Global (119), continental (118), regional (143), national (26), local (229), site (42).

TABLE 3.1

Continued

Descriptor Categories	Variables Included in Analysis	Number of Examples for Which Data Were Obtained	Data Categories and Frequency of Attempts in Each Category (n)
	Cause of decline (many taxa have >1 reason behind declines).	1,021	Urban and industrial development (130); agriculture when not specifically grazing (186); competition from other plants, including invasives, alien or native (136); grazing, including stock, goats, rabbits, and native herbivores (161); fire (26); climate change (10); overexploitation or collection (56); habitat loss when not specifically any other reason (97); flooding for reservoir construction or other water course engineering including draining (91); succession including disturbance suppression (63); disease (16); pollution (10); trampling or erosion (31).
	Cause of decline present at reintroduction site?	235 attempts*	No, primary threats causing species decline have been prevented from operating at this site (190) or, yes, primary threats still present (45).
	IUCN threat level assessed using the IUCN Red List 2009.1.	706 taxa	Extinct in the wild (4), critically endangered (41), endangered (12), vulnerable (15), lower risk (2), least concern (14), not evaluated (618).

Intervention descriptors	Site designation.	155	No site designation (14), designated site (141).
	Status of site within distribution.	278 attempts*	Previously extant site (39), within historic range (228), outside historic range (11).
	Provenance.	269 attempts*	Ex situ (167), direct translocation from wild population (102).
	Single or multiple donor populations.	140 attempts*	Single donor (100), multiple donor populations (40).
	Life stage of propagules.	299 attempts*	Seed (47), juvenile (132), adult (115), mixed (5).
Outcome: abundance	Surviving individuals.	301 attempts*	Taken from last reported survey and includes progeny.
	Number of time points.	301 attempts*	Does not include time 0, when reintroduction undertaken, i.e., time points = number of surveys since translocation. Mean = 1.69 ± 0.09 (1 SE).
Outcome: recruitment	Reproductive potential.	116 attempts	Individual plants with reproductive structures (89), no reproductive structures (27).
	Recruitment.	66 attempts	Recruitment reported (55), no recruitment (11).
	If recruitment evident, vegetative or sexual?	55 attempts	Vegetative (12), sexual (42), mixed (1).

Variables marked with an asterisk were used in the quantitative meta-analysis. All other variables are presented for summary purposes only. Units are either taxa or attempts to reintroduce a taxon, with the exception of cited causes of decline, which were collected by taxon; many taxa had more than one reported cause of decline.

America (thirty-eight of which were in response to a single reservoir creation project in Brazil) and twelve taxa in Asia, and reintroductions of two taxa have been documented in Africa.

The identification of threats that contribute, at least in part, to the decline of taxa in our sample reveals that agricultural intensification threatens 186 taxa, grazing threatens 161, competition from invasive plant species threatens 136 taxa, and urban or industrial development threatens 130 taxa (table 3.1). Threats were identified in the literature (including online sources such as the IUCN Red List) for 525 taxa, of which 62% were associated with more than one threat. It is noteworthy that climate change is infrequently mentioned as a threat to the survival of taxa identified for reintroduction programs even though it may exacerbate other threats. It is expected that given the prominence of this issue in the scientific literature and emergence of stronger climate signals over time that may reveal climate changes on a local scale, the threat from climate change will probably be emphasized more strongly in future threatened plant conservation literature.

Practitioners did not always report locality details that would enable us to characterize the type of reintroduction. For the majority of taxa in this data set, the information available was insufficient to describe reliably whether practitioners intended to undertake projects to restore extirpated populations or whether populations were intended to be created within the historic range or outside known range boundaries. Similarly, for the majority of taxa it was not possible to distinguish when reintroductions could also be classed as augmentations.

How Successful Are Reintroductions?

The data set was reduced to 128 taxa, which are represented as 304 attempts to create populations, because information on key variables needed for the analysis was not available. From this point onward, analyses treat each venture to reintroduce a taxon to a site and create a population at that locality as an individual reintroduction attempt.

To be included in the meta-analysis, a study must have reported propagule survival (where each reintroduction event constitutes an attempt to establish a population) and additionally may have reported attaining the life cycle's reproductive stages and achieving in situ recruitment of an offspring generation. Propagule survival is based on reported numbers of individuals surviving after a time period specified by the study authors. Because this is often reported as population size, it can be a number greater than one when the population is increasing and may include the number of outplanted propagules and their progeny. Proportional propagule survival and progeny recruitment must therefore be aggregated into a combined measure referred to as "population size expressed as a percentage of propagule input." This measure of population size therefore can provide an indicator of population growth or decline. Population survival is taken to mean that at least one of the reintroduced propagules, or their offspring, are still extant. The latter two measures (reproductive maturity and in situ recruitment) required only that either mature reproductive state or recruitment of at least one individual per attempted reintroduction was reported. The presence of flowering or fruiting bodies denoted reproductive maturity. Recruitment of an offspring generation included individuals resulting from vegetative and sexual reproduction. Although these are coarse measures of success, they were identified as being key outcomes with which to evaluate reintroductions (table 3.2).

Seed-based reintroductions used the highest mean number of propagules (5,640 ± 2,007) and had the lowest mean propagule survival at only 5% (averaged across all 47 attempts) compared to juvenile or adult plant reintroductions (appendix 1, #16). Seeds carry an inherent risk of mortality and are expected to perform worst when measured in this way. However, despite this very low propagule survival and low population survival, a much higher proportion of attempts using seed reached reproductive maturity (49%) and produced an offspring generation

TABLE 3.2

Summary statistics describing key parameters used to assess effectiveness of reintroductions.

Summary Parameters	Seeds $n = 47$	Juveniles $n = 134$	Adults $n = 115$
Mean monitoring period (months)	34.34 ± 7.93	41.16 ± 3.66	36.89 ± 4.23
Monitoring period range from point of reintroduction (months)	3–384	1–120	2.5–384
Mean number of surveys (range in parentheses)	1.38 (1–4)	1.38 (1–5)	2.23 (1–12)
Mean number of propagules	5,640.62 ± 2,007.51	157.30 ± 30.85	111.17 ± 21.55
Mean percentage propagule survival	4.6% ± 1.4%	65.0% ± 4.7%	998.5% ± 730.7%
Number of attempts to reintroduce annuals	25	2	3
Number of attempts to reintroduce biennials	0	0	3
Number of attempts to reintroduce perennials	22	132	109
Percentage of unsuccessful attempts (extinct at last survey)	36.1%	9.0%	15.7%
Percentage of "successful" attempts (extant at last survey)	63.8%	91.0%	84.3%
Percentage achieved reproductive maturity	48.9%	18.7%	34.8%
Percentage of attempts where offspring recruited	46.8%	5.2%	20.9%

The *n* value is the number of attempts, categorized by propagule type (seeds, juvenile or adult plants). Means are shown ±1 standard error.

(47%) than projects using whole plants. This result is explained partly by the high proportion of annuals that are introduced to a site as seed. If the practitioner has selected favorable microsite conditions that allow persistence for just a few months, there is a high probability that the resulting plants can reach reproductive maturity and set seed, leading to recruitment of an offspring generation in the following year. Of the attempts using seed that resulted in recruitment of offspring, only two were attempts to reintroduce perennials; all others were annual species.

Juvenile plants, defined as those not yet achieving reproductive maturity, included propagules described by the study authors as seedlings, saplings, and in some cases cuttings (depending on the part of the plant). The mean number of juvenile propagules used (157.3 ± 30.9) was much lower than that of seed-based reintroductions. However, this reflects both the greater resource requirement to produce juvenile plants through propagation and the lowered expected risk of mortality in individuals that have developed beyond a seed. Propagules of juvenile plants do indeed have a very promising survival of 65%, implying that they overcome much of the mortality experienced by seeds. The juvenile-based projects also had the lowest population mortality (9%) while achieving the longest mean monitoring periods. However, monitoring was still only 41 months, and it is not possible to say whether this encouraging survival rate would continue if monitoring data were available over longer timescales. The short monitoring times are a probable explanation for why reproductive measures are so low. Only 18.7% of the 134 attempts resulted in individuals producing flowers or fruits, and 5% showed in situ recruitment.

Adult plants used for reintroductions have the lowest mean number of propagules per reintroduction attempt (111.2 ± 21.6) and the highest mean propagule survival, which increased by several orders of magnitude. However, this measure of reintroduction success was strongly influenced by attempts to reintroduce *Aldrovanda vesiculosa*, a vegetatively spreading aquatic species that in one extreme case was able to increase from 60 propagules to 50,000 in 6 years (Adamec and Lev 1999; Adamec 2005). If that species is removed from the data set, propagule survival decreased to 85% ± 24%; although this is still higher survival than was recorded for other propagule types, the standard error value indicates that a great deal of variance exists in this data set. Within reported timescales, reintroductions using adult propagules have a high population survival rate; 84% of attempts are still extant at the last survey. Reintroductions initiated with adult propagules attained reproductive maturity and recruited offspring in 34.8% and 20.9% of attempts, respectively. Although this is low and might suggest that many attempts do not result in the creation of self-sustaining populations, these are higher than the equivalent figures for juvenile propagules, suggesting that within the short timescales reported by study authors, adult plants are further along in reproductive development, leading to a higher proportion recruiting into the next generation. Again, the monitoring period reported before publication was quite short, approximately 3 years on average. Despite the propagules being outplanted at a developmental stage, which should ensure reproductive maturity, the actual time needed for acclimation to new conditions and the length of reproductive cycle typical for each taxon will ultimately affect how soon recruitment can be achieved. Therefore, we hope that the proportions of attempts achieving reproductive maturity and progeny recruitment might increase if reintroductions were reported after longer monitoring timescales. If this were not the case, it could be taken that habitat selection was not good enough to ensure continued propagule health and subsequent recruitment, but because this lies in the realm of speculation, this indicator of success will have to be interpreted in future reviews.

Population Survival over Time

Efforts to determine reintroduction success are influenced by whether the population was extant, was extinct, or had an unknown fate at time of last survey. We examined reintroductions based on their installation date and time of last survey (fig. 3.1). In addition to demonstrating that attempts to reintroduce threatened plants have been undertaken largely in the last 10 years, we showed that the status of most attempts was unknown because we were unable to update the outcomes since the authors published their studies. This information was more useful than showing population survival at last survey alone (table 3.2), acting as a warning that using published results without follow-up surveys should be done with caution; they cannot convey whether reintroductions are a reliable tool for mitigating plant declines because timescales for publication are normally much shorter than the lifespan of the species of interest.

Unsurprisingly, after 6 months since outplanting most attempts were still extant (301 attempts; fig. 3.1). However, a small proportion (41 attempts) are classed as unknown because the last survey was undertaken within 6 months. Many of these studies focused on in vitro propagation methods and reported the success of outplanting only as a final stage of their project. Projects more than 10 years in duration tended to be few and have a higher proportion of project success classified as unknown, probably because of a lack of monitoring. An important threshold is crossed at the transition between 5 and 10 years: At 5 years since reintroduction, 46 attempts are still extant and 26 have gone extinct, whereas at 10 years only 20 are extant and 30 are recorded as being extinct. It could be argued that at some point between 5 and 10 years, reintroductions go from being a successful intervention to an unsuccessful intervention, assuming that we use only population existence as our measure of success. The small size of the final bar showing the

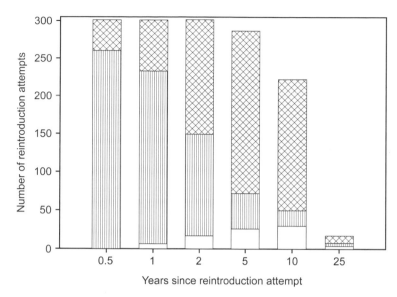

FIGURE 3.1. Survivorship of attempted reintroductions in relation to years since propagules were transplanted to site (initial $n = 301$). White bars represent failed attempts (all transplants and any resulting progeny dead), striped bars represent extant attempts, and cross-hatched bars represent attempts of unknown fate. Decreasing total number of attempts reflects how many attempts were undertaken in each time period preceding analysis in 2009.

outcome of attempted reintroduction at 25 years reflects the fact that very few reintroductions were initiated more than 25 years ago. At this point the number of extant and extinct attempts is equal, but there are only four in each category. Because small sample size and large variation limited our ability to make conclusive statements on the effectiveness of reintroductions, we applied further meta-analyses to explore possible explanations for heterogeneity in the data.

How Do Ex Situ–Derived Propagules Compare with Wild Propagules in Reintroduction Attempts?

The involvement of ex situ institutions such as botanical gardens, seedbanks, and horticultural nurseries has long been a feature of many reintroductions, whereby seeds or whole plants are propagated before being outplanted into recipient sites. There are potential problems with this approach because of possible genetic bottlenecks or genetic drift in populations held in cultivation for several generations (Neale, this volume). There may also be problems with outplanting technique (Guerrant, Haskins and Pence, this volume). Finally, risks are posed to the recipient community if pests or pathogens are introduced along with the propagules (e.g., in the cultivation substrate; Haskins and Pence, this volume).

We analyzed the pooled risk ratios of subgroupings of reintroduction attempts classified by seeds, juveniles, and adults and (1) whether the attempt had involved direct translocation of seeds or whole plants from wild populations or (2) whether the propagules had been sourced from ex situ institutions. No relationship existed between propagule mortality of wild-sourced as opposed to ex situ–sourced propagules, indicating that if ex situ techniques confer risk, it is not detected in propagule survival within the reported timescales. However, there were indications that wild-sourced propagules may achieve higher recruitment levels (fig. 3.2), but with the caveat that there are very small sample sizes for some subgroups. These findings must be treated within the context of rare plant conservation; when practitioners are faced with protecting a threatened plant species, it is imperative that wild populations are not harmed. Ex situ conservation institutions offer facilities for bulking up seed of rare plants and cultivating juvenile or adult plant

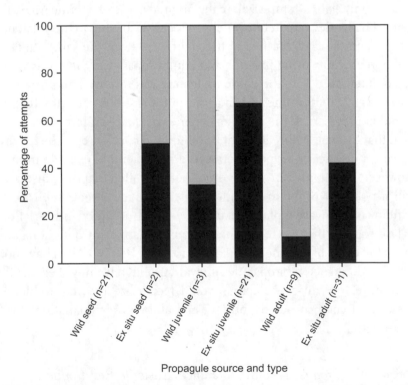

Figure 3.2. Percentage of attempts showing evidence of recruitment of progeny in situ (gray bars) or no evidence of recruitment (black bars) from propagules sourced from either wild populations or ex situ sources. Percentages used for purposes of comparison. Note the very small sample sizes for some categories.

propagules in order to improve each individual's chances of survival. Subsequently, we are not advocating the abandonment of ex situ approaches but instead recommend that practitioners follow guidelines for good practice set out by Guerrant and colleagues (2004b) and appendix 1.

If We Remove the Cause of Decline, Does This Improve Propagule Survival?

Subgroup analyses were undertaken to investigate whether heterogeneity can be explained by the presence or absence of the cause of the target taxon's decline. The analysis of reintroductions using seeds indicated that risk ratios are lower and therefore reintroduction success is higher where the cause of decline is no longer present. However, there is a large overlap of confidence intervals ($RR_{threat\ absent}$ = 1.911, $CI_{.95}$ = 1.867, 1.955; $RR_{threat\ present}$ = 1.937, $CI_{.95}$ = 1.906, 1.968). There is significant heterogeneity within the former subgroup, but not the latter (χ^2 = 298.50, df = 38, p < 0.001, I^2 87.3%). The former includes thirty-nine reintroduction attempts where the cause of decline was removed before outplanting; the latter includes only eight attempts where the original cause of decline was still acting at the site. The low level of heterogeneity may simply be due to the small numbers of attempts and the fact that five of the eight were from the same study. For reintroduction attempts using juvenile and adult propagules, there was similar imbalance of attempts where the cause of decline was removed and those where it was not. This fact undermines the analysis, but it is still noteworthy that the subgroups did not separate. If the cause of decline has not been removed, it might be expected that the type of threat might be difficult to control (e.g., the effects of climate change, disease, or air pollution would be difficult to exclude from a locality). However, attempts where the study authors admitted that the cause of decline was still a potential problem included quite localized threats such as grazing, competition from invasive plants, exploitation, and urban or industrial development. This suggests that the recommendations of many reintroduction guidelines to remove the cause of decline (e.g., Falk et al. 1996; IUCN 1998a) have not made a discernible difference to propagule survival. Although we might expect that failure to remove the cause of decline would jeopardize the reintroduction, this analysis cannot confirm our hypothesis, and alternative explanations for failure must be sought.

Are Species with Broader Habitat Tolerances Easier to Reintroduce?

Levels of endemism explained some variation in the survival of juvenile propagules, with regional endemics (i.e., those with subcontinental distributions) experiencing higher mortality than local endemics (fig. 3.3). This did not conform to

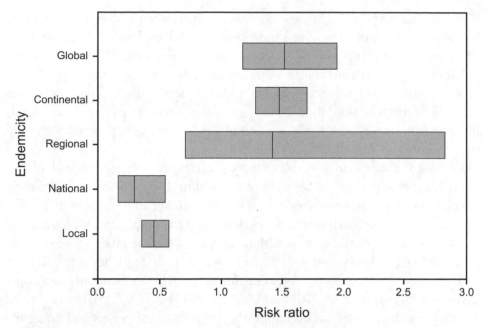

FIGURE 3.3. Pooled risk ratios (central vertical line in each box) and 95% confidence intervals for each attempted reintroduction using juvenile plants as propagules in subgroups categorized by level of endemicity. Numbers of attempts in each subgroup varied widely (global $n = 8$, continental $n = 10$, regional $n = 4$, national $n = 16$, local $n = 94$).

expectations and apparently has no biological significance because endemism was used as a range size indicator. We expected that species with very narrow ranges might have specific habitat needs, making it more difficult to select suitable sites for reintroduction. However, regional endemics were associated with higher mortality than the narrow endemics, indicating that assumptions of broader habitat tolerances making for simpler reintroduction site selection may be ill founded. As practitioners, we might assume that more widely distributed species could tolerate more variable conditions, but the results from the meta-analysis suggest that we may have overlooked key factors related to low success rates in reintroduction projects.

Does It Matter How Many Populations Are Used as Propagule Donors?

The number of donor populations (i.e., the wild populations that were the source for propagules) was included in the subgroup analysis to test whether using single or multiple populations is better in a reintroduction context. Using a single donor population has advantages, because threatened plant species often persist in isolated habitat fragments and become adapted to those specific conditions. If the reintroduction site is thought to be ecologically similar to the donor site, it is prudent to avoid disrupting coevolved and adaptive traits (Stockwell et al. 2003). Multiple donor populations may offer benefits in some situations for the very reason that they may be detrimental in others; propagules from multiple populations would be expected to have much higher pooled genetic diversity, and in situations where the reintroduction site was very heterogeneous or suspected to be slightly different from the conditions at the donor sites, greater genetic diversity would maximize survival (Neale, this volume).

Of the 134 attempts to reintroduce threatened species using juvenile plants, we could reliably classify only 66 attempts according to whether single-donor or multiple-donor populations were used. This is in part because very few authors described where the original material was sourced, if the propagules for reintroduction were raised in ex situ facilities. Meta-analysis of the risk ratios of the two subgroups showed that using multiple donor populations led to higher survival (RR = 0.253, $CI_{.95}$ = 0.129, 0.498), whereas relying on propagules from only one population had higher mortality (RR = 0.472, $CI_{.95}$ = 0.323, 0.691). However, given that the confidence intervals overlapped between these subgroups and no significant relationship with pooled risk ratios was discerned for seed- or adult plant–based attempts, we cannot conclusively make any recommendations that practitioners use single or multiple donors. Instead, it highlights the need for more research into (1) appropriate strategies for reintroduction accounting for ecological similarity of donor and reintroduction sites, (2) reproductive biology of the species of concern, and (3) isolation and subsequent genetic divergence of existing wild populations (and potential for inbreeding or outbreeding depression when mixing propagules from different donors).

Are Reintroductions within the Historic Range More Successful Than Introductions outside Range Boundaries?

We predicted that attempts to establish a population at a site outside the historic range would incur higher propagule mortality, because the habitat was more likely to be unsuitable. Instead, projects that used sites outside a species' historic range had a remarkably low risk ratio (RR = 0.177, $CI_{.95}$ = 0.053, 0.588, n = 7) compared with reintroductions to sites confirmed as supporting previously extant populations or those within the species' historic range, which showed very similar propagule survival ($RR_{previously\ extant\ sites}$ = 0.827, $CI_{.95}$ = 0.646, 1.059, n = 23; $RR_{sites\ within\ historic\ range}$ = 0.665, $CI_{.95}$ = = 0.578, 0.763, n = 99). It should be emphasized that the confidence intervals of all subgroups overlapped, and the number of attempts in each subgroup was, again, very unbalanced. Many more studies detailing attempts to move a species outside its historic range would be needed to test adequately whether this intervention is more successful than reintroductions within range.

How Do Reintroduction Practitioners Explain Failures?

In cases where attempted reintroductions resulted in very few or no surviving individuals, many authors offered explanations for the failure of their projects. Although these are often speculative, the insight gained from practitioners with experience working with different species and in different situations was a valuable component of amassing an evidence base on which to inform future use of a technique. Speculation can easily be translated into hypotheses, therefore, offering suggestions for productive research opportunities. If these hypotheses were tested rigorously (see Kennedy et al., this volume) and disseminated effectively, the reintroduction community could make major advances in filling the knowledge gaps that are still so evident.

According to practitioner accounts, reintroductions often failed because of unfavorable habitat conditions despite using conditions associated with extant, wild populations to select the recipient site. Specifically, causes such as drought, particularly in the first few years after outplanting (Jusaitis 2005; Batty et al. 2006b); inappropriate disturbance regimes, including too much and not enough disturbance (Drayton and Primack 2000; Leonard 2006a, 2006b; Maschinski and Dusquenel 2007); and unsuitable substrate texture (Fiedler and Laven 1996) have been cited. In some studies where authors deliberately included marginal habitat types, these were unsurprisingly shown to be less suitable for introduced propagules (Arnold et al. 2005; Jusaitis 2005). Competition from invasive plant species confounded several reintroduction attempts (e.g., Jusaitis 2005; comparison of establishment of propagules in weeded and nonweeded plots); in one case this was because nonnative weeds responded more positively to post-outplanting management than the target species (Mehrhoff 1996).

Other reported common causes for failure were linked to the species' developmental and reproductive biology, including propagules being outplanted at too early a stage in their development (Ruth Aguraiuja, personal communication; Batty et al. 2006b). In some cases the authors admitted that too few propagules had been introduced to overcome demographic and environmental stochasticity, leading to loss of all individuals (Dalrymple and Broome 2010). In other studies the transplanted individuals survived the duration of the reported monitoring period, but pollen limitation of flowering adults was cited as the reason for an absence of subsequent recruitment (Drayton and Primack 2000).

Preparing for Reintroductions under a Climate Change Scenario

Reintroduction practitioners must accept that for many species of conservation concern, remaining suitable habitat within the historic range may have changed in inapparent ways. This is due to the ecological lag in response to drivers causing habitat degradation. This means that reintroduction attempts are destined to failure unless the drivers have been identified and reconciled. MR is often criticized for being too costly in time and resources to assess wild population status and suitability of recipient sites in detail, and too risky given the inherent uncertainty in range shift predictions. However, we argue that current environmental change (resulting from pollution, climate, and land use changes) undermines any notion that historic range can be upheld as some sort of gold standard for reintroductions. This change means that the detail of research and monitoring, which is quite rightly thought to be appropriate for MR, must also be adopted for reintroduction attempts. We recognize that the distinction between true reintroductions (i.e., those to sites of extirpated populations) and MR to sites close to the target taxon's range (or within the extent of occurrence but to a site with no historic record of occupation) is increasingly artificial as we continue to alter our environment. Of course, there are many questions associated with the use of MR, including the target organism's invasiveness in a novel habitat (see Reichard et al., this volume) and how to reliably choose recipient sites based on future climate extrapolations (see Krause and Pennington, Maschinski et al. [chap. 7], this volume). However, we propose that the two techniques are separated only by historical definition, and as our review has begun to demonstrate, there is little ecological justification for using reintroductions throughout a species' historic range while ignoring the potential of close-range MR to mitigate against threatened plant declines.

Summary

This review highlights many aspects of reintroduction practice that require further research to inform the future use of the technique. We have been unable to state conclusively that the parameters commonly identified as being critical to reintroduction success actually made a difference to the longevity of reintroduced populations. However, although the absence of evidence of an effect is not the same as evidence of absence, these findings can direct us to areas that warrant attention in the future. For reasons that are expanded upon later, we strongly recommend that practitioners undertake (1) improved evaluation of wild population demographics to identify regions of decline and growth; (2) surveys that attempt to discern the specific causes of declines in wild populations; (3) surveys that identify conditions supporting population growth at candidate recipient sites; and (4) more detailed monitoring for more than 10 years after propagule reintroduction (appendix 1, #22, 23, 36, 40).

Disseminating detailed information about reintroductions and the consequent opportunities for experiential learning across the practitioner community might be facilitated by the existence of a single cataloging system. The Center for Plant Conservation has already initiated such a database: the *CPC International Reintroduction Registry* (Center for Plant Conservation 2009). We strongly recommend that practitioners contribute their reintroduction experiences, whether they are successful or unsuccessful, to this project. We encourage practitioners to realize the value of their data beyond any single project requirements and to publish their results in a national or international database. Although there are reasons to report results only to contracting agencies, sharing information in a database will make it easier to identify practitioners and conduct a review such as this in the future (appendix 1, #43).

Although we acknowledge that a limitation of this review is the lack of causality that can be attributed to the tested parameters, the one parameter that showed clear subgroup separation was the endemism or taxa range size of juvenile plants reintroduced to sites. As discussed earlier, this was counter to our expectations; species with larger range sizes had higher mortality than endemic species. This suggests that species that were once widespread but have declined in part of their range are being reintroduced to habitat that is not currently suitable. In effect, taxa with apparently broad habitat needs at species level may be lulling practitioners into a false sense of security with regard to site selection for reintroduction. In addition, the fact that introductions to sites outside the species' historic range are not generating high risk ratios suggests that our conventional notions of putting species back into their "core" habitat might be misguided.

Most recipient reintroduction sites are selected based on habitat surveys using fairly coarse indicators, such as community type. We hypothesize that as a community of practitioners, we have not successfully identified critical parameters describing the target taxon's niche. In addition, environmental change may alter key parameters in the intervening period between population extirpation and reintroduction and may be a key factor explaining propagule mortality or failure to recruit progeny in recipient sites (see Maschinski et al. [chap. 7], this volume).

Essentially, practitioners must take very great care to identify what specific factors determine the target species' decline; this must be more than just identifying the main threats (appendix 1, #2). For example, many species are cited as undergoing a decline due to agricultural intensification, but the actual underlying mechanism causing the decline is not always known. The species may be intolerant to shading from species that have positively responded to increased nutrient availability. Alternatively, the application of herbicides may have extirpated populations. In another example, Dalrymple and Broome (2010) relied on "community matching" to select recipient sites for *Melampyrum sylvaticum*, but they did not adequately address the potential for recent climate change. Introducing this species to sites at the warmer and drier margin of its UK range despite evidence that wild populations are moisture dependent resulted in failed reintroductions (Dalrymple et al. 2008). Myriad other examples could be given, but the key message we would like to communicate is that if the mechanism of decline has not been accurately discerned or has not been properly incorporated into reintroduction plans, there is little chance of selecting locations where the species can be restored or introduced successfully. Using the extant wild populations as indicators of habitat suitability is the most obvious way to achieve this (Maschinski et al. [chap. 7], this volume), but being sure to select habitat attributes where populations are growing, rather than declining, is essential (Knight, this volume). Admittedly, acquiring these data is time and resource intensive, but the failure of reintroduction attempts bears costs, which are difficult to defend and may potentially damage the conservation community as a whole.

Because of the short timescales of reintroductions reported in the conservation literature, it was not possible for us to describe decisively the efficacy of reintroductions here. However, there is little doubt that there is a great deal of room for improvement and that without a move toward more rigorous investigation and evaluation of reintroductions, we cannot expect to protect threatened species adequately or to learn how to defend biodiversity from increasing anthropogenic threat in the future.

This work was supported by the British Ecological Society (Ecology Into Policy Grant no. 921/1146) and benefited from unpublished data supplied by the Botanical Society of the British Isles.

PART II

Reintroduction Science and Practice

"All things are difficult before they are easy."

—Thomas Fuller

The tools available to conservation practitioners are improving, multiplying, and becoming more practical. This section addresses common tools and overlooked tools that are used in plant reintroductions and provides examples of how they are used for conserving some of the world's rarest plant species. As reintroduction practice increases, so do the opportunities for testing ecological and restoration theory, for improving methods, and ultimately for increasing our chances to recover endangered species.

All practitioners know that plant reintroductions require the help of many hands, as well as political and financial support. Integral to the preparation, implementation, care, and social acceptance of reintroductions is public support. Chapter 4 reveals the important role volunteers play in plant reintroductions and rare species recovery. Public participation is a conduit for rare plant education and conservation advocacy. Without these, none of the work described in this section could be sustained.

Plant reintroduction efforts have benefited from attention to plant reintroduction science. Chapters 5 through 12 are presented as components to consider while planning a reintroduction. Throughout the chapters, important findings are linked to specific recommendations in the *CPC Best Reintroduction Practice Guidelines* (appendix 1). The chapters represent diverse topics demonstrating multidisciplinary collaborations, which will play an increasingly important role in this growing field.

Chapters 5 and 6 examine propagule selection and handling. Chapter 5 addresses genetic considerations when collecting or choosing propagules for reintroduction projects. Chapter 6 reviews acclimatization of propagules from laboratory to greenhouse or from greenhouse to field, focusing on soil preparation and symbiotic microbes.

Site selection for reintroduced populations is critical to reintroduction success. Chapter 7 provides theory behind and tools for selecting optimal sites in which to reintroduce rare plants, and chapter 8 uses species distribution modeling and geographic information system software to predict current and future sites for reintroduction.

Once optimal sites are selected, the number of propagules and the most appropriate life stage of the propagule must be chosen for reintroduction. Chapter 9 examines the use of population viability analysis from natural populations to determine numbers and life stage of the species to introduce. Based on literature and the *CPC International Reintroduction Registry*, chapter 10 evaluates the evidence that founding population size and stage of plants with different life histories influence the establishment and demography of reintroduced populations.

Some of the most difficult cases associated with rare plant reintroduction are species that are long-lived and species with fewer than fifty individuals remaining in the wild. Because long-lived plants often have slow maturation and lifespans greater than funding cycles or tenures of individual researchers, Monks and colleagues (chap. 11) develop unique criteria for determining the success of long-lived species reintroductions using illustrations from long-lived Australian species. Chapter 12 looks at another set of extreme reintroduction cases: species that have few living individuals. Although many of the rare Hawaiian species face dire threats, Kawelo and colleagues give examples that highlight how science and best practice are preventing extinction.

Chapter 4

The Critical Role of the Public: Plant Conservation through Volunteer and Community Outreach Projects

JOYCE MASCHINSKI, SAMUEL J. WRIGHT, AND CAROLINE LEWIS

Universally, plant conservation practitioners know that it would be impossible to accomplish effective long-term species recovery without broad public support. Because more than 20% of plant species are estimated to be threatened with extinction worldwide (IUCN Species Survival Commission 2004; Gilbert 2010; Pennisi 2010), the task of conserving species and habitats is too large for the few existing professional practitioners to accomplish alone. The public actively supports plant conservation through participation as volunteers, through financial contributions to programs, and by influencing community and government actions related to rare species and habitat protection. The ever-increasing need to preserve species requires more public involvement daily. In this chapter we examine the impact of volunteerism on rare plant recovery. We provide quantitative evidence that public participation has reduced extinction risk of rare species. Furthermore, we provide examples of programs that have successfully engaged the public in plant conservation activities.

According to Parker (2008), meaningful community participation should be viewed as one of the primary outputs of any reintroduction effort. An important first step for building public support for plant conservation is to raise awareness of community opportunities and call attention to the plight of rare species and their habitats. This is followed by a phase of retaining commitment to the cause that can finally lead to permanent changes in public attitudes and choices made about the environment. The long-term welfare of reintroduced rare populations may depend on stewardship of human communities adjacent to reintroduction sites.

Sociologists and psychologists have shown that caring for nature promotes human well-being (Clayton and Myers 2009). Among the psychological benefits gained from being in nature are recovery from mental fatigue (cognitive restoration), opportunity for reflection, a sense of identity, and emotional attachment (Fuller et al. 2007). Volunteers have reported that conservation activities give them a richer quality of life, a sense of being in an alternate reality from their work or school lives (Bell et al. 2008) that are often overly structured, time-deficient, urban, electronic, and sedentary (Miller 2005; Lewis 2009).

Active participation also imparts social and intellectual benefits. Group activities with like-minded people tend to create a synergistic circumstance for learning; group members can learn from leaders and from one another while socializing and supporting a cause (Bell et al. 2008). At times when work is difficult or conditions are unpleasant, being in a group can make a daunting task palatable and increase opportunities for fun exchange (J. Maschinski, personal observation). Cooperation within groups can lead to committed activism (Olli et al. 2001). Establishing a personal connection with a natural area or a species while working with plants can promote advocacy for the conservation of species and their environments (Schultz 2000; Williams and Cary 2002; Primack 2006). Fortunately, people willingly participate in activities that they believe matter (Hines et al. 1987) and hunger for an opportunity to become part of a solution to join the good fight.

Personal, social, environmental, and spiritual aspects of involvement synergistically and simultaneously connect people to nature. Furthermore, experiences in nature have long-term impacts on human values, actions, and decisions. Fortunately, this is truly a mutualism, as rare species and their habitats also benefit (fig. 4.1).

Measurable Conservation Gains: Public Contributions to Endangered Species Recovery

From the plant's perspective, what are the benefits of volunteer contributions? Are volunteer efforts truly reducing extinction risk of rare species? We addressed these questions using information from our own reintroduction research with five endangered species (Maschinski and Wright 2006; Wendelberger et al. 2008; Possley et al. 2009; Maschinski et al. 2010; table 4.1). To examine how volunteers affect the status of the endangered species, we compared the total number of wild populations and individuals for each species with the numbers of new populations, new plants, and seedlings in the volunteer-established reintroduced populations. In addition, we documented the numbers of volunteers who participated in the reintroductions, including the immediate preparation for each reintroduction, their total hours of effort, and the total number of organizations that participated. All our target species have few populations and few individuals in the wild, so reintroduction was a critical next step for species recovery. Four species have had

Benefits of Public Involvement	Short-term Impacts	Long-term Impacts
The Plant Perspective	Increase individuals in wild	Species recovery
	Increase the number of populations	No extinction
	Increase reproduction	
	Increase recruitment	
	Increase probability for sustainability	
	Increase opportunities for funding	
The Human Perspective	Personal - Intellectual	Knowledge of species
	Environmental	Stewardship
	Social	Political activism - Actions to benefit rare plants
	Spiritual	Connection to where they live
		Hope

FIGURE 4.1. Short- and long-term impacts of public involvement in plant conservation from the plants perspective and the human perspective. Although this is displayed in a linear format, the aspects operate as a web. Examples of personal, social, environmental, and spiritual benefits to humans are noted in the text.

multiple reintroductions into multiple sites over several years. The exception is *Tephrosia angustissima* var. *corallicola*, which had two experimental reintroductions done at a single location on the same day. Collectively, the reintroductions in this example have doubled the total populations and increased numbers of extant individuals 4.6-fold while engaging 466 volunteers from eighteen organizations in 2006 hours of activities (table 4.1). All but one of the species has had documented next-generation recruitment. Thus, the reintroductions that were made possible by volunteer assistance are benefiting the endangered species in several measurable ways.

The Australian Trust for Conservation Volunteers reported in 2010 that teams of volunteers planted more than 1 million trees for habitat and land restoration, collected more than 907 metric tons of native seed for revegetation and reforestation projects, built and restored more than 300 kilometers of walking tracks and boardwalks, installed more than 80 kilometers of conservation fencing to protect vulnerable areas, and completed 500 wildlife surveys to assist with threatened species management.

It is important to note that large and small contributions make a difference. Small organizations collecting grams of seed and restoring meters can take pride

TABLE 4.1

The impact of volunteers on endangered species recovery.

Species	Total Wild Plants	Total Wild Pops	First Date of Intro	Last Date of Intro	Total New Pops	Total Augmentations	Total Plants Installed	Live Intro by 2010	New Sdlgs in Intro Pops	Volunteers	Hr	Orgs
Jacquemontia reclinata	846	8	Sep 2001	Jul 2009	11	1	1,656	1,166	96	339	1,376	17
Amorpha herbacea var. crenulata	303	4	Aug 1995	Aug 2007	3	0	658	368	302	73	350	3
Lantana canescens	2	1	Jul 2005	Jun 2007	3	0	367	55	4,934	31	160	3
Passiflora sexflora	41	3	Jun 2006	Nov 2006	0	3	136	22	?	12	60	1
Tephrosia angustissima var. corallicola	351	1	Jun 2003	Jun 2003	1	0	191	2	161	11	60	1
5 species	1,543	17			18	4	3,008	1,613	5,493	466	2,006	25

Total Wild Plants = total number of documented individuals of the species in the wild; Total Wild Pops = total number of natural populations; First Date of Intro = earliest date of introduction; Last Date of Intro = latest date of introduction; Total New Pops = total number of new populations created with introduction efforts; Total Augmentations = total installations of plants into an existing population; Total Plants Installed = number of whole plants transplanted into the site; Live Intro by 2010 = number of transplants alive by 2010; New Sdlgs in Intro Pops = number of documented new seedlings observed in introduced populations by 2010; Volunteers = number of volunteers who participated in the immediate preparation and installation of reintroductions (does not include long-term care for plants before reintroduction); Hr = number of volunteer hours; Orgs = number of organizations participating in the introductions. Organizations involved included Fairchild Volunteers, Fairchild Palms, Boy Scouts, Americorps, Florida Native Plant Society, City of Miami, Surfriders, Hands on Miami, Homestead Job Corps, Jupiter High School Environmental Research and Field Studies Academy, Palm Beach County Stewards, Florida International University, Miami Dade College, Palmetto Senior High School, Kids Ecology Corps, City of Hollywood Summer Kids Program, and University of Miami Law Students.

in their contributions just as well as larger organizations. One person at a time, the synergistic effort is benefiting us all and the rare plants in our care.

Outreach Programs for Children and Teachers

Reconnecting people with nature should begin with our children (Kahn 2002; Lewis 2009) because they are the stewards of the future. Programs for children are intended to inspire, engage, build confidence, develop critical thinking skills, instill the importance of biodiversity and conservation, and foster the recognition that individuals do make a difference. Alumni of such programs report that they were strongly influenced and motivated to make changes in their personal lives (Lewis 2009). This is the route to becoming informed, compassionate citizens.

The Fairchild Challenge, a competitive, interdisciplinary environmental education outreach program for middle and high school students, is influencing students and their community (Lewis 2009; table 4.2). Comprised of multidisciplinary competitions with an environmental emphasis, this free annual program open to all schools in the Greater Miami area in Florida is designed to promote engagement in environmental issues for students with diverse interests, abilities, talents, and backgrounds. In 2009, more than 55,000 students and 1,800 teachers from 100 schools participated. Using environmental themes, students composed rap songs, created postcards (box 4.1), wrote letters, helped restore habitat, planted endangered species, and created native plant gardens in their schoolyards. Students who helped reintroduce an endangered species had an opportunity to participate directly in species recovery and will be able to revisit the restoration site in the future, perhaps with their grandchildren. Reintroductions and gardens have given students an opportunity to gain real experience in ecology, habitat restoration, and conservation, while their teachers have had opportunities to incorporate garden practices in math and science curricula. Student and teacher enthusiasm has been powerful.

Several projects provide close supervision and training by professional conservation scientists followed by place-based service learning using reintroduction and restoration activities over the course of several weeks or months. For example, in Corvallis, Oregon, researchers at the Institute for Applied Ecology (IAE) work with local schools to enlist student participation in Restoration and Reintroduction Education. Meeting weekly with an IAE conservation biologist throughout the school year, students learn to germinate seeds of rare species and cultivate them in their school greenhouses. Students participate in outplanting the rare species into an appropriate site near the school after a reintroduction and educational plan that has been customized to site and species characteristics, student abilities, and teacher needs (Cramer 2008). Similar programs exist for restoring native plant communities (table 4.2).

TABLE 4.2

Programs designed to engage the public in plant conservation activities.

Programs	Program Attributes									Website
	CT	ST	M	S	I	F	R	Priv	Pub	
Australian Trust for Conservation Volunteers			x	x			x		x	http://www.conservationvolunteers.com.au/
Back to Natives	x						x		x	http://www.backtonatives.org/volunteer.shtml
Botanical Guardians		x	x					x	x	http://www.uga.edu/gpca/project6.html
Botanists in Training Program	x				x			x		http://archive.rbg.ca/cbcn/en/cbcn+kids/kid_main.htm
Budding Botanist Program of the Plant Atlas Project of Arizona	x						x		x	http://www.gcvolunteers.org/trainings_botanists.html
Channel Islands Restoration		x	x	x	x	x	x	x	x	http://www.channelislandsrestoration.com/
Connect to Protect	x			x	x		x	x	x	http://www.fairchildgarden.org/centerfortropicalplant conservation/connecttoprotect/
EcoHelpers	x	x		x			x		x	http://www.nps.gov/samo/forteachers/ecohelpers.htm
Endangered Plants Stewardship Network	x						x		x	http://www.epsn.org/Teachers/index.asp
EU-Wide Monitoring			x					x	x	http://eumon.ckff.si/volunteers.php
Gondwana Link			x	x	x	x	x	x	x	http://www.gondwanalink.org/
Greening Australia		x	x	x	x	x	x	x	x	http://www.greeningaustralia.org.au/
Native Plant Corps	x		x	x	x		x	x	x	http://fieldstaff.thesca.org/index.php?option=com _content&task=view&id=51&Itemid=24
New England Wild Flower Society Plant Conservation Volunteers	x	x	x	x				x	x	http://www.newfs.org/protect/rare-plants-and -conservation/Volunteer

Program	CT	ST	M	S	I	r (Priv)	r (Pub)	URL
Partners for Colorado Native Plants and Rare Plant Monitoring Stewards	x	x	x				x	http://www.botanicgardens.org/content/conservation-and-research-get-involved
Restoration and Reintroduction Education	x	x			x		x	http://appliedeco.org/ecological-education/programs/default-page
Roots and Shoots	x	x			x		x	http://www.rootsandshoots.org/?gclid=CImjiqyznJ0CFQZinAodrEw72g
San Cristobal Biological Reserve	x	x	x		x		x	http://www.volunteer.org.nz/ecuador/reserves/san cristobal.php
The Fairchild Challenge	x		x		x		x	http://www.fairchildgarden.org/education/fairchild challenge/
The Washington Rare Plant Care and Conservation Program	x	x	x		x		x	http://courses.washington.edu/rarecare/
Victorian Environment Friends Network	x	x	x	x	x		x	http://home.vicnet.net.au/~friends/index.html
Volunteer South America		x	x					http://www.volunteersam.com/what.htm
Wildlands Restoration Volunteers	x	x			x		x	http://www.wlrv.org/index.html
Woodland Watch/Healthy Ecosystems		x	x		x		x	http://www.wwf.org.au/ourwork/land/woodlandwatch/

Program attributes include activities for children and teachers (CT), specialized training (ST), monitoring (M), seed collection (S), invasive species removal (I), and restoration (r) on private (Priv) and public lands (Pub).

BOX 4.1. FUTURE STEWARDS

Contributed by: Samuel J. Wright and Caroline Lewis

In 2008 Fairchild Challenge featured the option "Postcard Design Contest and Political Action," in which middle school students created postcards that brought attention to wildlife living in South Florida's pine rocklands and wrote a short letter to a county commissioner asking him to help conserve pine rockland areas. The winning entry is shown in figure 4.2. This collaboration between Fairchild's education and research departments helps conservation programs.

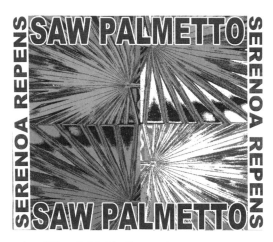

FIGURE 4.2. Example of Fairchild Challenge Competition Winning Design for a pine rockland postcard and letter to a commissioner encouraging the protection of this globally endangered ecosystem.

Dear Commissioner Gimenez,

My name is Madison Bec and I am writing to encourage you to conserve the Pine Rockland areas. I would like for you to encourage the conservation of the Saw Palmetto in the Pine Rocklands. The Saw Palmetto has been around for ages. It was a food for the Native Americans and it is used for medication for various illnesses. The Saw Palmetto is a significant part of the environment. Without it, the food chain of the eco-system would be missing an important link. I am enrolled at South Miami K–8 Center and study Computer Art Technology (C@T) in the secondary school magnet. As part

---

Box 4.1. Continued

of the Fairchild Challenge, I have created a postcard about the Saw Palmetto that I would like to share with you. I made this postcard using Photoshop and Printshop Programs. Learning about the environment while using my artistic vision makes learning fun!

Please take a minute to write back to me and tell me what you think.

Sincerely,

Madison Bec, South Miami K8

Roots and Shoots, a program of the Jane Goodall Institute, is a powerful, youth-driven, global network of more than 8,000 groups in almost 100 countries. Youth of all ages are taking action to improve our world through service learning projects that promote care and concern for animals, the environment, and the human community. For example, in Tanzania, Roots and Shoots Youth Leadership Councils is creating five new tree nurseries for Rebirth the Earth: Trees for Tomorrow, a project to restore trees in communities (table 4.2).

As important as hands-on activities are, there is also a need to provide support to teachers who want to incorporate conservation themes into curricula. A great example of such support is that given by the State Botanical Garden of Georgia, whose research, conservation, and education staff developed the Endangered Plants Stewardship Network. This program centered on plants, threatened habitats, and conservation biology provides training and materials to K–12 teachers. Documented links to state education standards allows these materials to be used easily in the classroom.

There are also opportunities for students to explore on their own via the Internet. A Botanists in Training Program for kids only, developed by the Canadian Botanical Conservation Network, explains biodiversity, conservation, and the wonders of botany to a virtual audience of children. Participants are given suggestions on conservation actions they may take.

Programs with Specialized Training

Throughout the world many opportunities exist for group volunteer activities in support of plant conservation. Among the most successful are those that have devoted paid personnel to coordinate, motivate, and train volunteers. Specialized

training for plant identification and scientific protocols for monitoring and recording observations enhances the education of the participants and increases the
value and accuracy of their data, which in turn expands the conservation impact
(e.g., Budding Botanists; table 4.2). Training programs led by organization professionals involve large institutional commitments of staff time and funds, and therefore many have investigated and made recommendations to increase volunteer retention. These include asking for signed commitments, providing opportunities
for group social networking and guidance by a professional, giving personal recognition, acknowledging contributions, and demonstrating that the contributions
are valued and effective (Martinez and McMullin 2004; Bell et al. 2008; Lewis
2009). Grades or prizes may motivate initial participation (Lewis 2009), but more
subtle internal factors permeate, indoctrinate, and solidify the experiences that
may lead to changed behavior. There is no substitute for enthusiastic individuals
and close personal interactions to increase volunteer commitment and retention
(Bell et al. 2008).

Regionally organized umbrella groups are effective means to accomplish plant
conservation goals in alignment with the "Think globally, act locally," philosophy.
Established in 1998, the New England Wild Flower Society Plant Conservation
Volunteers program has been revered and emulated as an effective conservation
strategy (table 4.2). Amateur field botanists are recruited and trained to perform
vital plant conservation work throughout New England. Each volunteer receives
training in plant identification, monitoring protocols, and recordkeeping, plus op-
tional participation in special classes, field trips, and symposia. Volunteers monitor rare plants, manage habitats, survey and control invasive species, and conduct
general botanical surveys. They provide invaluable data to state heritage programs
and plant conservation professionals and advocate for native plants region-wide.
Education, networking, and a yearly evaluation process ensure the reliability of
volunteer work and participant enjoyment. Similar programs operate in the Pacific Northwest (e.g., the Washington Rare Plant Care and Conservation Program), the Rocky Mountains (e.g., Partners for Colorado Native Plants and Rare
Plant Monitoring Stewards), and southeast regions of the United States (e.g.,
Georgia Plant Conservation Alliance and Botanical Guardians), where volunteers
participate in diverse activities from demographic monitoring to relocating "lost"
populations of rare plants. Participatory monitoring networks have arisen around
the world (Bell et al. 2008; Evans and Guariguata 2008). Amateur naturalists help
assess the abundance, distribution, and conservation status of species and their
habitats (Danielsen et al. 2005). Informing participants how their data are used or
linking the activity to a conservation goal helps to retain volunteers in these programs (Bell et al. 2008).

For research or monitoring programs that include volunteers, a critical concern is the quality of the information gathered. In an analysis of the quality of mammalian monitoring data collected by 155 volunteers, Newman et al. (2003) found that the reliability and accuracy of the data were directly related to the technical difficulty of gathering it, the amount of time devoted to specialized training with a professional, and the physical fitness of the volunteer. Clear, unambiguous protocols also help improve the accuracy and reliability of data collected by volunteers (Foster-Smith and Evans 2003).

Some volunteer groups concentrate efforts on preserves. A particularly successful program, Victorian Environment Friends Network, is a network of 278 groups of volunteers working with management authorities in natural areas throughout Victoria, Australia. Groups are independent and autonomous, but they all provide support for parks, reserves, or species of interest, assist with special projects, gather kindred spirits, foster public awareness of the reserve or species, and support effective management of native flora and fauna. After training, they participate in community-based monitoring of threatened species, weed mapping, fixed-point photography and faunal monitoring, sampling water quality, and ecological burns, all of which assist park rangers with ongoing management (Cooke 2008). Similar efforts are under way at San Cristobal Biological Reserve in the Galápagos Islands.

Examples of Restoration Programs

Habitat restoration must often begin with invasive species removal followed by native species planting. With little training, volunteers of all ages can learn to identify and remove most invasive species, especially if a single species is targeted for a session. Children taught to identify invasive weeds will continue pulling them along trails and roadsides as they grow to be teenagers and young adults (Maschinski, personal observation).

Having locally adapted seed available for restoration requires coordinated efforts for species selection, collecting at the appropriate season, and seed processing and storage. Native seed collection is a fine activity for volunteers of all ages and is a critical activity in preparing for climate change (Vitt et al. 2010). Additionally, the action of bringing the public out into the field to make wild collections rebuilds the connection between nature and people.

Groups may participate in multiple activities to accomplish plant conservation objectives. Broad habitat restoration often requires operating at large scales; groups such as the Australian Trust for Conservation Volunteers and Greening Australia (Cochrane 2004), Volunteer South America, and Wildlands Restoration

Volunteers, Channel Islands Restoration (table 4.2) have helped collect native seed for propagation and restoration of native vegetation. Sometimes this entails building fences to exclude herbivores. Other restoration activities may have the goal of supporting animal populations, such as the Back to Natives Restoration Project, which reinstates native food plants for native arthropods in protected areas (table 4.2).

Connecting fragmented natural areas has been endorsed as a proactive strategy for ameliorating the impacts of climate change. To increase the area and diversity of globally endangered pine rocklands in South Florida, the Fairchild Tropical Botanic Garden's Connect to Protect Network is creating corridors and stepping-stones between isolated pine rockland remnants (table 4.2). County land managers, schools, private landowners, and businesses have joined the effort to reduce extinction risk of pine rockland rare endemic species by removing invasive species or restoring natives into lands that once were pine rocklands. Within the first 3 years of the program, forty-three parcels were restored with native species. Participants helped collect more than 170,000 seeds of eighty-three pine rockland species to use for native plantings along public trails and roadways. Private landowners are fostering rare species by establishing experimental populations into their own pine rocklands. Through the Fairchild Challenge students from local schools have repeatedly participated in Connect to Protect Network activities over the past 5 years (fig. 4.2).

An even more ambitious and broader scale project to connect rare ecosystems is Gondwana Link, which has the goal of linking 1,000 kilometers between the Stirling Range and Fitzgerald Rivers national parks in Western Australia, a biodiversity hotspot with more than 2,500 species at risk of extinction (table 4.2). With guidance of the indigenous Noongar people and financial and logistical support from Bush Heritage Australia, Greening Australia, the Wilderness Society, The Nature Conservancy, Fitzgerald Biosphere Group, Friends of the Fitzgerald River National Park, and Green Skills, properties have been successfully purchased or protected with covenants. Large (70- to 250-hectare) restoration projects in dry woodlands and wet forests have shown some initial promise.

Outreach Programs for Private Landowners

An important public constituency to reach for effective plant conservation is private landowners. At times, soliciting private landowner support for plant conservation requires gently coaxing a relationship of mutual trust and respect (Curnow et al. 2008). Reluctant landowners may have fears about perceived legal constraints of having rare species growing on their property or may think that the only good plant is one that can be eaten by livestock (Janssen and Williamson 1996). With

Box 4.2. Continued

square meters). No seedlings or saplings were found surrounding the investigated trees or in their natural habitat, indicating that the species has difficulty regenerating naturally. Its poor natural regeneration may be due to habitat destruction, low fruit production, difficulty of seed germination, and seed overcollection.

FIGURE 4.3. (a) A big *Magnolia sinica* in the wild and (b) a small tree growing at Kunming Botanic Garden.

Goals of Population Augmentation and Progress: Urgently needed ex situ and in situ conservation actions have been undertaken by Kunming Botanic Garden and other organizations. Some 6,000 seedlings and saplings have been raised from seeds collected from extant individuals. In July 2007, a multidisciplinary working team of Flora and Fauna International China, Kunming Botanic Garden, and the local forestry bureau and nature reserve in Wenshan planned a strategic program to introduce the species to historic locations and develop practical restoration methods for threatened magnolias. This was the first attempt to follow IUCN Species Survival Commission reintroduction guidelines (IUCN 1998a) for endangered tree species in China. Based on field surveys, two augmentation sites within the historic

Box 4.2. Continued

range were selected. Before planting, Sun held a workshop to train twenty local people and discuss problems. To facilitate communication and to befriend locals, Sun made an effort to follow local customs, drink local beverages, and eat local cuisine. These local people were trained to weed, water, collect data, and plant the trees.

In November 2007 and July 2008, 400 saplings were hauled up a mountain, planted, mapped, and measured (fig. 4.4). Local participants were paid a stipend by the government to monitor and patrol plantings. Public education about the importance of the program followed the outplanting event. By March 2008, 89% and 92% of transplants survived at two sites and had excellent growth.

Figure 4.4. Tagged 2-year-old *Magnolia sinica* sapling returned to natural habitat.

A 26-year-old *M. sinica* (more than 8 meters tall) growing at Kunming Botanic Garden has not yet flowered (fig. 4.3b), so it certainly is too early to predict the future fate of these individuals. The excellent implementation of the program and establishment of the multidisciplinary working team has great significance for rescuing critically endangered *M. sinica* and can provide an effective model or guidelines for reintroducing other critically endangered trees in China.

personal attention, dedicated time to visit and listen, patience, compassion, honesty, and follow-through, private landowner skepticism can be overcome (Janssen and Williamson 1996).

In Western Australia, the Woodland Watch/Healthy Ecosystems Project is helping landholders, communities, and local government authorities protect priority threatened ecosystems (including eucalypt woodlands) on private and non-state-managed lands (table 4.2). Enthusiastic extension officers discuss the natural values of bushland with landowners one on one. Gaining permission to conduct floral surveys and providing reports to landowners enables extension officers to discuss the value of the land, its biodiversity, and its habitat value for fauna while providing technical advice about land management. Financial support for fencing bushland for protection from herbivores has improved project success (Curnow et al. 2008), as has voluntary land management agreements. More than 150 landowners are participating in this partnership with regional Natural Resource Management and World Wildlife Fund Australia. These efforts are helping to protect rare flora and fauna.

Practical Implications for Traditional Restoration Work

To promote a collective consciousness for plant conservation, plant conservation work with volunteer groups and engaged citizens is effective. Restoration work is labor intensive at every stage and is often underfunded at both state and federal levels. Although the work involved differs by target species, people of all ages with minimal training can usually assist with many tasks, including removing invasive species, collecting seed, and planting. With specialized training, important demographic monitoring can be accomplished.

Without support from the public, much of the ongoing plant conservation work throughout the world could not be accomplished. The need for conservation is far outstripping the capacity of professionals to complete the work. Professionals who manage a great deal of territory or many species may find that they can accomplish much more by training and supervising a volunteer workforce. Careful cultivation of volunteers can lead to long-lasting relationships. Although private citizens may begin their participation in a conservation project as volunteers, this role often expands to include the role of financial donor.

In return for their participation in conservation activities, the public has much to gain. While contributing to restoring biodiversity, volunteers may also achieve personal satisfaction, meet like-minded people, learn new skills from trained professionals, and discover that the implications of their work expand beyond one moment or one species to benefit the world community.

Preparing for Climate Change

In our rapidly changing world with an exponentially growing human population, even widespread species are becoming threatened by habitat destruction, overexploitation, pollution, invasive alien species, and climate change (Secretariat of the Convention on Biological Diversity 2010). The greatest legacy of public involvement is a realization that human action is the underlying cause of climate change, that biodiversity loss is linked to these drivers, and that personal changes in behavior will be needed to change the current trajectory.

Citizen cooperation and endorsement will be essential for any translocations or managed relocations of species outside their historic range (Parker 2008). Particularly for communities adjacent to relocation sites, there is a need for discussion, rather than one-sided or hierarchical explanations from land managers to citizens (Parker 2008). With open communication and public involvement, translocations can have logistical, financial, and political success.

Summary

Clearly it takes a village to accomplish conservation goals, and without public involvement plant conservation will not succeed. With the help of the public, rare species can benefit and their extinction risk can be reduced. Conservation organizations engage citizens of all ages to engender caring for the planet and its species. Focusing human energy on solutions creates healthier people and healthier ecosystems.

We are grateful to Arlene Ferris, Sandy Schoenfeldt, and Stephanie Bott for their expertise, cordiality, and finesse for coordinating the Fairchild volunteer program and to our volunteers, who have made our work possible and enjoyable.

Chapter 5

Genetic Considerations in Rare Plant Reintroduction: Practical Applications (or How Are We Doing?)

Jennifer Ramp Neale

The importance of considering genetic variation in conservation of rare plants is well understood (Falk and Holsinger 1991; Hamrick and Godt 1996; Rieseberg and Swensen 1996). In addition, the importance of considering genetic diversity in plant reintroduction has been well argued (Fenster and Dudash 1994; Mistretta 1994; Havens 1998; Falk et al. 2001, 2006; Hufford and Mazer 2003). As a result, various guidelines for collecting propagules (primarily seed) for reintroduction based on genetic theory have been put forth (Knapp and Rice 1994; Center for Plant Conservation 1996; McKay et al. 2005). Here I review the tenets of restoration genetics before discussing the means of measuring genetic diversity in species slated for reintroduction. Current collection guidelines for reintroduction and examples of empirical studies that have measured genetic variation before, during, and after reintroduction are presented and evaluated. Lastly, further arguments for the inclusion of genetic data in reintroduction practice are presented, with a brief discussion of the issue of managed relocation and how genetic data may inform the future of plant reintroduction in a changing climate.

Genetic guidelines for plant reintroductions suggest collecting locally and not mixing population sources in outplanting if possible (Knapp and Rice 1994; Krauss and Koch 2004; Sanders and McGraw 2005). It appears that in practice these guidelines have been generally followed, but are such strict guidelines really necessary (Wilkinson 2001; Broadhurst et al. 2008)? Empirical studies evaluating levels and patterns of genetic diversity of reintroduced species remain few (but see Smulders et al. 2000; Ramp et al. 2006; Fant et al. 2008; Liu et al. 2008 for exceptions), making the evaluation of guidelines for appropriate reintroduction source material difficult.

Incorporating genetic data (or theory) in reintroduction of rare species has been well covered in other sources and will not be covered in detail here (Fenster and Dudash 1994; Hufford and Mazer 2003; Falk et al. 2006). Genetic diversity in reintroductions should be considered in an evolutionary context because genetic diversity provides the basis for adaptation to changing environments and can prevent, or lessen, the effects of inbreeding depression or other genetic consequences of small population size (Ellstrand and Elam 1993; Newman and Pilson 1997; Reed and Frankham 2003). When establishing new populations, we determine a population's evolutionary potential (ability to adapt) by choosing the founding propagules (or individuals) that will be outplanted. By considering genetic diversity of founding propagules, we can mimic diversity seen in natural populations. Conversely, when populations are degraded and suffering from inbreeding or outbreeding depression, we can establish populations with greater diversity than the natural populations to provide greater evolutionary potential. Inbreeding depression, a reduction in the vigor and fitness of offspring, is caused by the breeding of closely related individuals, whereas outbreeding depression similarly results in reduced vigor and fitness but as a result of distantly related individuals interbreeding (see Glossary). Both inbreeding and outbreeding depression are well known as detrimental and counteracting forces acting within populations (Hufford and Mazer 2003; Edmands 2007) that can be especially detrimental in small populations such as those established through reintroduction efforts (Barrett and Kohn 1991; Ellstrand and Elam 1993; but see Bouzat 2010). Without carefully considering propagule source in reintroduction, we may be inadvertently establishing populations that are less fit as a result of inbreeding or outbreeding depression. Therefore, understanding levels and patterns of genetic diversity in both source and reintroduced populations is vital to establishing populations with high levels of fitness and the greatest chance of long-term success.

Methods for Assessing Genetic Diversity

Ideally, population genetic diversity and distribution would be well understood before reintroduction of any species. However, this is not always feasible. Measuring genetic diversity and distribution of the focal species before establishing a new population is ideal, yet we often work with what is available from data on life history traits and genetic data from closely related species. Several means of measuring genetic diversity exist, ranging from protein-level analyses (isozyme variation), to neutral DNA markers (amplified fragment length polymorphism [AFLPs], inter–simple sequence repeats [ISSRs], random amplified polymorphic DNA [RAPDs], microsatellites), to sequencing of coding genes (a fairly new practice), to quantitative trait analyses (common garden and reciprocal transplant studies)

TABLE 5.1

Summary of genetic methods used to analyze population genetic diversity.

Feature	Allozymes	AFLP, ISSR, RAPD	Microsatellite, SNP	Quantitative Traits (Garden Studies)
Source of marker information	Protein	Genomic DNA (anonymous)	Genomic DNA (anonymous)	Morphology, physiology
Dominance	Codominant	Dominant	Codominant	Variable
Transferability between species	High	High	Variable	High
Type of information	Molecular phenotype	Molecular phenotype	Genotype	Phenotype
Ease of development	Moderate	Moderate	Technically difficult	Logistically difficult
Time and expense	Substantial time, moderately expensive	Moderate time, low to moderately expensive	Substantial time, expensive	Substantial time, moderate to expensive

Modified with permission from Falk et al. (2006). AFLP = amplified fragment length polymorphism, ISSR = inter–simple sequence repeats, RAPD = random amplified polymorphic DNA, SNP = single nucleotide polymorphism.

(table 5.1). Different genetic metrics for analyzing diversity within and between populations include measures of genetic diversity (Nei's or Shannon's), measures of differentiation between populations (Wright's *F* statistics, analysis of molecular variance), and Bayesian estimates (see Hartl 2000; Frankham et al. 2002 for general overview of genetic metrics).

The most widely used molecular markers typically measure "neutral" genetic variation, or genetic variation in noncoding regions of the genome. This means that the adaptive significance of the measured variation is not known. Despite not being able to measure directly the adaptive effects of neutral genetic diversity, several studies have shown strong positive correlations between fitness and genetic diversity (Fischer et al. 2003; Reed and Frankham 2003; Dostálek et al. 2010). Positive correlations have also been demonstrated between community diversity and genetic diversity (Reusch et al. 2005; Crutsinger et al. 2006; Johnson et al. 2006). Markers that measure neutral genetic variation have become more accessible to researchers in recent years and offer the benefits of providing data quickly, inexpensively, and with minimal invasive sampling. For example, in plants data can be acquired from a few leaves or flower heads (see Wolfe and Liston 1998; and Falk et al. 2006 for reviews). Although many restoration practitioners may not have direct access to genetic facilities, advances in genetic technologies have made it easier to find collaborators with genetic expertise, such as Genetic Identification Systems in California, the Nevada Genomics Center, and Genewiz (see URLs for these companies at the end of the chapter). Thus, although the adaptive significance of neutral genetic markers is not always known, they can provide broad data sets on fitness potential.

In the absence of genetic data, using information on species life history traits to inform seed collection methods for reintroductions is valuable. For example, collections of an obligate outcrossing species should be made more thoroughly within a population and between populations than collections for a species that is an obligate selfer or reproduces primarily via clonal reproduction, where fewer collections could be made from each population. Several reviews have examined the effects of life history traits including habit (annual versus perennial) and breeding system (self-compatible versus self-incompatible) on levels and patterns of population genetic diversity (Hamrick and Godt 1996; Nybom and Bartish 2000; Nybom 2004). Incorporating life history data (e.g., germination needs) for the target or other species with similar life history traits can prove invaluable (Primack 1996; Broadhurst et al. 2008). Following a set of guidelines for conducting a reintroduction will help ensure that the reintroduced population has the highest possible chance of success (Falk et al. 1996; Guerrant et al. 2004; appendix 1, this volume).

In addition to using life history data to inform the choice of source materials for reintroduction, quantitative studies (e.g., common garden and reciprocal transplant) can be invaluable (Hufford and Mazer 2003; Hereford 2009). Quantitative traits (viability, fecundity) that are measured through common garden or reciprocal transplant studies often provide great insight into locally adapted traits within and between populations (Raabová et al. 2007; Hereford 2009). Local adaptation, where local genotypes are better adapted to their local environment than are foreign genotypes, has been widely demonstrated in plants (McKay et al. 2001; Becker et al. 2006; Hereford 2009). Local adaptation is not always directly correlated with geographic distance, and ecology and climate are also key factors influencing adaptation (Raabová et al. 2007; Hereford 2009). Knowing whether species are locally adapted is essential for guiding reintroductions, because locally adapted species will probably perform poorly in novel environments.

Sampling Guidelines

Over the last two decades, many guidelines for seed collection have been set forth in order to assist practitioners in choosing source locations for seed conservation collections (e.g., no more than 10% of the seed from 10% of the flowering individuals, one seed from each of fifty individuals; Center for Plant Conservation 1991; Lawrence et al. 1995a, 1995b; Guerrant et al. 2004a) and reintroductions (e.g., collect from nearby populations, do not mix sources; Knapp and Rice 1994;

Linhart 1995; Center for Plant Conservation 1996; Jones and Johnson 1998). In addition to these broad recommendations, many authors provide guidelines for restoration collections based on species-specific studies (Millar and Libby 1989; Sanders and McGraw 2005; Sinclair and Hobbs 2009). It is important to note that collection goals will vary between projects, and protocols will vary with differing conservation and reintroduction goals. For instance, collections made for reintroduction in a known location will differ from collections held in long-term ex situ collections intended to serve as source material for future reintroduction projects at unknown locations. In addition, the number of individual propagules collected will vary with end goals. If the end goal is simply to conserve the genetic diversity of a species, then collection protocols will require far fewer seeds than those for establishing a new population that contains genetic representation for the species (Center for Plant Conservation 1991; Brown and Marshall 1995; Lawrence et al. 1995a, 1995b; Falk et al. 2006; Lockwood et al. 2007). However, simply maintaining genetic diversity in a few seeds in a freezer does not equate to conserving the species, so collections should always aim to conserve diversity and numbers sufficient to reestablish populations. The general recommendation for reintroduction projects is to use local seed, collected from a similar habitat, keeping natural levels of gene flow in mind (reviewed in Rogers and Montalvo 2004; Schaal and Leverich 2004; McKay et al. 2005).

Given the broad recommendations typically presented in guidelines, can we assess whether certain propagule collection guidelines or seeding techniques have measurable effects on the genetic diversity of reintroduced populations? How often are particular guidelines followed in practice, and how often are genetic considerations explicitly considered in reintroductions? Answering these questions has proven incredibly difficult because few reintroduction projects have examined genetic diversity directly (see exceptions later in this chapter). Despite the existence of guidelines for maintenance of genetic diversity and integrity for more than 20 years, it was difficult to find empirical investigations into their genetic impacts. Here studies with considerations of genetic diversity in current reintroduction practice are presented.

Examples Involving Nursery or Stock Analyses

Reintroduction efforts, particularly those at large scales, are often conducted with commercially provided, nursery-increased seeds or plants. Few studies have directly examined genetic diversity in stock or nursery propagules to be used for restoration. A growing number of studies suggest that reintroduction success is greater for whole plants than for seeds (Albrecht and Maschinski, Dalrymple et al., this volume), and therefore it is important to consider the effect of nursery cultivation on the genetic diversity of a species (Menges et al. 2004; McKay et al. 2005). Given that fitness in greenhouse or nursery conditions differs from that seen in the field, using individuals reared in a nursery may produce individuals that will grow well but may not be able to withstand natural conditions (Primack 1996; Raabová et al. 2007). In addition, the source of nursery or stock propagules plays a large role in the amount of genetic diversity available for restoration.

Fant and colleagues (2008) examined genetic diversity in natural, reintroduced, and stock propagules of the threatened American beachgrass (*Ammophila breviligulata* Fern. [Poaceae]) in the Great Lakes region of the United States of America. This perennial species initiates dune formation along lakeshores. It is primarily an outcrossing species that is largely self-incompatible but also reproduces clonally through rhizome development. Because of its importance in dune stabilization, it is widely used for restoration throughout the Great Lakes region. Fant and colleagues (2008) measured ISSR diversity within and between well-established populations, newly arisen populations, reintroduced populations, and nursery stock to examine the genetic diversity of the population types. They found that whereas the native populations held high diversity, the two commercial cultivars (nursery stocks) lacked diversity across the three primers used in this study (Fant et al. 2008). In addition, reintroduced populations derived from the commercial cultivars all had the same genotype, and the diversity in the reintroduced populations differed from that found in nearby local populations. The repercussions of low diversity are not yet known, but presumably the populations at the reintroduced sites will be less likely to adapt to future change. Furthermore, the reintroduced populations may interbreed with nearby natural populations, thereby increasing the risk of outbreeding depression in those natural populations. Here the examination of genetic diversity in natural, stock, and reintroduced populations revealed that the reintroductions are not mimicking the natural populations, and supplementing stock with additional genetic variation may be warranted.

In Japan, microsatellite and chloroplast DNA data were used to identify possible source populations for twenty-nine cultivated stocks of the rare Japanese sakurasoh (*Primula sieboldii* E. Morren [Primulaceae]) (Honjo et al. 2008). Samples taken from cultivated stocks maintained in personal gardens, botanic gardens, and agricultural research centers were compared with natural populations. For most stocks, a presumed original population (or location) was known. In some cases, the stocks had been rescued before population destruction. All known local populations (within 30 kilometers of presumed origins) were sampled for comparison. Assignment test methodology was applied to the microsatellite data to determine which local populations were the purported origins of the cultivated stock individuals. Eleven of the stocks were assigned to populations near the source area, indicating that they were closely related to those populations and could be used for restoration in the area. The remaining stocks were assigned either to other local populations or to natural populations that do not exist. This indicated that the assumed origin was incorrect or that the genetic diversity captured in the stock populations differed from those in the sampled natural populations. In addition, diversity was detected in stock populations, which was not detected in natural populations. Stock population genetic diversity could then be used as a future source of diversity if natural populations were deemed inadequate for restoration.

Genetic variation of four species was compared across natural, nursery, and restored prairie habitat at Kankakee Sands in Indiana (Dolan et al. 2008). Under the recommendations of Millar and Libby (1989), seed for restoration was collected from natural remnant populations located within 80 kilometers of reintroduction sites and was sown in an on-site nursery. To maximize the genetic diversity captured, seeds were collected from many plants of varying size across the remnants, with some collections occurring over multiple years. Reintroduced populations were seeded with both seed obtained from nursery plants and seed collected directly from remnant populations. Four species (*Asclepias incarnata* L. [Asclepiadaceae], *Baptisia leucantha* Torr. & Gray [Fabaceae], *Coreopsis tripteris* L. [Asteraceae], and *Zizia aurea* L. [Apiaceae]) were chosen for allozyme analysis to compare diversity levels across remnant, nursery, and restored populations. Despite decreased allelic diversity for three of the species from remnant to nursery to restored populations, reintroduced sites contained 88.9% of the alleles detected in the remnant populations. The majority of the variation lost was that of alleles present at a frequency of 1% or less, whereas approximately 90% of the alleles present in remnants at a frequency greater than 1% were retained in the nursery and restored populations. Overall, the restoration techniques applied at Kankakee Sands capture the majority of variation present in the remnant populations and could be applied in other prairie restorations.

Each of these studies examined genetic diversity in a nursery or stock population as well as source populations. In studies where multiple sources fed into the nursery (Dolan et al. 2008) or where multiple cultivated populations existed (Honjo et al. 2008), genetic diversity was maintained and would provide sufficient diversity for future restorations. In the case where a commercial stock was sampled and used for restoration, no genetic diversity was detected, indicating a discrepancy between the natural and stock populations, which could have led to a loss of fitness in the restored populations or susceptibility to environmental change over time.

Pre-Reintroduction Analyses

Most conservation genetic studies could be considered evaluations of genetic diversity before reintroduction. Results from these studies should be applied to reintroduction programs, but the frequency with which these results are actually used to inform reintroduction is unknown. However, in several published examples genetic data directly informed reintroduction of a rare species.

Many reintroduction projects of rare species result from field observations and studies. The eastern four-nerved daisy (*Tetraneuris herbacea* Greene [Asteraceae]; *Hymenoxys acaulis* var. *glabra*) is a federally listed species limited to the Great Lakes region of the United States and Canada. Ten years of pollinator observations coupled with 6 years of hand pollinations did not produce any viable seed in members of one remnant Illinois population. The lack of viable seed led to an investigation of the breeding system with the aim of establishing recovery criteria for the species (DeMauro 1993). Common garden studies revealed that the species has a sporophytic self-incompatibility system and that reproduction was inhibited by a lack of compatible mating types within the remnant individuals. In light of this genetic knowledge, recovery goals for the species were established. To increase outcrossing potential for individuals, researchers conducted reintroductions with individuals from multiple populations to increase the likelihood of capturing different mating types (DeMauro 1994). Additionally, the provenance of transplanted individuals was identified, and transplant locations were chosen carefully to maximize the potential for outcrossing. Reproductive success was evident in all reintroduced populations in the years immediately after restoration. Here, genetic data directly guided the reintroduction and recovery of a rare species.

In the Australian *Zieria prostrata* Armstrong (Rutaceae), genetic data informed a planned reintroduction (Hogbin and Peakall 1999). The endangered plant has only four known populations along the eastern coast of Australia. Individuals from the four populations and outplantings of a fifth "rescued" population were sampled for levels and patterns of genetic diversity to inform collections for additional plantings at the fifth population. RAPD data indicated a large divergence between populations and disclosed that the individuals of the fifth population were probably collected from one of the other known populations. When interviews and further investigation failed to confirm the occurrence of this fifth population, the researchers negated the existence of the fifth population. Because of the results of the genetic study, additional planned reintroductions at the fifth population were abandoned and conservation efforts focused on the four confirmed populations.

RAPD data guided efforts to restore the federally endangered beach cluster-vine (*Jacquemontia reclinata* House [Convolvulaceae]) in Florida (Thornton et al. 2008). Samples collected from eight of the ten known populations across the species range represented approximately 20% of all known individuals. A genetic analysis indicated that one population, located inland from the coast and found in a different habitat than the remaining populations, was highly divergent from the other seven populations sampled. The analysis suggested that two genetically distinct groups are represented within the species and that they should remain distinct in future reintroductions. Introductions carried out after the RAPD study maintained the two groups, and initial analyses indicate that the introduced individuals are persisting and reproducing.

Ideally, land managers and conservation practitioners will have genetic data (molecular or quantitative) before conducting a reintroduction. The aforementioned examples provide case studies where genetic data greatly informed, altered, or directed management actions based on the results.

Post-Reintroduction Analyses

In recent years, several published studies have examined genetic diversity after a reintroduction has occurred. These studies specifically examined levels and patterns in reintroduced populations as compared with natural or source populations. Knowledge of the source populations of reintroduction propagules allows examination of the seeding and planting protocols over time.

Population genetic diversity levels of reintroduced and natural populations of eelgrass (*Zostera marina* L. [Zosteraceae]) were examined through allozyme analysis (Williams and Davis 1996; Williams and Orth 1998). The two studies found disparate results where restored (transplanted) seagrass beds in California showed lower diversity than natural beds. Alternatively, comparable levels of genetic diversity were found in restored and natural bed types in the Chesapeake Bay region. Three hypotheses may explain the differences in diversity levels: (1) The transplantation process affected diversity levels, (2) initial diversity levels were lower in the source areas providing transplants in California than those in the Chesapeake Bay region, and (3) inputs from sexual reproduction may be greater in the Chesapeake Bay transplants (Williams and Davis 1996; Williams and Orth 1998). Transplants in the Chesapeake Bay region were from multiple sources over multiple years as well as additional seed inputs (Williams and Orth 1998). In contrast, transplants in California were from the closest natural population, which was often disturbed by dredging and thus may have exhibited lower genetic diversity (Williams and Davis 1996). Thus, in the reintroduction of eelgrass populations, frequent transplants from multiple locations and sources (vegetative and seed inputs) resulted in restored populations with higher levels of diversity than when transplants were taken from disturbed natural populations.

Seeding protocols used to establish several reintroduced populations of the vernal pool annual Contra Costa goldfields (*Lasthenia conjugens* Greene [Asteraceae]) at Travis Air Force Base, California successfully captured the genetic diversity of the source populations (Ramp et al. 2006). Ten seeds collected from each of ten source pools were placed into each reintroduced pool for a total of 100 seeds. Genetic analyses using ISSR markers were conducted in three consecutive years following reintroduction to determine whether and how genetic diversity changed over time. In each year sampled, the majority of the genetic diversity was distributed within pools (89%), with no significant differences between population types detected over time. For this annual, self-incompatible species, the seeding protocol successfully captured and distributed the genetic diversity present at the site and has been recommended for additional reintroductions of the species.

Post-restoration analysis of reintroduced and natural populations of the Chinese *Cyclobalanopsis myrsinaefolia* (Blume.) Oersted (Fagaceae) indicated low levels of diversity in the reintroduced populations of this long-lived tree (Liu et al. 2008). Seeds for reintroduction were collected within a 10-square-kilometer area, germinated in a greenhouse, and outplanted into two areas. Genetic samples were compared from the donor population, the two reintroduction sites, and two additional sites using microsatellite analysis. The reintroduced populations showed lower levels of genetic diversity on three of the four parameters measured. In addition, the seed source for the restoration efforts had a lower level of diversity than other natural populations, so it was not the best choice as the source for restoration. Although seeds were collected haphazardly across a large area, they did not successfully capture the diversity present in the donor population. Future seed collection efforts should take these results into consideration, and collections should be made from populations with higher levels of diversity or from multiple populations.

Smulders and colleagues (2000) examined post-reintroduction levels of genetic diversity in two prairie species (*Cirsium dissectum* L. [Asteraceae] and *Succisa pratensis* [Dipsacaceae]). They found that although reintroduced populations had less diversity than their remnant source population, the diversity levels of the two populations as a whole were not significantly different. Gustafson and colleagues (2002) examined diversity levels in remnant and reintroduced populations of the purple prairie clover (*Dalea purpurea* Vent. [Fabaceae]). Their results indicated that reintroduced sites established from seed within 80 kilometers maintained diversity and displayed greater allozyme diversity than individual remnant populations, possibly because of multiple sources for the reintroductions. Gustafson and colleagues (2002) also provided empirical evidence that using seeds from multiple sources reduced the chance that founder effects will decrease diversity in restored populations.

Whereas several studies have detected lower genetic diversity in reintroduced populations (Friar et al. 2000; Li et al. 2005), additional studies have found comparable levels of genetic diversity in restored and natural populations (McGlaughlin et al. 2002; Gustafson et al. 2004b; Travis and Sheridan 2006). Results submitted to the *CPC International Reintroduction Registry* (Center for Plant Conservation 2009) that considered genetic source of reintroductions showed few effects of source on reintroduction success (although no submitted study explicitly measured genetic diversity [E. O. Guerrant, personal communication]).

Evaluation of Seeding Techniques

The post-reintroduction studies described earlier demonstrate that using multiple sources for reintroduction results in higher levels of diversity in reintroduced populations than when reintroduction occurs from a single source (see table 5.2 for a summary of studies). Another finding from these studies is that the greater the number of seeds or plants used to initiate a population, the greater the chance that unfit individuals will be outcompeted by more fit individuals in the population, thus reducing the risk that outbreeding depression will have a negative impact on the population (Schaal and Leverich 2004). Despite the results illustrated here, whether or not local seeds are best to use for reintroductions has yet to be resolved because researchers have generally followed the recommended guidelines of using local seed from a single source. An exception is Kephart (2004), who demonstrated that reintroduced populations originating from outbred progeny showed significantly greater survival and reproduction than inbred progeny. This body of literature argues for the inclusion of propagules from many locations or those that are definitively outbred as compared with inbred individuals.

Selecting the appropriate propagule type is a critical step in all reintroduction projects. Evidence has shown that reintroductions conducted with individuals rather than seeds are more successful, at least in the short term (Albrecht and Maschinski, Dalrymple et al., this volume). However, individuals produced in the greenhouse and seeds derived from "seed increasing trials" in the nursery experience different selection pressures than they would face in a natural setting (Jones and Johnson 1998; Menges et al. 2004; McKay et al. 2005). Reintroductions conducted with individuals rather than seeds should be done with careful attention to diversifying source location so as to start with a genetically diverse population that may be more adaptable to changing environmental conditions (Gustafson et al. 2004a). Reintroductions conducted with high numbers of propagules and those conducted over multiple years have been more likely to result in success (van Andel 1998; Kirchner et al. 2006; Collinge and Ray 2009).

Practical Implications for Traditional Reintroduction Work

Evaluation of the few published studies examining genetic diversity in reintroduced populations as compared with natural populations provided contrasting

TABLE 5.2

Summary of reintroduction studies measuring genetic diversity in reintroduced populations.

Objectives	Reintroduction Protocol	Markers or Analysis	Years after Reintroduction	Results	Management Implications	Reference
Compare genetic diversity and structure of transplanted and "natural" seagrass beds in California and Mexico.	Transplant from nearest neighboring population, up to a few km away.	Allozymes	Sampled 3–16 yr after transplant	Most genetic diversity is within populations, low diversity in transplanted beds.	Source material may be low in diversity and is vegetative only.	Williams and Davis (1996)
Compare genetic diversity and differentiation of transplanted and "natural" seagrass beds in the Northeast.	Transplant and seed additions over multiple years. Seed from local populations.	Allozymes	Sampled 1–12 yr after transplant or seeding	Comparable levels of genetic diversity in both population types.	Reintroduction with both vegetative material and seed is useful.	Williams and Orth (1998)
Assess genetic consequences of a population bottleneck and subsequent reintroduction on Mauna Kea silversword population.	Seed collected from 2 individuals has generated >1,500 plants in 30+ yr.	Microsatellites	Initial outplanting in 1973, intermittent within a 20-yr period	Outplanted population has significantly less variation than the natural population. Outplanted is not genetically depauperate compared with natural.	Populations suffering a genetic bottleneck may not be genetically depauperate; reintroduction plans should carefully consider the genetic impacts of reintroduction.	Friar et al. (2000)

Objective	Description	Marker	Time	Results	Recommendation	Reference
Examine differences between source and reintroduced populations of two meadow species.	Seed from 200–500 individuals collected across 4 or 5 populations. Reintroductions consisted of 2,000 seeds outplanted into 16 plots (source tracked).	AFLP	1 yr	Reintroduced populations have less diversity than sources and more closely resemble each other than their source.	Maintenance of variation in reintroductions is essential for maintenance of adaptive potential. Introductions from >1 source may provide for greater diversity maintenance in reintroduced populations.	Smulders et al. (2000)
Compare genetic relationships between remnant and reintroduced populations of a grassland species.	Reintroduced populations established with seed from remnant populations within 80 km. Multiple sources used for seeding.	Allozyme and RAPD	Approximately 15–38 yr	Comparable diversity between remnant and reintroduced prairies. Diversity measures are not related to prairie size.	Local seed from multiple sources established new populations with adequate diversity.	Gustafson et al. (2002)
Maintain initial levels of genetic diversity in introduced populations.	Tissue culture was used to create clones from 10 individuals, which encompassed 87% of measured diversity. Additional individuals introduced from seed.	AFLP	At least 4 yr after collection of ramets for cloning	Detected a 22% increase in inbreeding from parental to F1 generation. Four clones generated more than 85% of the offspring, resulting in an effective population size of 2.	Continued monitoring is needed to prevent further loss of genetic variation. New translocations should aim to increase diversity.	Krauss et al. (2002)

TABLE 5.2

Continued

Objectives	Reintroduction Protocol	Markers or Analysis	Years after Reintroduction	Results	Management Implications	Reference
Assess genetic diversity in source and reintroduced populations. Investigate relationship between population size and genetic variability in reintroduced populations.	Single source sampled to establish four reintroduced populations. Seeds collected from 20 or more individuals pooled. Initial seeding with 5,000 seeds resulted in few individuals, subsequent seeding with 50,000 seeds led to established populations.	ISSR	1–3 yr	Largest reintroduced populations most closely resemble source. Positive correlation between population size and genetic diversity. Suggest 90% of source variation can be maintained in reintroduced populations of approximately 1,000 individuals.	Advise seeding with large numbers of seeds and across multiple generations.	McGlaughlin et al. (2002)
Assess genetic diversity and genetic relationships between remnant and restored prairies and cultivars.	Remnants established with seed from multiple sources within 20–150 km of the site.	RAPD	1 yr	Reintroduced populations as genetically diverse as remnants. Some geographic structure to diversity. Cultivars are not less diverse than remnants but are genetically distinct.	Multiple sources result in diverse reintroduced populations, but local sources are best because of the geographic structure of remnant diversity.	Gustafson et al. (2004a)

Objective	Methods	Marker	Time	Results	Recommendations	Reference
Compare genetic diversity and differentiation of natural and reintroduced populations.	No detailed information available. Trees outplanted by various groups to make up artificial populations, probably from multiple sources.	RAPD	Data not available	Artificial populations more closely resemble each other than wild. Artificial populations have less diversity than wild.	Additional artificial populations should be established to restore wild populations. Multiple reserves should be established from different sources and should be maintained independently.	Li et al. (2005)
Determine whether seeding protocol used for reintroductions captured genetic diversity present in natural populations.	Collected seed from on-site natural populations. Introduced 10 seeds from each of 10 plants from different pools to each reintroduced site.	ISSR	1–3 yr	Genetic diversity sampled in natural populations captured and distributed within reintroduced populations.	Seeding technique was successful in creating populations that genetically mimic the natural populations for this species.	Ramp et al. (2006)
Assess genetic diversity in natural and restored populations (restored range in age from 2 to 7 yr).	Restored sites were established with material from 5 or 3 donor sites.	AFLP	2–7 yr after restoration (1 site planted more than 1 time)	Restored populations contained similar diversity to donor populations.	Recommend multiple donors for each restored site.	Travis and Sheridan (2006)
Examine diversity levels in natural and restored populations of a dominant tree species.	Seeds collected from a single population grown to seedlings in a greenhouse and outplanted into two locations.	Microsatellites	At least 4 yr	Restored populations show less diversity than natural populations. Source population had lowest measured diversity.	Suggest investigating genetic diversity of donor populations before restorations.	Liu et al. (2008)

AFLP = amplified fragment length polymorphism, ISSR = inter–simple sequence repeats, RAPD = random amplified polymorphic DNA.

results, probably because of the low number of published studies. Given the variety of techniques used to collect propagules and conduct outplantings in these reintroductions, it is difficult to determine whether one practice is better than another, which precludes any definitive recommendations for future reintroduction efforts. A combination of measured genetic data from existing populations, information on life history traits, and ecology of the species should be considered in any reintroduction program.

The studies reviewed here also indicate that seeding from multiple sources and over multiple years should accomplish the capture of genetic diversity present within the species and lead to a high chance of success for the reintroduced population. This suggestion is supported by a recent review of plant reintroductions by Godefroid and colleagues (2011). This evaluation is in contrast to the generally presented guidelines of local seed collection without population mixing, furthering the argument for more study into post-reintroduction success per Wilkinson (2001) and Broadhurst and colleagues (2008). No matter which propagule collection (local versus broad) and outplanting techniques (seed versus nursery-grown plants) are applied, detailed documentation of every step of the process should be kept and all reintroductions should be established as scientific experiments. Additionally, more studies are needed that examine fitness directly and follow potential phenotypic shifts in the newly established populations. Through in-depth examinations, we will gain greater knowledge of how to establish populations with the greatest chance of success.

Areas of Need and Research Opportunities

The phenomenon of outbreeding depression in species with restricted ranges lacks empirical research that is sorely needed to identify risks. Because outbreeding depression has been documented primarily at a large scale (generally more than 200 kilometers) (see Waser and Price 1994 for an exception; Pelabon et al. 2005; Becker et al. 2006), this phenomenon may not be a serious threat to rare species reintroductions. Although local adaptation can occur across small spatial scales (Linhart and Grant 1996), it seems that the risk of inbreeding depression may be more serious for some of these rare species than the risk of outbreeding depression. Studies such as Kephart's (2004) and Pinto-Torres and Koptur's (2009) that show outbred individuals to be more fit than inbred individuals, and the suggestion that most sexually reproducing plants should be able to tolerate a small reduction in fitness, indicate that inbreeding and outbreeding depression may have minimal impact on reintroduced populations of rare species (Schaal and Leverich 2005). However, more empirical research on these phenomena in rare species is needed. Explicit testing of seed collection guidelines also needs to be conducted to test whether using local seed is always the best default protocol or whether collecting seed from diverse habitats is more likely to lead to successful reintroduction. A review of published literature indicates that collections from more disparate populations may not adversely affect genetic diversity in reintroduced populations. One thing to consider is that our rarest plants generally have highly reduced ranges, a few square miles or less (Kawelo et al., this volume), so collections across the species range, specifically along the periphery or unique habitats, might prove vital for capturing the genetic diversity in the species. In order to examine these issues rigorously, all reintroductions must be established as experiments with testable hypotheses (McKay et al. 2005; Bottin et al. 2007). Improved collaborations between conservation practitioners and researchers will help ensure that reintroductions are set up in a way that will immediately address some of these questions. Lastly, we need to publish reintroduction successes and failures in the scientific literature. Even though many reintroductions are not designed as experiments, the observations resulting from these projects can still be highly valuable learning devices. Publishing these projects will greatly increase our general knowledge about rare species reintroductions. If nothing else, increasing communication between practitioners and researchers and gaining a better understanding of what occurs in practice versus in theory will advance the conservation of many rare species (Young et al. 2005; Smith et al. 2007).

Prospects and Cautions for Appropriate Use of Managed Relocation

As the global climate changes, organisms will be forced to adapt or shift ranges or face extinction as selective pressures increase (Rice and Emery 2003; Jump and Peñuelas 2005; Kramer and Havens 2009). As discussed earlier, genetic variation is a critical component of an organism's ability to adapt to changing environments. If the climate changes faster than most species are able to adapt (Etterson and Shaw 2001) or reasonably disperse (Primack and Miao 1992), then land managers will be faced with fewer options to conserve the native species found on their lands. One tool gaining attention in recent years is that of managed relocation (Haskins and Keel, this volume). Managed relocation involves moving individuals to facilitate or mimic a natural range expansion for a species (Hoegh-Guldberg et al. 2008). Genetic issues to be considered in managed relocation are generally the same as those considered in reintroducing a population. Practitioners will want to establish populations with high levels of genetic diversity to maximize their adaptive potential. High levels of diversity will also defend against potential genetic pitfalls of small populations such as founder effects and inbreeding depression. Reusch and colleagues (2005) found that communities with higher levels of genetic diversity showed greater recovery from climatic extremes (temperature) than communities with less diversity. Depending on distance from the nearest natural population, if interbreeding with natural populations seems likely, the possibility of inbreeding and outbreeding depression should be considered. However, although local seeds are most commonly recommended for reintroductions, seeds from diverse habitats and the periphery of the species' range may provide genetic variation, which affords new populations greater adaptive potential (Lesica and Allendorf 1999; Broadhurst et al. 2008). As in conducting reintroductions within a species' range, consideration of life history traits (Kramer and Havens 2009) and microevolutionary processes including genetic drift and selection (Rice and Emery 2003) should be considered along with habitat suitability in any managed relocation.

Summary

For rare plants, genetic diversity is essential for reintroduction success and offers the greatest chance of long-term survival. When considered with other ecological and biological features such as breeding system, habitat, and symbiotic relationships (pollinators, seed dispersers, and microbial mutualists), genetic data can add an additional layer of knowledge to a practitioner's toolbox. Practitioners should aim to obtain genetic data ahead of reintroduction activity, if warranted (appendix 1). Studies on single-population propagule collections are more numerous than studies using multiple propagule collection sites. Remaining unbiased is important because our rarest species may benefit from more diverse propagule collection. Collecting from multiple local sources and seeding over multiple years may lead to greater success than when single sources and single years are used for reintroductions. When local adaptation is assumed or anticipated, reciprocal transplants or common garden studies can inform seed collection strategies for reintroductions. Considering reintroduction as a study of evolutionary processes beyond simple levels and patterns of genetic variation will also continue to broaden the lessons learned from ongoing work (Latta 2008). Above all, reintroductions should be conducted as experiments with close collaboration between researchers and practitioners when possible (Schaal and Leverich 2005; Young et al. 2005; Bottin et al. 2007).

Genetic Facilities

Genetic Identification Systems: http://www.genetic-id-services.com/
The Nevada Genomics Center: http://www.ag.unr.edu/genomics/
Genewiz: http://www.genewiz.com/

Transitioning Plants to New Environments: Beneficial Applications of Soil Microbes

KRISTIN E. HASKINS AND VALERIE PENCE

An important goal of ex situ conservation and conservation science is to reintroduce rare plants to nature. A crucial step, which often begins in the laboratory or greenhouse, is the preparation of propagules. Whether you are using in vitro (tissue culture) or seed germination techniques, a major hurdle in propagule production is overcoming issues with acclimatization. Generally speaking, acclimatization is the process of adapting propagules to new environments. The use of in vitro techniques is gaining popularity for some of the rarest plant species (Fay 1992), but the development of cuticles, stomata, and photosynthetic functions are maladapted to the soil environment when grown in this manner. Acclimatization aims to facilitate those transitions, but the process has been plagued with high mortality rates (Kapoor et al. 2008). These high rates of failure occur whether plant material is moved from culture containers to soil (Pospíšilová et al. 1999) or from greenhouse to field (Estrada-Luna et al. 2000).

The primary goal of this chapter is to elucidate the challenges associated with acclimatizing tissue-cultured and greenhouse-grown propagules. We review the current knowledge of soil microbe–root interactions for conservation purposes (table 6.1) and emphasize the use of microbes in acclimatization. We highlight a case study of a rare lupine (Fabaceae) that examines a new tissue culture technique. Finally, we discuss different applications of these methods, suggest critical areas for future research, and provide prospects and cautions for these methods in managed relocation.

Aboveground plant tissues experience changes during in vitro and ex situ propagation that create challenges for acclimatization, and although these will be addressed, the bulk of this chapter focuses on belowground plant tissues. Root systems anchor plants to the soil, provide a means for water and nutrient absorption, and

TABLE 6.1

Studies in which beneficial soil microbes were applied to rare plant species during conservation practices or research.

Species	Family	Microbes	Application	Response	Reference
Astragalus applegatei	Fabaceae	AMF (N)	Propagation	+	Barroetavena et al. (1998)
Plantago atrata, Pulsatilla slavica, Senecio umbrosus	Plantaginaceae, Ranunculaceae, Asteraceae	AMF, PGPR (B)	Propagation	+	Zubek et al. (2009)
Platanthera leucophaea	Orchidaceae	OMF (N)	Propagation	+	Zettler et al. (2001)
Curculigo orchiodes	Hypoxidaceae	AMF (N)	Ex vitro transfer	+	Sharma et al. (2008)
Platanthera praeclara	Orchidaceae	OMF (N)	Propagation	+	Sharma et al. (2003)
Mitragyna parvifolia, Withania coagulans, Leptadenia reticulata	Rubiaceae, Solanaceae, Asclepiadaceae	AMF (N)	Outplanted	0 Survey	Panwar and Tarafdar (2006)
Pulsatilla patens	Ranunculaceae	AMF (N)	Propagation	+	Moora et al. (2004)
Jacquemontia reclinata	Convolvulaceae	AMF (N)	Propagation, outplanted	+	Maschinski et al. (2003)
Astragalus tennesseensis	Fabaceae	Nitrogen-fixing bacteria (E)	Propagation	−	Bowles et al. (1988)
Tephrosia angustissima var. corallicola	Fabaceae	AMF (N)	Propagation, outplanted	0	Fisher and Jayachandran (2006), data unpublished
Jacquemontia reclinata, Amorpha crenulata	Convolvulaceae; Fabaceae	AMF (N)	Propagation	+	Fisher and Jayachandran (2002)
Viola elatior, V. pumila, V. stagnina	Violaceae	PGPR	Propagation	−	Eckstein and Otte (2005)
Quercus euboica	Fagaceae	Whole soil (N)	Propagation	+	Kartsonas and Papafotiou (2007)
Moringa concanensis	Moringaceae	AMF (N)	Propagation	+	Panwar and Vyas (2002)
Diuris micrantha, Thelymitra manginiorum	Orchidaceae	OMF (N)	Propagation, outplanted	+	Batty et al. (2006b)
Astragalus cremnophylax var. cremnophylax	Fabaceae	Nitrogen-fixing bacteria (E) Whole soil (N)	Propagation	+	Haskins and Murray, unpublished data

AMF = arbuscular mycorrhizal fungi; PGPR = plant growth promoting rhizobacteria; OMF = orchid mycorrhizal fungi; B = both native and exotic; N = native; E = exotic. Responses of rare plants to soil microbes were either positive (+) or negative (−) in regard to growth. A zero (0) indicates that no effect was detected or that the study was survey only.

host many important soil microbial mutualists; thus, it is clear that roots are important. The paucity of attention in the literature regarding the care and treatment of rare plant roots before reintroduction was revealing. The exception to this finding is the work that has been done in the family Orchidaceae; here you find a long history of integrating soil microbes with propagating techniques (Dixon et al. 2003 and references therein). New and improved methods focusing on minimizing acclimatization-induced mortality are needed to ensure reintroduction success.

The Challenge of Acclimatization

Plant propagules prepared ex situ for restoration projects are, by definition, not fully adapted to their natural habitat. Whether they are produced through the germination of seeds, by cuttings, or through in vitro propagation, the conditions in which they are produced are generally less stressful than those in the wild. Such plants need a period of acclimatization, during which they can adapt to ambient conditions (appendix 1, #21). This period of transition allows the plant to develop natural physiological and morphological traits for dealing with environmental stress, traits that often do not develop fully under the conditions used for propagation.

Plants produced in vitro generally exhibit the most extreme deviation from natural conditions and pose the greatest challenge to the acclimatization process. Tissue culture propagation takes place in containers that create a highly humid environment, reducing the need for mechanisms and structures in the plant tissues that aid in retaining water. Tissue culture media generally contain sugar, which reduces the need for photosynthesis. Thus, when plants are removed from culture, they often cannot survive ambient conditions without treatments, before or after transfer, that can stimulate the development of normal water-retaining tissues and photosynthetic function.

Survival through acclimatization is correlated with normalizing several morphological and physiological traits related to transpiration and photosynthesis. In vitro–grown plants often exhibit abnormal stomatal frequency, morphology, and function, with excess, rounded, and nonfunctional stomata (Preece and West 2006; Khan et al. 2009) or cuticles with reduced thickness or abnormal composition (Brutti et al. 2002; Louro et al. 2003). In addition, leaf parenchyma may not form normally (Apostolo et al. 2005), and normal photosynthetic activity may not become established in vitro (Fila et al. 2006). These factors may be significant not only between species but also between genotypes of a given species (Aracama et al. 2008).

Treatments that have contributed to successful acclimatization center on decreasing humidity, increasing CO_2, or decreasing sugar levels in cultures and

Contributed by: Cheryl Peterson and Pilar Maul
Species Name: *Lupinus aridorum*
Common Name(s): Scrub lupine, McFarlin's lupine
Family: Fabaceae
Legal Status: US, endangered; Florida, endangered
Location: Scrub habitat on the Winter Haven Ridge in Polk County and
the Mt. Dora Ridge in Orange County, central Florida

Scrub lupine is a short-lived perennial with pink pea flowers and silvery simple leaves lacking stipules. The well-drained sandy soils of central Florida's ridges support endemic scrub species such as scrub lupine and are ideal for orange groves and development, so very little habitat remains. In 1990, forty populations were documented; these declined to nineteen in 1998 and only eight by 2007. Most of the remaining sites have few plants, are on private property, and are highly threatened by development. Germplasm collection is vital for preserving this species, yet transplantation has been mostly unsuccessful and vegetative propagation not possible. Because few seeds are produced in the wild, efforts have turned to tissue culture micropropagation for germplasm preservation.

Successful micropropagation protocols have been developed by the Cincinnati Zoo and Botanical Garden's Center for Conservation and Research of Endangered Wildlife. However, transitioning the plants into potted conditions has met with limited success. This "hardening off" step is typically stressful for plants as they adjust to the presence of pathogens and lower humidity, obtain nutrients from different types of substrate, and develop a protective waxy leaf cuticle.

Mortality during acclimatization has been much higher for scrub lupine than for other scrub species acclimatized by Bok Tower Gardens, with few plants surviving past 6 months, possibly because of its sensitivity to any form of root disturbance. Transferring plants into pots requires that all traces of the agar medium be removed from the roots in order to prevent microorganism infection, and some breakage or disturbance of the root system is unavoidable. To eliminate the need for mechanical agar removal from the roots and to minimize root disturbance, scientists at St. Thomas University have developed a liquid culture system for this species.

Methods and Findings

In a preliminary study, we germinated sterilized seeds in half-strength Murashige and Skoog (MS) (Murashige and Skoog 1962) medium on either

Box 6.1. Continued

0.8% agar or on bridges made from Whatman® No. 1 filter paper suspended over liquid media (fig. 6.1). The two culture systems produced plants that were similar in leaf number, growth rate, and height. However, root morphology was distinctly different: multibranched roots formed over liquid media, whereas agar cultures developed one main root with a small number of thin lateral roots.

FIGURE 6.1. In vitro seedlings on filter paper over liquid media (left) and on solid agar (right).

Using 103 plants grown in liquid media composed of either half-strength MS salts or half-strength MS salts supplemented with +0.5 mg/L benzyl-aminopurine (BAP), we tested the effect of this plant growth regulator on plant architecture. After 3 months, we transferred plants into peat pots containing a 1:1 mixture of native, nonsterilized, dried and sifted sand (collected from the site of a wild scrub lupine population) and either Fafard™ professional nursery soil, Miracle-Gro® Seed Starting potting mix, or coco-fiber. We placed plants into trays covered with plastic protective domes and maintained them in one of three conditions: a "winter" growth chamber (21°C day, 10°C night, 13 hours light), a "summer" growth chamber (27°C day, 18°C night, 16 hours light), or a greenhouse under 50% shade. We watered them with filtered rainwater, gradually reducing soil moisture levels and removing domes after 6 months.

In all treatments, 44% of plants survived after 6 months and 20.4% after 12 months (the highest we have obtained to date for this species). Plants had

Box 6.1. Continued

highest survival (fig. 6.2) in media without BAP, when grown using seed starter mix in "winter" conditions. Surviving plants appeared healthy and had grown noticeably (fig. 6.3). Low soil moisture and black spot disease caused mortality. The results of this trial suggest that liquid cultures can be successful for in vitro growth and subsequent acclimatization of scrub lupine.

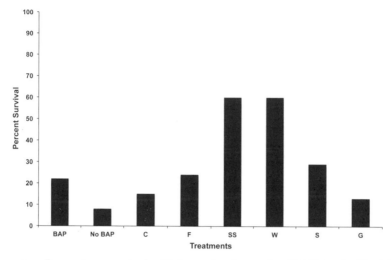

Figure 6.2. Percentage survival with benzylaminopurine (BAP) in the liquid media and no-BAP controls; cocofiber (C), commercial soil mix (F), or seed starter mix (SS); and "winter" conditions (W), "summer" conditions (S), or greenhouse conditions (G).

Figure 6.3. Plant grown without BAP and acclimatized using seed starter mix in "summer" conditions (a) after removal from test tube and (b) 12 months after transfer.

increasing stress tolerance of tissues. Humidity levels in tissue culture containers are generally close to 100%. Thus, before plants are transferred to soil, increasing stress tolerance can be accomplished through air exchange or treatments with growth regulators or related substances (Lee and Kim 2009; Mills et al. 2009; Pospíšilová et al. 2009). Alternatively, plants can be transferred to soil and maintained at high humidity initially, before gradually decreasing relative humidity (Gilly et al. 1997). Increasing CO_2 levels has also been effective for increasing survival (Siano et al. 2007), and often this is done in combination with decreasing or eliminating sucrose in the medium (Shim et al. 2003) or with increasing light intensity (Batagin et al. 2009). Photoautotrophic cultures have been shown to have higher acclimatization survival than heterotrophic or even mixotrophic cultures (Siano et al. 2007). Whereas in vitro–propagated plants may need more acclimatization, similar approaches have been taken with plants propagated or grafted ex vitro (Fordham et al. 2001; Nobuoka et al. 2005).

Failure of micropropagated plants to survive is often attributed to poor root development (Wang et al. 1993), which deprives the developing plant of strong structural support and nutrient acquisition capability. Because beneficial soil microbes have been shown to promote strong root systems and functioning (Smith and Read 1997), it is logical that these organisms be used to improve acclimatization methods for tissue-cultured plants. Mycorrhizal fungi have been established as a necessary component for in vitro culturing of orchids since the early 1970s (Warcup 1971, 1973). Similarly, Padilla and colleagues (2006) showed that inoculating micropropagated lucumo plants with either mycorrhizal fungal inoculum or with a soil bacterial wash significantly increased plant growth 4 months after ex vitro transfer to soil. Mycorrhizal fungi and soilborne bacteria are two large and diverse taxonomic groups, members of which can provide numerous benefits to rare plant propagation techniques.

Mycorrhizal Fungi

The beneficial attributes of mycorrhizal fungi to plant growth and survivorship have been known for well over 100 years (Frank 1894), and their use in field applications has been ongoing since at least the early twentieth century (Kessell 1927; Anonymous 1931). Currently, research investigating the use of important soil microbes to improve plant growth, plant production, and outplanting success is being done in the fields of agriculture (crops and forest products) and horticulture. Conservation practitioners are increasingly using soil microbes to propagate and restore rare species (table 6.1) but still lag behind the progress made by horticulturists (see Kapoor et al. 2008 for a reference list).

Of the seven recognized types of mycorrhizal fungi (Smith and Read 1997), three will be considered here: arbuscular mycorrhizal fungi (AMF), ectomycorrhizal fungi (EMF), and orchid mycorrhizal fungi (OMF). Generally, in all mycorrhizal relationships the host plant provides the fungus with a food source, and the fungal partner provides the host plant with increased nutrient and water uptake. Recent studies have revealed even more benefits to hosts of fungal partners, including protection from shoot and root herbivores (Rabin and Pacovsky 1985; Jones and Last 1991; Gange and West 1993), resistance to root pathogens (Kapoor et al. 2008), drought resistance (Allen and Allen 1986), improved photosynthetic efficiency (Augé 2001), and increased survivorship (Allen 1991).

Mycorrhizal relationships extend far beyond flowering plant taxa, with which most people are familiar, to include gymnosperms (e.g., conifers and cycads), pteridophytes (e.g., spore-bearing ferns and horsetails), and some gametophytes (e.g., mosses) (Smith and Read 1997). Arbuscular mycorrhizal relationships are formed by a number of plant species, including most of our crop species and many rare species. AMF associate primarily with herbaceous species, but associations with woody plants have been described (Smith and Read 1997). Thus, most of our crop species and horticultural varieties use AMF. A vast body of research has firmly established that AMF can benefit plant growth (table 6.1).

Woody plant species typically form associations with EMF, with a few exceptions that dually colonize with AMF and EMF (e.g., members of the Fagaceae and Salicaceae). Research in the use of EMF to improve plant survivorship and production is also substantial and has been pioneered by the forestry industry. There are more than five thousand recognized species of EMF (Allen 1991), but our collective knowledge of their distribution, life histories, common plant associates, and function as mutualists in terms of plant establishment, survival, and growth (Cairney 1999) is still very limited.

Some of the rarest plant species in the world are orchids (Dixon et al. 2003). Orchid species form very unique relationships with mycorrhizal fungi of the form-genus *Rhizoctonia*. Orchid seeds are dustlike, enabling them to travel far but not to survive for long without assistance from fungal symbionts. Many orchid species are highly dependent on OMF at some point in their life, often starting at the seed stage (Rasmussen 1995). Practitioners have been using this relationship to improve seed germination and propagation in the laboratory and greenhouse with great success. Symbiotically generated seedlings are more vigorous and exhibit higher survival rates when transferred from in vitro conditions to soil (Ramsay and Dixon 2003; Batty et al. 2006b). Orchid mycorrhizas represent an extreme example of mycorrhizal dependency.

Other Beneficial Fungi

In 1998, a new fungal species was described, *Piriformospora indica* (Verma et al. 1998). This species was accidentally discovered in association with AMF spores in the rhizosphere of desert-dwelling plants in India (Verma et al. 1998). Now known to be a root endophyte in the division Basidiomycota (a fungal taxon that hosts many EMF species), *P. indica* is showing great promise as a beneficial plant symbiont (Verma et al. 1998; Varma et al. 1999). This species, unlike members of the Glomeromycota (the fungal taxon that hosts most AMF species), is easy to cultivate and exhibits strong growth-promoting properties with tobacco, parsley, poplar, and more (Varma et al. 1999). Because this species was discovered in an arid environment and is probably drought adapted, it may have significant potential to benefit plants whose environment is becoming warmer and drier due to climate change.

Named for their physical appearance, dark septate endophytes (DSEs) are members of the fungal division Ascomycota, which also hosts many EMF species of fungi, and are being examined as beneficial plant symbionts (Jumpponen 2001). These fungi do not form mycorrhizas, but like mycorrhizas they are incredibly widespread across plant taxa and geography (Jumpponen and Trappe 1998). The fungal taxa that comprise DSEs have an advantage in that they are easy to cultivate relative to AMF and thus hold potential for large-scale production. However, as with mycorrhizal fungal species, the nature of the partnership exists along a continuum from highly beneficial to parasitic (Johnson et al. 1997) and varies with host species and applied fungal species. Research in DSE application will probably yield important contributions to horticulture practices.

Root Nodule–Forming Bacteria in Legumes

The legume family (Fabaceae) consists of about eighteen thousand species (Polhill and Raven 1981), and 12% of these species are listed as endangered according to the IUCN Red List (IUCN 2010). One key morphological characteristic of this family is the formation of nodules on the root systems that house nitrogen-fixing bacteria. The bacterial member of the symbiosis (*Rhizobium, Azorhizobium, Sinorhizobium, Bradyrhizobium,* and *Mesorhizobium*) is a member of the family Rhizobiaceae (Crespi and Gálvez 2000). Sequestering nitrogen is an important advantage for plants, particularly in nitrogen-limited environments.

Root nodulation is controlled by a set of bacterial *nod* genes that produce Nod factors (Long 1996) that act as signals to specific hosts to initiate nodulation. Determining and characterizing the presence of *nod* genes enables researchers to identify species of rhizobia to ensure appropriate host–bacteria pairing. It should be noted that bacterial symbionts of fewer than 10% of the 750 legume genera have been fully characterized despite more than a hundred years of study (Moulin et al. 2001), and thus the potential for discovering more efficient rhizobial partners for rare species is wide open.

Root Nodule–Forming Bacteria in Nonlegumes

A diverse group of nonleguminous plants can also form root nodules for nitrogen-fixing bacteria. The bacteria in these associations are Actinomycetes (e.g., *Frankia*), and therefore their host plants are called actinorhizal plants, of which there are approximately twenty-five genera and 220 species (Wall 2000). Actinorhizal plants are ecologically important because they are often pioneer species that can live in marginal habitats. The symbioses of *Frankia* and actinorhizal plants have similar structures and outcomes to rhizobia–legume symbioses, yet our knowledge of how they form lags behind their better-studied counterparts, probably because of difficulties associated with extracting *Frankia* from some plant roots and culturing these bacteria ex situ (Wall 2000). Although *Frankia* can occur as free-living bacteria in the soil, whether or not they can benefit plants through nitrogen fixation in this state is unknown (Wall 2000).

Plant Growth–Promoting Rhizobacteria

Certain free-living soil bacteria can improve plant growth through different mechanisms. *Azospirillum* is the genus of plant growth–promoting rhizobacteria (PGPR) that is best characterized. Other free-living diazotrophs (organisms that do not need fixed nitrogen to grow) repeatedly detected in association with plant roots include *Acetobacter diazotrophicus*, *Herbaspirillum seropedicae*, *Azoarcus* spp., and *Azotobacter* (Steenhoudt and Vanderleyden 2000). Bowen and Theodorou (1979) have suggested that soilborne bacteria associated with the mycorrhizosphere of plants enhance mycorrhizal colonization; these bacteria have hence become known as mycorrhiza helper bacteria (MHB). This relationship is generic in that many species of bacteria perform similar functional roles with all types of mycorrhizas, including AMF and EMF (see Frey-Klett et al. 2007 for a review). Like mycorrhizal fungi, MHB also exhibit variation along a continuum in their beneficial effects (Johnson et al. 1997) depending on the partner species (Garbaye and Duponnois 1992).

Cyanobacteria are photoautotrophs, many of which can also fix atmospheric nitrogen, like the rhizobial bacteria mentioned earlier. The cyanobacterial symbionts of which we are most knowledgeable are found in the order Nostocales (*Anabaena*, *Nostoc*, *Calothrix*, *Scytonema*) and form associations with fungi. These associations are better known as lichens. Additionally, this group of nitrogen-fixing cyanobacteria form beneficial associations with four major groups of plants: (1) spore-producing, nonvascular bryophytes; (2) spore-producing, vascularized ferns; (3) cycads (gymnosperms); and (4) members of the genus *Gunnera* (herbaceous angiosperm; Rai 1990). Studies on the relationship between *Nostoc* species and *Anthoceros punctatus*, a hornwort, are revealing the importance of these relationships for host plant nitrogen acquisition (Meeks 1998). Symbiotic *Nostoc* species typically release 40–95% of their fixed nitrogen as ammonium to their plant partners (Meeks 1998), which provides a huge advantage to plant species, possibly including endangered edaphic endemics growing in nitrogen-limited habitats.

Combinations of Soil Microbes

Studies focusing on single-microbe inoculations are numerous (Cairney 1999; Cavallazzi et al. 2007). However, many plant species are known to form dual colonizations; for example, legumes are typically mycorrhizal with AMF and are also colonized by nodule-forming bacteria. In these cases symbiont co-presence may be crucial to plant performance. Asai (1944) found that some legumes do not form root nodules unless they are already mycorrhizal. Furthermore, by alleviating phosphorus stress, mycorrhizal fungi can indirectly improve the nitrogen status of a legume by affecting the root nodule bacteria (Hayman 1986).

A few researchers have begun investigating the merits of inoculation with multiple soil microbes and are exploring the effects of these combinations on plant emergence, survivorship, and growth. For example, Requena and colleagues (1997) explored the effects of AMF, *Rhizobium*, and PGPR in different combinations on *Anthyllis cytisoides*, a woody legume. The results from the microbial addition treatments varied depending on the other microbial soil inhabitants present. By propagating their target plant in both sterile and nonsterile soil, they were able to track how the microbial treatments differed upon outplanting. Second, inoculating with strains of *Rhizobium* that naturally occur in the reintroduction site can improve treatment efficacy in environments with varied levels of phosphorus and moisture (Paau 1989; Requena et al. 1997). Third, plant performance outcomes depend on the specific combination of soil microbes selected (Requena et al. 1997). Selection of species and strains native to the reintroduction site sometimes resulted in better plant performance; additionally, problems with introducing foreign microbes were avoided.

Application of Soil Microbes in the Laboratory

In vitro propagation can be used to produce rare plants for reintroductions and is particularly useful when species produce few or no seeds or are reduced to very low numbers. In vitro plant materials must be maintained under sterile conditions to prevent pathogen growth, and therefore introducing whole soil inoculum to provide symbionts to growing plantlets is not feasible. Augmenting potting soil with the microbial inoculum at the time of transfer is much more feasible and has met with success (Morte et al. 1996). However, co-culturing plants and isolated microorganisms in vitro is increasingly being explored.

During in vitro culture, the nutrients in plant tissue culture media can replace the nutrients that would be supplied in nature by beneficial microorganisms. This method has long been used for seed germination in orchids, a group known to have mycorrhizal associations in the wild. However, orchid propagation is routinely done aseptically, in vitro (Butcher and Marlowe 1989), and when the propagated orchids are acclimatized, it is assumed that they must reestablish a mycorrhizal relationship. In some cases, adding fungi has aided successful acclimatization (Gutiarrez-Miceli et al. 2008), but techniques have also been developed for symbiotic germination of orchid seeds, in which an appropriate fungus and seeds are co-cultured (Beardmore and Pegg 1981). Although this method requires isolation of the fungus and different culturing procedures, it has been shown that symbiotic seedling growth is often more robust than that of seedlings germinated without a symbiont (Huynh et al. 2002).

Many species are naturally associated with beneficial microorganisms, which can improve growth of micropropagated plants when growing in vitro or during acclimatization. The benefits of mycorrhizal fungi for survival and growth of acclimatized plants have been demonstrated with temperate and tropical fruit trees, forest trees, and several other species (e.g., Dolcet-Sanjuan et al. 1996; Estrada-Luna and Davies 2003; Khade and Rodrigues 2008). Co-culture with fungi has improved rooting and growth in vitro (Grange et al. 1997; Martins et al. 1997), and a mixture of fungi and bacteria can also be beneficial (Rodriguez-Romero et al. 2005). Although the methods for co-culturing are somewhat different from those of traditional plant tissue culture, the potential for improving growth and the ability of plants to survive acclimatization open the door to broadening the usefulness of in vitro methods for propagating endangered plants for successful reintroduction projects.

Application of Soil Microbes in the Greenhouse

Beneficial microbes have been applied in greenhouse propagation for years (Schwartz et al. 2006) with great results, particularly in the agriculture and horticulture industries. Sources of inoculum range from field-collected soil to self-cultured inoculants to mass-produced commercial sources. There are pros and cons associated with each of these inoculum sources.

Field soil is clearly the least expensive source of microbial inoculum and can provide an entire suite of microbes: mycorrhizal fungi, root nodule–forming bacteria, and plant growth–promoting bacteria, which may be beneficial to plant survivorship and growth. Kartsonas and Papafotiou (2007) used native whole soil to improve survivorship and growth significantly during greenhouse acclimatization of *Quercus euboica* propagules, a rare oak from Greece. In native whole soil, oak propagule survival ranged from 79% to 93%, as compared to 21–36% for propagules grown in a compost–perlite mixture (Kartsonas and Papafotiou 2007). In some cases, however, a field soil inoculum can also contain sources of pathogens (Packer and Clay 2000) and other undesirable propagules (e.g., spores or weed seeds), which would be detrimental to introduce into the greenhouse.

Studies on plant–soil feedback systems have significant implications for understanding plant performance, species distributions, community structure, and ecosystem functioning (Bever 1994; Kardol et al. 2007). Home soils can host many native pathogens, and high pathogen loads could override the merits of beneficial microbe populations. For example, Packer and Clay (2000) found that an accumulation of a pathogen (*Pythium* spp.) underneath *Prunus serotina* trees led to greater mortality when juveniles were grown in soil collected from under the trees. However, if that soil was sterilized, then lower mortality occurred. Results of other studies have corroborated these findings (Nijjer et al. 2007; Brenes-Arguedas et al. 2008). On the other hand, home soils can also offer a suite of preadapted beneficial microbes. This may be particularly important for edaphic endemic taxa (Taylor and Levy 2002). A number of studies support the home soil advantage (Montalvo and Ellstrand 2000; Grøndahl and Ehlers 2008).

One of the big questions with using inoculants is whether the applied inoculum actually ends up in the root systems of the target plants once they are outplanted. An early study of EMF inoculum on *Pinus* species tried to follow pure species inoculum treatments from the culture container into the greenhouse and out into the field (Riffle and Tinus 1982). Using reisolation and culturing techniques, the researchers were able to detect their original EMF inoculum based on morphological culture traits after 2 years in the field in some cases. After 2 years in the field, other presumably native EMF root tips were also observed in addition to the inoculum but were not identified (Riffle and Tinus 1982). Since the early 1980s, modern molecular analyses, such as restriction fragment-length polymorphism and DNA sequencing, have helped confirm the transfer of inoculum from culture to host plant (Redecker 2000; Dickie and FitzJohn 2007; Mummey and Rillig 2007). These tools provide a strong boost to our confidence that providing ex situ–propagated plants with an inoculum source before outplanting is effective and beneficial.

Some practitioners have chosen an intermediate technique between whole soil application and purchased inoculum. They have selected, isolated, and cultured known or hypothesized inoculum species from the target plant's rhizosphere. Growing your own inoculum requires an additional skill set and more time and resources, which can be costly, but should greatly reduce the probability of introducing pathogens.

More than thirty companies worldwide are producing mycorrhizal fungal inoculum, and these products go by various brand names, but they are generally marketed as growth promoters (Gianinazzi and Vosatka 2004). Mycorrhizal fungal inoculum products typically consist of one to a few fungal species, and they may also contain certain PGPR species. One drawback to using purchased inoculum is the expense, but more important is the reduced diversity of microbes to which your plants will be exposed. Most fungal mixes are based on a few species that are known to be excellent colonizers, thus indicating their potential aggressive nature. The aggressive behavior of these mycorrhizal fungal species makes them good early colonizers but can also make them good competitors for native fungi. This is something to keep in mind.

Methods for Applying Microbes in the Greenhouse

Once inoculum has been acquired, application to the target plants is quite simple. Rosbrook (1990) examined three ways of introducing *Frankia* to *Casuarina* seedlings using dried, ground root nodules. Acquiring *Frankia* nodules involved collecting fresh nodules, surface sterilizing, air-drying for 24 hours, and grinding. Three application treatments ensued: (1) watering a slurry of *Frankia*, suspended in a 1% sucrose solution, onto potting soil; (2) applying *Frankia* slurry directly to the root zone using a syringe; and (3) adding dried, ground nodules to potting soil. All treatments achieved nodulation, but syringe-inoculated plants nodulated faster and the plants grew bigger.

Applying AMF and EMF is also simple. AMF inoculum is generally acquired in the form of spores (see the International Culture Collection of Vesicular Arbuscular Mycorrhizal Fungi [INVAM] website at the end of this chapter). Spores can be added to soil as slurry. EMF inoculum can be from fungal spores collected from the mushroom fruit body or from colonized plant root tips. Both of these inocula can be directly added to potting soil. When using whole soil as an inoculum source, a ratio of one part whole soil to four parts potting soil is generally sufficient. Because inoculum potential varies, it is recommended that inoculum levels be assayed before use. Usage instructions are generally provided with purchased inoculum.

Unlike mycorrhizal fungi, PGPR are more difficult to acquire. Products containing PGPR may be commercially available (e.g., Plant Growth Activator; Organica, Norristown, Pennsylvania). However, native bacterial washes can be made from native whole soil. Blending about 25 grams of soil in about 400 milliliters of water, then filtering through Whatman® No. 1 filter paper will effectively remove mycorrhizal spores and inoculum while retaining the bacterial community. This wash can be directly applied to potting soil.

Methods for Applying Microbes in the Field

The application of beneficial soil microbes in crop species has been well researched for more than 35 years (Okon and Labandera-Gonzalez 1994). In 1981, Hall and Kelson developed a technique for dispersing AMF fungi in a pellet form that could be spread like a solid fertilizer. Soil pellets are infused with AMF inoculum and can even be attached to seeds using gum arabic before dispersal (Kapoor et al. 2008). This application method has the potential to cover large areas, but the longevity of viable AMF inoculum in the soil once dispersed is still unknown.

Far fewer studies have tracked the fate of rare plants and their beneficial microbes after introduction to the field, but one study on sea oats (*Uniola paniculata*) showed improved growth of AMF-inoculated plants up to 19 months after outplanting (Sylvia 1989). Alterations that may have occurred in the microbial community of these plants are unknown, but several scenarios could occur. At one extreme, populations of greenhouse-introduced microbes may explode, potentially becoming invasive. Second, the introduced and native microbes could coexist. And third, the introduced microbes could disappear or become drastically reduced by competition from local soil microbes in a swamping effect; there is more evidence for this outcome (Weinbaum et al. 1996).

Practical Implications for Traditional Restoration Work

The least expensive and probably the most appropriate source of microbes will be via whole soil collection at the propagule source site or from the recipient site. Whole soil will typically contain the full complement of beneficial microbes (mycorrhizal fungi, root nodule–forming bacteria, PGPR), but may also include some pathogens. Before collection, acquire permits. When acquiring the sample, collect from the top 20–30 centimeters of soil, where the majority of microbes that are beneficial to a seed or seedling will be found.

If whole soil collection is not an option, then beneficial microbes can be cultured or purchased. For example, INVAM offers a large selection of different AMF species and online instructions for culturing fungi. A detailed explanation of OMF extraction for culturing purposes is available in Dixon and colleagues' (2003) pivotal work on orchid conservation. It is essential to know the types of beneficial relationships formed between the plant species to be reintroduced and soil microbes as well as the identity of those microbes. If such information is not available or cannot be found in the literature, information from closely related species may be used. Although the microbe species may differ, the type of relationship (e.g., ectomycorrhizal vs. arbuscular mycorrhizal) is often consistent within genera.

Areas of Need and Research Opportunities

This chapter reveals only a portion of the primary literature covering the topic of beneficial soil microbes, but what is already apparent is the need for additional research. Most important, a clear understanding of the fate of inoculated microbes once they are outplanted in the wild is needed. Some have tried tracking inoculum after outplanting, but these monitoring efforts have generally been very short (Riffle and Tinus 1982). Short-term results suggest that inoculum is usually overtaken by microbes already present in the outplanting environment (Weinbaum et al. 1996). Often, ex situ plants provided with inoculum outperformed plants without inoculum, providing an advantage upon outplanting (Weinbaum et al. 1996). More conclusive research on the longevity of microbial inoculum would also help allay fears about microbes "escaping" and becoming invasive species themselves.

One of the goals of this chapter has been to develop awareness of beneficial soil microbes. A promising area of research is in examining different microbial combinations to determine which mixes will maximize the beneficial effects. For example, Jäderlund and colleagues (2008) examined different combinations of two species of PGPR (*Pseudomonas fluorescens* SBW25 and *Paenibacillus brasilensis* PB177), two AMF species (*Glomus mosseae* and *G. intraradices*), and one pathogenic fungus (*Microdochium nivale*) on growth of winter wheat (*Triticum aestivum* cultivar Tarso). In a greenhouse trial they found that different combinations of PGPR and AMF either enhanced the growth of wheat in the presence of the pathogen or had no effect on wheat growth. These findings support the need to examine fully the choice of microbes that are used and how they interact to promote plant performance.

Developing new tools for rare plant work will be imperative for conservation progress. Brundrett and colleagues (2003) made advances in work with orchids by developing "baiting" techniques for OMF to use in situ (Rasmussen and Whigham 1993) and ex situ. These techniques can be used to help determine OMF availability in potential recipient soils. Similarly, Batty and colleagues (2006b) developed a new protocol to improve the acclimatization success of orchid seedling transfer from axenic cultures to soil. Furthermore, the horticulture and agriculture industries have been working for centuries to learn how to grow large, healthy plants for retail and reforestation. Rather than reinventing the wheel, conservation practitioners may be able to use some of that knowledge base to develop techniques and tools that will apply to rare species.

Prospects and Cautions for Appropriate Use of Managed Relocation

Managed relocation of rare plants is controversial; while one hand is full of hope, the other is full of doubt. An important consideration for managed relocation involves the ability to prove that population declines are caused by climate change factors (McLaughlin et al. 2002). Macel and colleagues (2007) used a reciprocal transplant experiment to separate soil factors from climate and seed origin to elucidate mechanisms of plant adaptation. Two very important findings emerged. First, different plant functional groups (e.g., grass vs. legume) responded very differently, suggesting that not all plant species can be analyzed in a similar manner. Second, climate factors, not soil or genetics, significantly affected measures of fecundity for the grass species studied, *Holcus lanatus*. Despite the extensive cost in time and resources to conduct reciprocal transplant experiments, more are needed to help determine whether managed relocations are warranted.

Moving soil around, as required by reciprocal transplants, introduces other concerns about managed relocation. The unintentional introduction of soil pathogens and the release of novel microbes into foreign environments are two strong reasons to doubt managed relocation as a safe practice. However, moving soil microbes as inoculum is currently happening and has been happening for quite some time. A review of EMF introductions traced the first recorded EMF introduction back to 1839 and reported more than 770 introductions from more than 190 publications (Vellinga et al. 2009). These findings beg the question of why we are not more concerned about this practice when we know that the introduction of novel plant species can have detrimental effects on recipient communities (see Reichard et al., this volume).

Interest is growing in the role of soil microbes in invasion ecology (see reviews by Schwartz et al. 2006; Desprez-Loustau et al. 2007; Vellinga et al. 2009). Although the benefits of soil microbes in acclimatization and horticultural practices are well supported, researchers and practitioners alike should still proceed with caution when conducting introductions. Schwartz and colleagues (2006) do an excellent job of outlining and supporting three recommendations when considering the application of mycorrhizal fungal inoculum: (1) Determine whether inoculum addition is necessary, because adequate microbe sources may be available at the recipient site; (2) use local soil microbe sources whenever possible; and (3) if nonlocal sources of inoculum are used, attempt to select microbes that do not exhibit invasive characteristics. These recommendations can easily be extended to include all beneficial soil microbes.

Summary

For a long time, the soil environment was frequently ignored during plant reintroduction planning and practice, to the point where soil environments could deserve the title "The Last Frontier." Reasons for this are becoming less abundant, but the fact remains that working with soils and plant roots is difficult. Collecting soil near rare plants is disruptive, and examining the actual roots of rare plants is often destructive. Because not every conservation biologist has the expertise to identify, quantify, and culture the microbes of interest, appropriate collaborations are advised. However, once soil microbes have been used for acclimatization improvement, then tracking the fate of the applied microbes at the reintroduction site becomes another challenge. Molecular genetic tools are typically needed to identify inoculum species precisely, and these are currently too expensive for most reintroduction trials. Obstacles such as these may be more easily overcome by collaborating with appropriate experts.

The available research describing how soil microbes can be applied to improve rare plant propagules' acclimatization to soil is growing (see table 6.1). Not only are the traditional mycorrhizal fungi and root nodule–forming bacteria being applied to rare plant propagation, acclimatization, and outplanting, but use of the lesser-known PGPR is growing also (Steenhoudt and Vanderleyden 2000; Adesemoye et al. 2008). The soil microbes mentioned in this chapter can provide a suite of benefits, including increased water and nutrient uptake, nitrogen fixation, resistance to herbivores, heavy metal and salt tolerance, pathogen resistance, and improved competitive abilities. The effort needed to achieve these benefits exists on a continuum from minimal effort, such as applying a pinch of native whole soil to rare plant seedlings, to maximum effort, such as identifying, culturing, and designing multispecies inoculum mixes. As with any conservation practice, using soil microbes for reintroduction purposes comes with a list of concerns such as cost-effectiveness, invasibility, and phytosanitation, and these must be considered. However, depending on the case, the potential benefits could outweigh the potential risks. Advancing our knowledge of beneficial soil microbe ecology and application is bringing us one step closer to ensuring reintroduction success.

Online Resources

AMF inoculum source and mycorrhizal techniques: http://invam.caf.wvu.edu
/index.html

AMF and EMF inoculum source: http://www.fungi.com/index.html
Rhizobia inoculum source: http://www.emdcropbioscience.com/homepage
   .cfm
PGPR inoculum source: http://homeharvest.com/beneficialmicroorganisms
   .htm

We thank Sheila Murray for assistance with the manuscript. Joyce Maschinski and two anonymous reviewers improved this chapter by providing insightful comments.

Chapter 7

Optimal Locations for Plant Reintroductions in a Changing World

Joyce Maschinski, Donald A. Falk, Samuel J. Wright,
Jennifer Possley, Julissa Roncal,
and Kristie S. Wendelberger

Of all conservation strategies currently practiced throughout the world, reintroductions require the most sophisticated understanding of species biology and ecology (Falk et al. 1996). Whether augmenting existing populations, reintroducing within a species' known range, or introducing to a location outside the known range, finding optimal sites for long-term survival, growth, reproduction, and establishment of new populations is often "not as self-evident as it might otherwise seem" (Fiedler and Laven 1996, p. 157). Identifying appropriate habitat is essential to establish sustainable populations in existing or new locations, and yet for many species of conservation concern habitat needs are unknown. This uncertainty takes on even more importance in the context of contemporary and projected near-term changes in landscape and regional climate (Giorgi and Francisco 2000; Millar et al. 2007).

Many factors operating across a range of spatial and temporal scales influence where plants can successfully establish and persist. At geographic scales, distributions are constrained broadly by physiological tolerances, dominant ecosystem processes, and the distribution of suitable biomes (Lambers et al. 1998; Antonelli et al. 2009). Climate, land form, and geology drive species distributions (MacArthur 1972; Woodward 1987). At the scale of populations, fine-scale processes become more important, including gap dynamics, interspecific interactions, availability of symbionts, and environmental heterogeneity. Between the biome and population scales lies the mesoscale realm of landscapes and ecoprovinces $(10-10^4$ square kilometers), within which shifts in species distributions can happen over 10–1,000 years when constraints of dispersal and episodic climate or species interactions are altered (Ohmann and Spies 1998; Huston 1999; McKenzie et al. 2010). Climate-driven adaptation is most active at mesoscale (Cornwell and Grubb 2003; Parmesan 2006).

To improve the success of establishing rare plant populations, it is essential to understand the factors and processes that govern where species grow. In this chapter we review some theories related to species' distributions, their conservation applications, and relevance to plant reintroductions. We propose a process-oriented perspective that emphasizes population growth and the factors that influence it as an indicator of reintroduction success. We then provide examples of experimental introductions that demonstrate how fine- and broad-scale environmental variation influences population persistence.

Reviewing practical implications of these findings, we recommend a recipient site assessment, statistical techniques, and experiments for determining appropriate potential sites for reintroductions. These can be extended to consideration of managed relocation (MR) outside a species' current range, where niche theory indicates that suitable habitats may exist (Colwell and Rangel 2009). Because current conditions may differ from those in which the species evolved and may undergo dramatic change, the implications of these approaches are significant for understanding where on a landscape reintroductions can succeed and how to maintain biological diversity in a rapidly changing world.

Factors Influencing Distributions of Plant Species

Niche theory is central to our understanding of species' distributions and provides the basis for predicting where to find suitable habitat for rare plant reintroductions. Because niche was historically defined differently by different authors and is currently used interchangeably (Soberon 2007), we provide some historical context here. Grinnell (1917) defined the niche as the range of environmental conditions needed for a species to carry out its life history, and Elton (1927) articulated the niche concept as the status of a species in its community that includes biotic and resource–consumer dynamics. These early niche concepts are linked to particular places (Colwell and Rangel 2009). Hutchinson (1957) described a niche concept that is a species or population attribute. For an environment with n properties that define potential habitat attributes, the Hutchinsonian *fundamental* niche can be conceptualized as the n-dimensional hypervolume. Mathematically it can be described in n-dimensional space along n axes corresponding to environmental variables that permit a species' population growth rate to be positive indefinitely. The *realized* niche Hutchinson (1957) described as the hypervolume remaining after competitive exclusion, but other authors modified this definition.

The Hutchinsonian niche is not equivalent to a species' range (fig. 7.1). Observed geographic distributions of species do not necessarily overlap entirely with

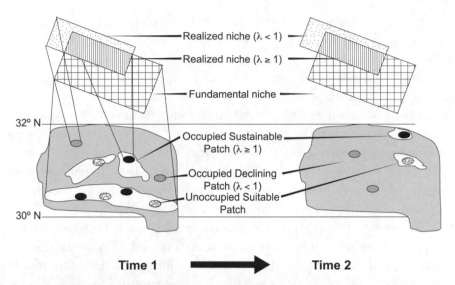

FIGURE 7.1. An illustration of the relationship between the geographic or physical space (latitude 30°N–32°N, with landmass shaded gray), the theoretical fundamental niche space (denoted by grid rectangle), and the realized niche, denoted by vertical hatch lines where conditions support stable or positive population growth, or $\lambda \geq 1$ (Hutchinson 1957), and denoted by speckled area where conditions do not sustain populations, or $\lambda < 1$ (Pulliam 2000). White areas indicate the possible geographic range related to measures of fundamental niche attributes. Extant populations with $\lambda \geq 1$ are indicated by black dots, and extant populations with negative growth $\lambda < 1$ are indicated by gray dots. Unoccupied but suitable patches are indicated by wave-patterned dots. At Time 1, there are only three sustainable populations of this rare species. Two populations with negative growth rates occur outside geographic areas correlated with the fundamental niche space. There are three suitable but unoccupied habitats in geographic areas correlated with the fundamental niche space. At Time 2, climate change (or other conditions) caused the size of geographic areas correlated with the fundamental niche space to shrink and shift northward. Remaining populations of the species are few; two exist outside geographic areas correlated with the fundamental niche space, and one colonized a new favorable patch and has positive population growth. The shape and size of the theoretical fundamental niche and realized niches at Time 2 are unpredictable at Time 1 because there may be adaptation or genetic drift, or they may be conserved and unchanged.

their fundamental niches. The discrepancies between distribution and fundamental niche space arise from dispersal limitations, interspecific interactions, the possibility that a portion of niche space is unexpressed in the physical environment (Jackson and Overpeck 2000; Colwell and Rangell 2009), and the stipulation in the definition of Hutchinsonian niche space that the conditions must permit populations to persist indefinitely (i.e., have population growth rates, or $\lambda \geq 1$). For species occupying habitat types that are inherently patchy, such as freshwater wetlands, edaphic anomalies, and mountaintops, dispersal is commonly limited across patches (Pulliam 2000). Suitable habitats with fundamental niche attributes may exist into which a species has never been able to migrate and colonize. Alternatively, competitive interactions (Case and Gilpin 1974) or predation pressure (Louda 1982; Dangremond et al. 2010) may limit a species' ability to colonize and maintain positive λ indefinitely. Suitable patches corresponding to environmental or resource niche values may exist but may be unoccupied or may occur outside the existing geographic range (fig. 7.1).

The fact that many populations exist in areas where they do not have positive or sustainable growth rates is explained in part by the additional concepts of source–sink theory, metapopulation theory, and dispersal limitation (Pulliam 2000). Source–sink theory posits that within a species' range there are source populations, where local reproduction exceeds mortality, and sink populations, where mortality exceeds reproduction (Pulliam 1988). Sink populations may arise from site-specific conditions leading to $\lambda < 1$ and are not sustainable unless they regularly receive propagules or immigration from a source population (Brown and Kodric-Brown 1977). Metapopulation theory incorporates colonization and extirpation of populations in space and time (Hanski 1999).

Population growth rates integrate multiple factors of species ecology, including reproductive effort and success, individual survivorship and growth, and biotic interactions (Knight, Monks et al., this volume) and are site-specific (Pulliam 2000). Thus, population growth rates provide a fundamental index for the dynamic condition of a population, whether naturally occurring or introduced, at a given site for a given time. Any local condition outside a species' niche space may limit population growth. A species' response to changes in local conditions could be extinction, adaptation, or emigration (Davis and Shaw 2001). If adaptation occurs, then there would be a subsequent change in the fundamental and realized niche, but it is not guaranteed that a species will be able to adapt to changes in the physical world, especially if conditions change rapidly (fig. 7.1).

Geographic Scale

On broad scales (e.g., ecoregions and biomes), plant distributions are correlated strongly with physiological tolerances to the abiotic template, including climate, soils, geology, hydrologic, and biogeochemical cycles. For example, cold tolerance, temperature optima for photosynthesis, and drought tolerance can influence where a species can persist (Woodward 1987). Climatic factors, such as annual precipitation and mean temperatures of the coldest and warmest months, have been combined with non–time varying factors, including major soil groups, and incorporated into models of climate envelopes to predict where species may be able to grow under future climate scenarios (Krause and Pennington, this volume). Contemporary distribution models use two approaches: (1) correlations between environmental variables and observed species distributions or (2) relationships between the functional ecology of the species and the environment (Morin and Lechowicz 2008).

To describe unambiguously the niche attributes operating at the scale of geographic range, Silvertown and colleagues (2006) suggested using the term *γ-niche*, following the Whittaker (1975) classification of species diversity. At this scale, species traits (γ-traits) influencing physiological tolerance are predicted to be evolutionarily conservative (Silvertown et al. 2006) and set the limits on species distributions (Morin et al. 2007). For example, a γ-trait such as frost tolerance could predict the ecoregion where a species can occur. Because defining a species' fundamental niche using actual occurrence data is very likely to underestimate niche space, experiments are often necessary to test γ-traits, especially for life stages involving seeds (e.g., Baskin and Baskin 1998). For example, seed germination trials done for some subtropical grasses indicated frost tolerance, an unexpected γ-trait considering their geographic distribution (Maschinski et al. 2009). Unfortunately, such data are generally unavailable for many rare species.

The proximal climatic controls on plant distributions are often linked to modalities in the climate system that operate over large scales of space and time. For example, in much of North America the El Niño Southern Oscillation (ENSO), Atlantic Multidecadal Oscillation (AMO), Pacific Decadal Oscillation (PDO), and other global circulatory processes in the ocean–atmosphere system contribute to changes in the frequency and intensity of storm events and seasonal precipitation (Beckage et al. 2003). These processes operate on a range of timescales, with temporal variance concentrated in interannual (ENSO) (Li and Kafatos 2000) to multidecadal (PDO, AMO) periods (MacDonald and Case 2005). Climate variation on these timescales plays a major role in the distribution of species and biomes by influencing temperature regimes, available moisture, and disturbance frequency and intensity. On long timescales (10^3–10^5 years) climate variation governs the global distribution of biomes, within which most species-level distribution processes occur.

Landscape Scales

At landscape levels or β-niche scales (Silvertown et al. 2006), plant communities are maintained by dynamic processes that can result in multiple stable states (Hobbs and Norton 1996; Suding and Gross 2006). These processes include abiotic filters (e.g., climate, substrate, and structure), biotic filters (e.g., competition, predation, mutualisms, parasitism, trophic interactions, dispersal, succession, disturbance, and site history), and socioeconomic filters (Hobbs and Norton 2004). For example, invasions of native or exotic species can result in species replacement, changes in nutrient cycling, and changes in community composition and structure (Suding and Gross 2006).

Across many terrestrial ecosystems, succession and disturbance are fundamental processes that shape plant communities (White and Jentsch 2004; Turner 2010). The residence time of any species in a particular place is related to its physiological tolerance, individual growth rate, and ability to extract resources in low and high densities (Tilman 2004). Disturbance processes often open space, altering competitive dynamics and redistributing light energy, nutrients, and carbon (McKenzie et al. 2010). Disturbance frequency, seasonality, magnitude, spatial extent, and synergisms (interactions and feedbacks) will influence how plants establish, persist, and disperse in any landscape (White and Jentsch 2004) and how the ecosystem will recover after the event (Holling 1996).

Anthropogenic activities also directly and indirectly influence species distributions. Altered ecosystem processes modify periodicity, frequency, and intensities of disturbances. For example, fires and insect outbreaks have modified the global distribution of major plant communities at the landscape scale (Bond and Keeley 2005).

Population and Patch Scales

At finer scales of population persistence and patch occupancy, or α-niche (Whittaker 1975; Silvertown et al. 2006), a different suite of factors influences where species occur. Fine-scale abiotic factors, including soil chemistry and nutrients, are particularly important determinants of patch occupancy. Local topography influences physical variables, such as solar radiation, soil water retention, and temperature regimes that in turn affect where plants will be able to colonize and persist. Biotic interactions with competitors (Keddy 2001), mutualists (Haskins and Pence, this volume), and herbivores (Maron and Crone 2006) also significantly constrain where a plant species will persist.

Conditions needed for early life stages may be different from those needed for later life stages. Germination, establishment, and early growth are the most critical phases of plant life history for regeneration, and these are influenced greatly by microenvironment (Grubb 1977; Veblen 1992; Wendelberger and Maschinski 2009) and the preexisting established species in the colonization microsite (Tilman 2004). Harper (1977) described this microenvironment as the *safe site* for germination, and Grubb (1977) subsequently identified the *regeneration niche*, which includes all the requirements for successful replacement of one generation with the next. These requirements are adequate seed production of the maternal plant, ability to disperse to a suitable microsite, seed germination, seedling establishment, and development to maturation. Heterogeneity in environmental conditions (e.g., light, litter, and soil moisture) can influence seed and seedling survival on scales less than 1 square meter (Molofsky and Augspurger 1992; Wendelberger and Maschinski 2009).

Periodic disturbance creates spatial and temporal heterogeneity or patch dynamics that allow coexistence among species that might otherwise be driven to competitive exclusion in stable environments (Pickett 1980). To regenerate successfully, some species need large-scale disturbance, such as wildfires, landslides, or blowdowns, whereas others with good dispersal and rapid colonization ability can use small-scale gaps of less than 250 square meters (Veblen 1992). Gap dynamics and time since disturbance also influence resource availability spatially and temporally (Tilman 1988); without disturbance, gap-dependent species would be driven to extinction (Pickett 1980).

Microsite variation can influence the strength of biotic interactions. For example, *Cercocarpus ledifolius* seeds have significantly greater predation in open microsites than in microsites with shrub cover. Predation levels coupled with secondary seed dispersal resulted in a structured spatial arrangement of seeds across microsites in northeastern Utah (Russell and Schupp 1998). Mutualisms are also influenced by microsite conditions. In the Sierra Nevada of northeastern Spain, where sunny open areas tend to have high temperatures in comparison to shaded areas, nurse shrubs increased evergreen and deciduous seedling growth and survival, particularly at low altitudes (Gomez-Aparicio et al. 2004).

Long-term demographic studies indicate that microsite variation can influence population growth and persistence (e.g., Kephart and Paladino 1997; Renison et al. 2005; see boxes 7.1 and 7.2). When short-term studies (2–3 years) have not found significant correlations between microhabitat parameters and survival (e.g., Baraloto and Goldberg 2004; Akasaka and Tsuyuzaki 2005), it may be a consequence of the short duration of the studies and the length of time needed to witness demographic effects. Alternatively, it is possible that out of the boundless possible factors influencing patch occupancy, the key microsite factor was not measured. Evidence from reintroductions indicates that it often takes many years to detect change in population demography (Maschinski 2006; Dalrymple et al., Albrecht and Maschinski, Monks et al., this volume); therefore, identifying optimal habitat also may require long-term studies (appendix 1, #22, #35, #38).

Conservation Applications of Niche Theory

A wealth of research and new theory has emerged from the original niche theories described earlier. Although much of it is beyond the scope of this chapter, within the past two decades niche theory has been invoked as a tool for conservation purposes. Here we briefly examine a few key uses of niche theory with particular relevance to rare plant reintroductions.

Using Niche Models to Improve Rare Species Surveys

Rare species are often difficult to detect using a strictly random sampling protocol. To increase efficiency and cost-effectiveness of surveys for rare species, Guisan and colleagues (2006) recommended using a niche-based model with stratified random sampling. The technique requires first fitting a model of climatic and topographic predictors (e.g., rock or soil type and vegetation type) onto a geographic information system map and then randomly selecting from areas with at least one patch with high habitat suitability (defined by the presence of predictors) as a starting point to search for rare species. Using this procedure, the researchers increased efficiency 1.8- to 4-fold; they detected more new occurrences of target rare species in less time.

Are Rare Species Rare Because Their Habitats Are Rare?

In his stochastic niche theory, Tilman (2004) suggested that within a community, common species are found in the most common habitat, whereas rare species, that is, those with low abundance and small range sizes (Gaston 1994), would be found in the rarest habitats or microsites within communities. Although this idea has theoretical support, broad empirical support is still needed. Alternatively, rare species may maintain positive population growth rates (λ) over a narrower range of values along one or more niche axes, in contrast to more common species. Coupled with the previously described sampling strategy (Guisan et al. 2006), these ideas provide a caution related to rare plant reintroductions. Suitable habitat for a rare species will not be randomly distributed and probably will not be common in any location; therefore, using an entirely random experimental design to place rare propagules in a reintroduction site is inefficient and doomed to have low success. We recommend using a stratified random sampling method within suitable microhabitat that has been determined by niche characteristics (appendix 1, #22).

Modeling Historic and Future Distributions

The niche concept forms the theoretical basis of species distribution models to predict future distributions (Colwell and Rangell 2009) and can be applied to locating and evaluating suitable reintroduction sites (Martinez-Meyer et al. 2006).

The fundamental niche can be approximated using species occurrence data, broad-scale climate or environmental data (e.g., Krause and Pennington, this volume), or field measurements taken where the species occurs (Collins and Good 1987). However, to define the realized niche adequately for each species requires experimentation, where interspecific interactions with environment can ensue (Soberon 2007), and in situ demographic analysis to confirm positive growth rates (Knight, this volume). Thus, broad categorizations of potential habitat may be developed from models, but actual outplantings will be needed to test the realized niche.

Practical Implications for Traditional Restoration Work

Because species have particular characteristics that allow them to occupy and persist in certain environments, both broad- and fine-scale environmental qualities must be considered when a reintroduction site is selected. Matching the environmental attributes at a potential reintroduction site to the species' niche requirements will help promote long-term population persistence (appendix 1, #22). Although it is not always easy to identify and quantify the critical environmental factors driving observed patterns of plant stage distributions (Baraloto and Goldberg 2004), it is still an essential first step for assessing reintroduction sites. Potentially suitable habitats may be identified based on apparent environmental factors, but a species' long-term persistence will be verified only after in situ trials where biotic interactions operate and where population growth rates can be estimated over multiple generations.

In the past 20 years a growing number of reintroduction studies have used experiments to test aspects of niche on successful establishment, survival, and reproduction of reintroduced species (Guerrant, Dalrymple et al., this volume). Of the 200 studies in the *CPC International Reintroduction Registry* (CPC 2009) that conducted experimental tests comparing factors influencing reintroduction success, 11% tested microsites, 13% tested broad-scale multiple sites, 6% tested abiotic aspects of niche space (e.g., light, soil, aspect, water table level), 7% tested biotic factors (e.g., competition, herbivory, weeds), and 1% tested an ecosystem process (burning). Several have shown that reintroduced plant performance and population growth depend on microsite (see boxes 7.1 and 7.2).

We recommend evaluating potential reintroduction sites using three different approaches. The first is a recipient site assessment, which is a criterion-ranking system that can be used to evaluate either single or multiple potential reintroduction sites (Wright and Thornton 2003; table 7.1). We recommend ranking factors influencing the species' ability to persist at a site related to logistics or ease of implementation, quality of habitat, and management. Because the percentage of

TABLE 7.1

Recipient site assessment based on ranking criteria related to logistics, habitat quality, and management.

Date: Observers:
Site: Description:
Criteria for prioritizing potential restoration site

	3	2	1	Score
Category 1: Logistics, implementation, management				
A) Status of relationship with landowner and management	None	Some	Good	
B) Commitment level of agency to protect introduced population	None	Some	Good	
C) Willingness of agency to manage habitat for target species	None	Some	Good	
D) Site preparation, threats removed	No	Partially	Completely	
E) Amount of public access or susceptibility to human disturbance	High	Medium	Low	
F) Accessibility for planting logistics and future monitoring	Poor	Fair	Good	
G) Water source present		No	Yes	
Category 1 Total				
Category 2: Habitat characteristics				
A) Percentage of associated species common with extant sites	0–40%	41–70%	71–100%	
B) Quantity and diversity of aggressive invasive plant species	High	Medium	Low	
C) Current and future impact of invasives	High	Medium	Low	
D) Size of potential reintroduction area	Small	Medium	Large	
E) Quality of adjacent habitat	Poor	Fair	Good	
F) Quantity of good-quality habitat adjacent to reintroduction site	None	Some	Abundant	
G) Soil texture similar to extant sites	No	Partial	Yes	
H) Soil nutrients similar to extant sites	No	Partial	Yes	
I) Canopy cover optimal for target species	No	Partial	Yes	
J) Hydrology similar to extant sites	No	Partial	Yes	
K) Topography similar to extant sites	No	Partial	Yes	
L) Target species presence at site	Never	Historic	Current	
M) Special needs of target species present	No	Partial	Yes	
N) Mutualists present	No	Some	Yes	
O) Herbivores present	Yes	Some	No	
P) Ecosystem processes functional	No	Partial	Yes	
Q) Number of potential translocation areas in site	1	2	>3	
R) Proximity to existing wild populations	>10 km	5–10 km	<5 km	
S) Natural disturbance regime	Excessively high or low	Moderate	Normal	

TABLE 7.1

Continued

Date:	Observers:			
Site:	Description:			
Criteria for prioritizing potential restoration site				
	3	2	1	Score

*				
Category 2 Total				
All Criteria Total				

*Additional habitat feature of interest for this species (e.g., seasonal flooding, salinity level, nurse plants present)

Notes:

For evaluating a single site: Add total scores. 27 is a perfect score.

Total scores of 27 to 54 are acceptable reintroduction sites.

Any criterion with score of 3 should be improved before moving forward.

For choosing between multiple sites: The best site has the lowest total score and no single criterion scoring 3.

The assessment can be used to score a single site or to prioritize multiple sites.

associated species common with extant populations, soil texture, soil nutrients, canopy cover, and so on may be easily quantified, we advise using quantitative data to determine the rank whenever possible. Categorical data are subjectively determined by the practitioner. Single sites achieving a total score of 27 are ideal reintroduction sites, but sites with total scores of 27–54 can be considered suitable recipient sites. However, any criterion scoring 3 should signal caution because it may limit reintroduction success. In some cases, it will be possible to improve the rank of this criterion before proceeding with the reintroduction. For example, negotiations may improve the commitment of the agency to protect the reintroduction. If multiple sites are being compared, the site with the lowest total score and without any criterion with a score of 3 should be considered as the best suitable recipient site. Wright and others used this process to identify thirteen potential reintroduction sites of thirty-two evaluated along the eastern coast of Florida for endangered *Jacquemontia reclinata* (Wright and Thornton 2003; Maschinski and Wright 2006) and two potential reintroduction sites in the Florida Keys for endangered *Pilosocereus robinii* (Goodman et al. 2007). In the case of *J. reclinata*, eleven reintroductions persist in the new locations properly identified by this assessment (table 4.1), and *P. robinii* reintroductions are in planning stages.

Many of the factors in the recipient site assessment are based on the assumption that similarity to extant populations is the best option for successful establishment of a reintroduced population. There are two important caveats to this assumption. First, climate change, novel disturbance regimes, and other disruptions

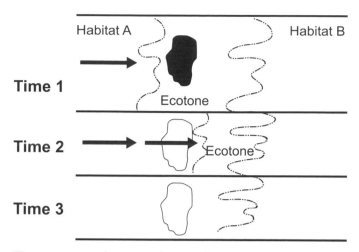

FIGURE 7.2. Time sequence depicting change of spatial location of a reintroduced population planted in an ecotone in response to a directional process, such as sea-level rise or temperature shift. At Time 1, a reintroduced population (black area) is planted into an ecotone between Habitat A and B. At Time 2, the process has shifted the boundary margin of the ecotone. At Time 3, the reintroduced population is left as a hollow polygon in inappropriate habitat. To persist, it would have to migrate eastward to new ecotone spatial boundaries. Succession or a disturbance event can also change boundaries over shorter timescales.

of ecosystem processes may shift the spatial footprint of suitable habitat (fig. 7.2). Second, some rare species exist solely in habitat fragments that no longer support positive or stable population growth, nor do they allow dispersal to new habitat as conditions change. In such cases, using occupied habitat as the reference condition for planning a reintroduction may fail to capture the optimal conditions for a species' positive population growth. For this reason, known historic range may not necessarily be the best basis for assessing optimal habitats for successful reintroduction. For example, Maschinski and Duquesnel (2007) found that *Pseudophoenix sargentii* was successfully reintroduced to only one of the two known historic US sites. Absence from a historic site may mean that the site is no longer suitable for the species.

Clearly, the more that is known about a species' ecology, the better the chance of reintroduction success. Ideally, quantitative in situ assessment of a target species' niche attributes is recommended. Although many practitioners may use a "gestalt" impression of the target species' growing requirements, measurements of key abiotic and biotic factors, as well as multiyear estimates of λ, can help bracket the range of conditions that allow persistence of the species. These conditions can then be used to compare or prioritize potential reintroduction sites.

Our second recommendation is to use a quantitative assessment to identify suitable recipient patches at a fine scale. Several assessments using different statistical approaches are available. Structural equation models (Iriondo et al. 2003), generalized logit models (Gomez-Aparicio et al. 2005), general linear models (Batllori et al. 2009), and principal component analysis combined with discriminant function analysis (Collins and Good 1987) have been used to elucidate important microsite factors related to plant reproduction, growth, seedling emergence, or abundance in natural wild populations. These, in turn, can be applied to selecting suitable recipient patches for reintroductions. For example, Wright (2003b) assessed soil chemical and community attributes associated with occupied patches of the endangered *J. reclinata*. Occupied *J. reclinata* patches had low salt concentrations and high graminoid and herbaceous diversity (Wright 2003b). These attributes indicated preferred placement of reintroduced plants on the landward side of foredunes, where they are buffered from salt spray (Wright 2003a; Maschinski and Wright 2006).

Experimental reintroductions can also be used to increase the probability of selecting suitable microsites for subsequent introductions, so our third recommendation supports previous authors, who suggested that reintroductions be conducted as experiments as a fine-scale assessment (Falk et al. 1996; appendix 1, #7). Essentially, reintroductions are bioassays (Maschinski et al. 2003; Maschinski and Wright 2006; Roncal et al. in press): Test plantings can reveal which microhabitat conditions are optimal for individual growth and survival as well as long-term population growth (see box 7.1). Even if the quantitative characters defining the regeneration niche are unknown, it is important that reintroduction sites provide a variety of microsites to meet requirements of all life stages of a species (appendix 1, #22 and #26).

Understanding a target species' tolerance for competition and disturbance can help inform spatial and temporal placement of any reintroduction. Evaluating the landscape from the perspective of topography, ecosystem dynamics, and patterns of possible restoration trajectories will help determine the locations with greatest likelihood of sustaining a reintroduced population (Larkin et al. 2006; Suding and Gross 2006). Dispersal pathways are an especially critical consideration; where suitable habitat is anthropogenically fragmented, the restoration strategy is more likely to require some degree of MR. Where species niche and landscape patterns and processes coincide will determine the potential domain for successful reintroduction.

Prospects and Cautions for Appropriate Use of MR

In landscapes that are rapidly changing through development or climate change, habitat options for rare species are often greatly diminished. Many rare species are left to occupy fragments of habitat that may not represent optimal habitat (i.e., where they do not have positive growth rates). This is especially true in landscapes

Contributed by: Jennifer Possley and Joyce Maschinski
Species Name: *Lantana canescens*
Common Name(s): Hammock shrubverbena
Family: Verbenaceae
Reintroductions Initiated: 2005
Location: Ecotone between pine rockland and rockland hammock in south Florida
Length of Monitoring: 2005–2009
Factors Tested: Modified habitat versus intact ecotone
Federal or State Status or IUCN Ranking: Florida endangered, G4

Species Description and Conservation Concern

In fragmented, fire-suppressed landscapes, optimal habitat for disturbance-dependent species may no longer exist. *Lantana canescens* Kunth (Verbenaceae) is a sprawling woody shrub native to South and Central America and the West Indies (Gann et al. 2002) that needs high light and low competition for growth. At the northern limits of its global range, *L. canescens* currently grows in one small south Florida park and in three southern counties of Texas. Two Florida populations disappeared this decade because of habitat degradation from hardwood and weedy species invasion and fire suppression. The single remaining wild population in Florida is growing in fragmented habitat with dysfunctional ecosystem processes (fig. 7.3a). The seri-

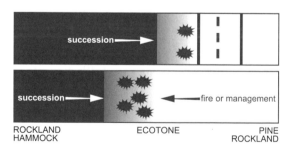

FIGURE 7.3. Depiction of *Lantana canescens* (denoted by black shapes) wild population habitat in south Florida. *L. canescens* grows in the ecotone between rockland hammock and pine rockland habitat and has been critically reduced. **(Top)** Rockland hammock plants have invaded the ecotone, and adjacent pine rockland habitat has been developed into a paved road, leaving little ecotone and little space for population expansion. **(Bottom)** In a functional ecosystem, *L. canescens* would exist within ecotone boundaries fluctuating between rockland hammock species succession and fire. To maintain an ecotone without fire would require ongoing management to thin encroaching hardwood species.

Box 7.1. Continued

ous population decline from sixty-six plants in 2004 to forty plants in 2006 spurred our action to increase the numbers of sites and individuals in the wild. Because the extant population was not growing in optimal habitat (wild population $\lambda < 1$), we tested the survival and persistence of populations reintroduced into two historic ecotones maintained by mechanical thinning and one modified habitat. The modified habitat was human altered. Once pine rockland forest, in the 1940s the site was converted to a lime grove, subsequently abandoned, invaded with 90% nonnative, invasive plants, and restored to a nonanalog community after invasive removal.

Reintroduction Methods and Findings

In 2005, we reintroduced *L. canescens* into three locations: Two locations were historic ecotones adjacent to rockland hammock but where substrate was similar to the existing wild population site (fig. 7.3a), and one was a modified habitat located distant from rockland hammock. After 18 months, 69% of transplants survived in the modified habitat and 65% and 84% in two historic ecotone sites. The modified habitat had significantly higher photosynthetically active radiation (PAR) (75%) than the historic ecotones (25–39%). Correspondingly, by 2007, 267 seedlings recruited into the modified habitat, whereas only 8 emerged at both historic ecotone sites. Seedling establishment was associated with higher PAR at the modified habitat than in the historic ecotone.

Ideal habitat for *L. canescens* would include fire balancing succession (fig. 7.3b) to maintain an ecotone with high light and low competition.

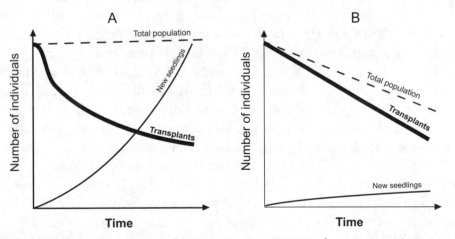

Figure 7.4. Comparative population trajectories in reintroduction sites (**a**) separated from rockland hammock versus (**b**) adjacent to rockland hammock. Higher light conditions account for higher recruitment in (**a**).

BOX 7.1. CONTINUED

Restoring fire could reduce woody species and maintain sunny conditions for the ecotone species, but if this is not possible, continued hardwood thinning would be needed to maintain structure and vegetation suitable for *L. canescens* (fig. 7.3b). When *L. canescens* was restored to a location physically distant from rockland hammock, the succession trajectory slowed, overall population sustainability increased (fig. 7.4), and maintenance costs decreased (Possley et al. 2009).

Lessons Learned

Historic habitat may not always be suitable to sustain rare populations.
If ecosystem processes are disrupted in wild habitat, modified habitat may prove to be a viable alternative for sustaining rare populations.

---

that have been subjected to anthropogenic change or that face fundamental modification driven by climate change. Thus, one might ask, "Where can plants go when their habitats are gone?" and "How will they reach new suitable habitat without assistance?"

Managed relocation into suitable unoccupied habitat across short distances from existing populations is a potentially useful conservation strategy. Following the same procedures outlined earlier for assessing suitable habitat, several studies provide evidence that rare species can persist when placed into suitable habitat outside their known range (Maschinski and Wright 2006; Wendelberger et al. 2008; Marsico and Hellmann 2009; Roncal et al. in press) or in modified habitat (see box 7.1; Possley et al. 2009). Experimental tests indicated that plant establishment or failure differed at fine scale (less than 10 square meters), yet the environmental factors measured did not always predict outcome (Roncal et al. in press). In addition, what was once appropriate habitat historically may not exist today, especially for species adapted to postdisturbance open habitats.

Using three *Lomatium* species, including one rare species, *L. dissectum*, Marsico and Hellmann (2009) tested whether short-distance (less than 25 kilometers) northern relocations outside the current range would be possible. Although the rare *L. dissectum* had lower germination and lower reemergence than the common species, all three species had equal or better survivorship in outside-range undisturbed treatment plots than they had in within-range plots. Climate of the outside-range plots had recently changed, but conditions were within the species' 100-year climate envelopes. The authors concluded that dispersal limitation is

Contributed by: Kristie S. Wendelberger and Joyce Maschinski
Species Name: *Tephrosia angustissima* var. *corallicola*
Common Name(s): Coral hoary pea
Family: Fabaceae
Reintroductions Initiated: 2003
Location: Pine rocklands of south Florida
Length of Monitoring: 2003–2009
Factors Tested: Microhabitat preference
Federal or State Status or IUCN Ranking: Florida endangered, G1 T1
Topic: What is the optimal microsite for species growth, establishment, and persistence?

Species Description and Conservation Concern

The Florida endangered *Tephrosia angustissima* var. *corallicola* is a prostrate, sprawling pine rockland forb in the Fabaceae family. Although herbarium specimens documented *T. angustissima* var. *corallicola* from pine rocklands throughout the Miami Rock Ridge from 1877 to 1927, today it is known only from a single location in the United States and eight populations of unknown size in Cuba (Beyra Matos 1998). The single US location is an agricultural field that was rock plowed and is currently mowed regularly. Neither herbarium specimens nor the reference extant site provided information on microhabitat (realized niche) requirements.

Reintroduction Methods and Findings

In June 2003, we experimentally introduced genetic clones of plants propagated from stem cuttings at Fairchild Tropical Botanic Garden into three different pine rockland microhabitats: fifty-seven plants into sunny habitat (*Serenoa*), twenty-seven plants into shady habitat (pine), and fifty-seven plants into disturbed habitat along the edge of the pine rockland (firebreak). Some plants had mature fruits at the time of transplant. We planted into locations where rebar penetrated the soil at least 20 centimeters, and we watered holes before and after planting. We watered every 3 days for 2 months, then occasionally for 3 months.

Tephrosia survival and persistence varied between microhabitats and life history stages. From fall 2003 through fall 2006, the greatest percentage

Box 7.2. Continued

adult survival occurred in the pine habitat; by 2009 only a single adult survived in firebreak habitat. Three months after transplanting, seedlings emerged. Greatest seedling recruitment occurred in the firebreak (2,067 seedlings), followed by pine (498 seedlings) and *Serenoa* (435 seedlings; fig. 7.5). Microsite significantly influenced maturation rate, survival, and population growth rate. By September 2009, recruited seedlings had become reproductive: fifty-nine in the *Serenoa*, ten in firebreak, and one in pine habitats. Seedling survival significantly differed across microhabitats ($\chi^2 = 179$, $p < .001$); seedlings had the highest chance of surviving to 2,000 days in the *Serenoa* (2%), but only 1% chance of surviving in the pine and no chance of surviving for 2,000 days in firebreak. Although there was low overall seedling survival in all microsites over 6 years, seedlings that germinated in the *Serenoa* had the greatest probability of surviving to reproductive age. Thus, although transplants established and seedlings recruited into all habitats, *Serenoa* habitat was the only one to have positive population growth ($\lambda > 1$) to F_2 generation within 2 years after installation (Maschinski et al. 2006).

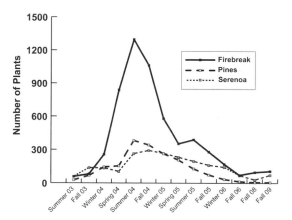

Figure 7.5. Numbers of *Tephrosia angustissima* var. *corallicola* growing in three microhabitat types from introduction in 2003 to 2009.

Lessons Learned

Provide heterogeneity in reintroduction sites to capture optimal conditions for different life stages.

Microsite significantly influenced vital rates of different life stages and species persistence at the site.

constraining current ranges of these species and could be overcome with human assistance.

The success of MR will be possible only if species are placed into habitats that encompass their niche space. It is unlikely that MR will be successful (in the sense of achieving ecological persistence) into habitat that is fundamentally unsuitable, where *unsuitable* indicates a significant mismatch with the species' niche space (Matyas and Yeatman 1992). As many traits are retained over evolutionary time (Ackerly 2003), broad geographic-scale MR would probably be outside the γ-niche space and thus unsuccessful.

As is advised for traditional reintroductions, MRs should be done as experiments and should be considered only if suitable recipient habitat is available. Each attempt can help define optimal habitat, factors limiting population persistence, and health of ecosystem processes. If we are to safeguard plant species against extinction from pervasive land conversion, climate change, and sea-level rise into the next century, an integrated strategy will be necessary combining ex situ and in situ conservation with introducing species to suitable protected locations outside their current or historic range (Falk 1990). What remains to be seen is whether there will be suitable habitat available anywhere for some of the rarest species.

Areas of Need and Research Opportunities

Can environment be manipulated to improve persistence of rare species populations? Is it possible to create microhabitat conditions that will favor positive population growth of rare species?

Most restoration ecologists agree that in situ conservation is the preferred conservation strategy whenever possible. For extant rare populations with negative population growth ($\lambda < 1$), conserving the species may entail finding ways to improve in situ population growth. For example, removing invasive species from rare plant habitats may increase λ. If done experimentally, invasive removal treatments coupled with measures of population dynamics could give insight into the ability to manage habitat to improve persistence of rare species (Pulliam 2000). Experimental manipulations are relevant and can inform future plant reintroductions, because creating or locating optimal microsites can be enhanced with this knowledge.

What Are the Best Ways to Define a Species' Habitat Needs Considering the Shifting Template of Climate Change?

Because of anthropogenic activities and their effects on future climate scenarios, optimal habitat conditions for plant species persistence may change in latitude and elevation (Thomson and Parker 2008). Experiments to determine whether species can withstand predicted temperature and precipitation change will shed light on physiological tolerance and the feasibility of natural migration or MR of endangered species. Future optimal locations will depend on the environmental factor that limits the range of a species, and therefore future studies should aim at revealing this factor first.

In addition, individual species are expected to react to these sweeping environmental changes in unique ways, potentially disassembling contemporary biotic communities while novel communities assemble (Temperton et al. 2004). Studies exploring contemporary mutualistic interactions of rare species with their pollinators, dispersers, soil symbionts, and microbial communities could help us understand the dynamics of these relationships. Experimental manipulations to examine the effects of predicted climatic shifts (e.g., increased rainfall, temperature, CO_2 levels) on coadapted species may provide insights about how these associations and habitat needs may change in the future.

Summary

Niche theory provides a foundation for understanding species' distributions at geographic, landscape, population, and patch scales. Several important conservation applications of niche theory are relevant to plant reintroductions: using niche models to refine surveys of rare species, heightening awareness of the degree of rarity of the species' habitat, modeling historic and future distributions, and defining suitable habitats and microsites for reintroductions. In the past 20 years, a number of reintroductions reported in the *CPC International Reintroduction Registry* confirm that microsite properties are especially critical for long-term persistence (i.e., nonnegative population growth rate) of plant reintroductions. Using a recipient site assessment along with quantitative analysis of occupied patches and experiments can help define suitable habitat for reintroductions. Because microhabitat strongly dictates success or failure of establishment and population growth, and because historic conditions where species were supported may no longer exist, we recommend that experimental approaches be used in reintroductions where more than one microhabitat is used as a trial. This is especially true in dynamic ecosystems that present unpredictable conditions associated with pervasive impacts of climate change, altered disturbance regimes, land cover change, and invasive species. Broad-scale plant distribution models that account only for physiological tolerance may overestimate potential future distributions of species if they do not incorporate microhabitat concerns, dispersal pathways, biotic limitations, and the availability of reliably protected natural areas. Population growth rates can be used as a barometer of reintroduction success across taxa and over ecologically realistic periods of time.

We gratefully acknowledge funding for reintroductions conducted by JM, SJW, JP, JR, and KSW from the US Fish and Wildlife Service, Florida Department of Agriculture and Consumer Services Division of Plant Industry, Miami–Dade County Environmentally Endangered Lands and Natural Areas Management, and Biscayne National Park. DAF was supported by the US Department of Agriculture McIntyre–Stennis Program support of the College of Agriculture and Life Sciences, University of Arizona. We thank Fairchild Tropical Botanic Garden and their volunteers for supporting rare plant conservation in south Florida. This chapter benefited from helpful comments and suggestions from Matthew Albrecht and reviewers.

Chapter 8

Strategic Decisions in Conservation: Using Species Distribution Modeling to Match Ecological Requirements to Available Habitat

CRYSTAL KRAUSE AND DEANA PENNINGTON

Identifying a species' geographic range is the first step in understanding whether the range is changing or being disturbed, fragmented, or modified by some other process. A species' range is constrained in part by specific environmental limiting factors that collectively define the environmental space of the species or its habitat needs. Other factors, such as dispersal and interspecific competition, further constrain a species' geographic distribution. However, environmental space creates a primary restriction on a species' current geographic distribution that is easily measured and analyzed. Understanding environmental conditions, how those conditions interact across geographic space, and how the spatial distribution of those conditions may change through time can be informative for making conservation decisions. This chapter addresses the following questions: (1) What environmental conditions make up suitable habitat for endemic species of the Colorado Plateau? (2) Where do those conditions currently exist on the landscape? (3) Will those conditions persist in the future?

There are many ways to identify a species' geographic range, from drawing circles on a map to intensive modeling procedures. Species distribution models (SDMs) are one approach to identifying suitable habitat and are widely regarded as the best available tools for producing species-specific information necessary for conservation planning (Hannah 2003). The results of SDMs have been used for conservation planning in many different ways, including (1) to guide field surveys to accelerate detection of unknown distributional areas and undiscovered species, (2) to project potential impacts of climate change, and (3) to predict species invasions (Peterson et al. 2003; Thomas et al. 2004; Bartel and Sexton 2009). Results have also been used for selecting reserves and guiding reintroductions of endangered species (Pearce and Lindenmayer 1998; Guisan and Thuiller 2005).

Over the last two decades, several multivariate techniques have been developed to predict species' distributions. These techniques were first built from presence and absence data. More recently, the need for presence-only models has become popular because of the enormous amounts of presence-only data becoming available through museum collections, herbaria, and other institutions providing online databases (Graham et al. 2004). These presence-only distribution models integrate a wide range of environmental data to project potential habitat for a species based on known occurrences and locations of specific habitat conditions (Phillips et al. 2009).

The biodiversity informatics community has invested substantially in shared repositories of species' occurrence data over the past decade and has developed numerous algorithms for SDMs, including genetic algorithms (Genetic Algorithm for Rule Set Production [GARP]), maximum entropy algorithms (Maxent), and neural networks. These algorithms generate models based on the species' locations and various environmental conditions. Each model is assessed for accuracy by measuring its performance on a set of reserved species locations (test data) that were not used to build the model. Once accurate models are developed, they can be projected onto past, current, and modeled future landscapes by substituting appropriate data. In the case of future landscapes, modeled climate change data are substituted for present-day climate data used to formulate the model. In addition to internal model evaluation, model output can be validated by comparison with expected outcomes based on other validated sources. Outcomes may then be used to inform conservation management decisions.

Among the conservation management decisions public land managers face is the prospect of losing biodiversity on public lands as a consequence of climate change. Through working groups, many federal agencies are addressing the topic of managed relocation (MR) as a potential tool to reduce the negative effects of climate change on biodiversity (see Glossary; Haskins and Keel, this volume). Along with traditional conservation strategies, some major land management units on the Colorado Plateau and in the western United States have developed policies that consider MR. For example, one policy allows MR if it is consistent with multiple-use objectives and the species is desirable, ecologically sound, and noninvasive (Camacho 2010). Another policy allows MR in wilderness areas only if it is consistent with preserving wilderness character and natural conditions (Camacho 2010).

Extensive protocols must be followed for an MR program to be approved. Identifying suitable habitat is part of this process and is critical for the species' success (Maschinski et al. [chap. 7], this volume). Although not all habitat components that are necessary for a species to survive can be relocated, many conditions can be identified with appropriate modeling. Understanding where suitable habitat

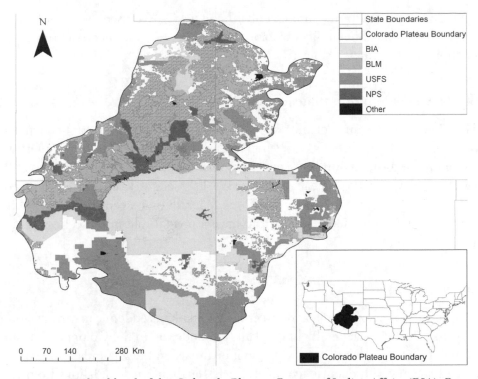

FIGURE 8.1. Federal land of the Colorado Plateau, Bureau of Indian Affairs (BIA), Bureau of Land Management (BLM), US Forest Service (USFS), National Park Service (NPS), and others, including the Department of Defense, Bureau of Reclamation, and Department of Energy.

currently occurs and where it might occur in the future can help land managers identify areas for MR.

In this study we conducted an SDM for twenty endangered or threatened plant species endemic to the Colorado Plateau (fig. 8.1) and used each species' output to understand the species' environmental space, the current habitat occupancy across geographic space, and potential changes in habitat occupancy due to spatial migration of environmental space linked to future climate change. We also identified areas of future habitat outside the species' current range for possible MR locations in addition to areas with constant suitable habitat (i.e., habitat that is suitable from present day to future time periods). Using current species' occurrence data, we identified relevant geologic and climatic conditions that provide suitable habitat for each species, henceforth called *relevant environmental space*. Our goals were (1) to identify environmental space needs for each species, (2) to define which environmental variables were most limiting to each species, (3) to examine model predictions of suitable habitat for current and future climate conditions, and (4) to measure potential habitat loss. We used potential habitat loss to assess the vulnerability of Colorado Plateau endemic species to risk of extinction due to climate change.

Study Area: Colorado Plateau

The Colorado Plateau ecoregion supports one of the highest levels of endemism in the United States; it ranks in the top three ecoregions on the North American continent for the total number of endemics in all taxonomic groups (Ricketts et al. 1999). The Colorado Plateau contains 373,128 square kilometers of land, of which 276,577 square kilometers (74%) is federally managed (fig. 8.1). Public land managers of the Colorado Plateau have a pivotal role in protecting species and ecosystems.

Species of Interest

Our study focused on threatened and endangered plants endemic to the Colorado Plateau. Thirty-four federally listed plants are endemic to the Colorado Plateau. Many of these plants are rare, with few individuals and few populations. We excluded species with occurrence points for fewer than ten populations from our models because prior studies have demonstrated that using fewer than ten populations will not be meaningful without extensive habitat requirement data and defined climate envelope constraints (Stockwell and Peterson 2002). For this study we modeled twenty taxa that met these guidelines.

The selected taxa include seven succulent perennials (*Pediocactus bradyi*, *Pediocactus despainii*, *Pediocactus peeblesianus* var. *fickeiseniae*, *Pediocactus winkleri*, *Sclerocactus glaucus*, *Sclerocactus mesae-verdae*, *Sclerocactus wrightiae*), twelve perennial herbs (*Astragalus humillimus*, *Carex specuicola*, *Cycladenia humilis* var. *jonesii*, *Erigeron maguirei*, *Erigeron rhizomatus*, *Gilia caespitosa*, *Glaucocarpum suffrutescens*, *Lesquerella congesta*, *Packera franciscana*, *Physaria obcordata*, *Schoenocrambe argillacea*, *Townsendia aprica*), and one perennial subshrub (*Eriogonum pelinophilum*). All taxa are listed as endangered or threatened with the US Fish and Wildlife Service.

Model Input Data: Predictor Variables

Our study area comprised federal land on the Colorado Plateau. All data are represented in a geographic information system (GIS) grid. A grid structure divides an image (map) into rows and columns of even size. Each point of data in a grid is known as a pixel or data cell. Each pixel contains location and other data according to the map information. For example, a map of temperature contains data of temperature at each grid pixel. For each grid cell of 1 square kilometer in the study area (fig. 8.1), we calculated values for twenty predictor variables, nineteen climatic and one geologic (table 8.1). We selected these variables because they are biologically meaningful to plants (Parolo et al. 2008). We did not incorporate other commonly used predictor variables such as topography because elevation is highly correlated with temperature and precipitation patterns. Although slope angle regulates soil wetness and erosion, the geologic data used instead provide much more detailed information.

We used the US Geological Survey preliminary integrated geologic map databases for the geologic layer. The database uses state geologic maps. We prepared the database maps using project standards for structure, fields, attribution, and data dictionaries so that each state map could be collated with those of other states. The geologic information consists of a unit link (this creates a unique identifier for every geologic unit), unit name, unit description, unit age, rock type (one is the dominant lithology of the unit, rock type, and two is the second most abundant lithology), province, and stratigraphic unit information (Ludington et al. 2007; Stoeser et al. 2007).

Worldclim, a 30–arc second global climate data model (http://www.worldclim .org), provided nineteen climatic variables that are thought to be biologically important (table 8.1). Derived from monthly averages of temperature and precipitation, these data are measured at weather stations from a large number of global, regional, national, and local sources (Hijmans et al. 2005). The Shuttle Radar Topography Mission (SRTM) elevation database provided elevation data (Hijmans

TABLE 8.1

Environmental predictor variables used in species distribution modeling.

BIO1 = Annual mean temperature (°C)	BIO11 = Mean temperature of coldest quarter (°C)
BIO2 = Mean diurnal temperature range (°C)	BIO12 = Annual precipitation (mm)
BIO3 = Isothermality (°C)	BIO13 = Precipitation of wettest month (mm)
BIO4 = Temperature seasonality (°C)	BIO14 = Precipitation of driest month (mm)
BIO5 = Maximum temperature of warmest month (°C)	BIO15 = Precipitation seasonality (mm)
BIO6 = Minimum temperature of coldest month (°C)	BIO16 = Precipitation of wettest quarter (mm)
BIO7 = Temperature annual range (°C)	BIO17 = Precipitation of driest quarter (mm)
BIO8 = Mean temperature of wettest quarter (°C)	BIO18 = Precipitation of warmest quarter (mm)
BIO9 = Mean temperature of driest quarter (°C)	BIO19 = Precipitation of coldest quarter (mm)
BIO10 = Mean temperature of warmest quarter (°C)	Geology = Landform description categorical units

et al. 2005). These data are interpolated using the thin-plate smoothing spline algorithm implemented in ANUSPLIN software (Hijmans et al. 2005). Latitude, longitude, and elevation were the independent variables used for cross-validation (Hijmans et al. 2005). We transformed monthly data into the nineteen bioclimatic variables. We tested predictor variables for correlation (see supplementary tables 8.S1–8.S3).

The Intergovernmental Panel on Climate Change (IPCC) database provided future climate data (Nakicenovic et al. 2000). We modeled four representative scenarios and three different climate models (BCCR, CSIRO, and NIES) at three time steps. The three selected models originated from different modeling centers to represent a range of future conditions: Bjerknes Centre for Climate Research (BCCR) is from the University of Bergen, Norway; CSIRO is from Australia's Commonwealth Scientific and Industrial Research Organization; and NIES is from Climate Risk Assessment Research Section of the Center for Global Environmental Research, National Institute for Environmental Studies, Japan.

For each modeling center, we chose and compared four scenarios related to future economic and environmental conditions provided by the IPCC Special Report on Emissions Scenarios; these have the abbreviations SRA1B, SRB1, SRA2, and 1PTO2X (BCCR 2005, 2006a, 2006b; Collier 2005a, 2005b, 2005c; Nozawa 2005a, 2005b, 2005c, 2005d, 2005e). The key assumptions for SRA1B are a future world of very rapid economic growth, low population growth, and rapid introduction of new and more efficient technology. Major underlying themes are economic and cultural convergence and capacity building, with a substantial reduction in regional differences in per capita income. In this world, people pursue personal wealth rather than environmental quality. Key assumptions of scenario SRB1 are a convergent world with the same global population as in the A1 scenario but with rapid changes in economic structures toward a service and information economy, with reductions in material intensity, and the introduction of clean and resource-efficient technologies. Key assumptions for SRA2 are a very heterogeneous world. The underlying theme is that of strengthening regional cultural identities, with an emphasis on family values and local traditions, high population growth, and less concern for rapid economic development. The 1PTO2X scenario is an experiment run with greenhouse gases increasing from preindustrial levels at a rate of 1% per year until the concentration has doubled and held constant thereafter.

The future climate scenarios we studied all show an increase in mean annual temperature; nearly all scenarios also predict a drier future. The NIES SRA2 2070–2099 scenario is the warmest, with mean annual temperature increasing by 6°C. The NIES SRA1B 2070–2099 scenario is the driest scenario, with a decrease in annual precipitation of 90 millimeters, and the BCCR SRB1 scenario is the wettest, with an increase in precipitation of 26 millimeters. For each model and scenario, we used minimum temperature, maximum temperature, mean temperature, and precipitation and transformed them into bioclimatic variables to match the Worldclim data (see supplementary tables for detailed information about each model and scenario).

It is standard practice to average climate simulation data into 30-year increments to remove year-to-year variation, yielding a "current" projection that is the average of 1960–1990 and future projections in 30-year increments beginning in 2010 (e.g., 2010–2039, 2040–2069, and 2070–2099). The modeled "current" projection is validated against observed climate data for the period of 1960–1990, which has been assimilated globally for all of the modeled climate parameters. General climate models (GCMs) typically have coarse spatial resolution (on the order of 300 square kilometers depending on the GCM). In order to generate finer spatial resolution projections needed for SDM, each future 30-year period is compared to the modeled current prediction, and a change value (anomaly) is calculated. The anomaly value is then added to the much finer spatial resolution (1 kilometer) observed climate data to generate high–spatial resolution future climate projections that are calibrated to the observation data (see Wilby et al. 2004 for details).

Model Input Data: Species Occurrence Points

Species location points are from Natural Heritage Programs of the four corner states and the Navajo Nation, regional herbaria, Bureau of Land Management (BLM), and National Park Service parks. We requested information from each of these institutions as point data with latitude and longitude, date of collection, and a general location description. We checked all points for quality by visual analysis in GIS. If a point was outside the plateau boundary, it was automatically removed. If a point was an outlier of other points, we used the general location description to cross-check the latitude and longitude. We used the date the species was collected or sampled to match up species location points with the same timeframe of the current-day climate variables. If the point date was before 1950 or after 2000, we omitted the points from the analysis. We eliminated points collected after 2000 because those populations may already be influenced by current-day climate change and may not coincide with climate variables, whereas points with a collection date before 1950 may represent extinct populations. We did not remove points with duplicate latitude and longitude because these locations provided an emphasis on suitable habitat needs.

Species Distribution Modeling

In this study we used Maxent as the modeling technique (Phillips et al. 2006). Maxent estimates a target probability distribution by finding the probability distribution of maximum entropy. Entropy is a measure of the uncertainty associated with a random variable, and maximum entropy ranges from the point that is closest to uniform to the point that is the most widespread (Phillips et al. 2006). The higher the entropy, the more choices or less constrained a model is. The constraints used in Maxent are the values of the environmental variables compared with the average environmental values from existing sites where species occur. The probability distribution produced is subjected to these constraints, and no unfound constraint is placed on the distribution (Phillips et al. 2006). The distribution then "agrees with everything we know but carefully avoids assuming anything we do not know" (Phillips et al. 2006, p. 236). An important reason for choosing Maxent was that it allowed us to use "presence-only" species data and continuous and categorical environmental variables. In addition, Maxent has been shown to perform better than other algorithms for modeling distributions with limited data points (Elith et al. 2006; Pearson et al. 2007).

For each species, we built a distribution model with the twenty predictor variables and projected models onto the four representative scenarios and three different climate models (BCCR, CSIRO, and NIES) at three time steps. After model completion, we applied a threshold value to each output grid to transform the data from a logistic output to a binary output. The threshold selected was equal training sensitivity and specificity. The application of the model influenced the threshold choice, because we were interested in identifying suitable habitat across the study area boundary, so low thresholds were sufficient. If our focus were to identify areas for reintroduction or another costly investment, a higher threshold may be needed. For a more complete review of threshold selection, see Liu and colleagues (2005).

Model Evaluation

It is important to use more than one test metric for performance testing because each metric quantifies a different aspect of the models' predictive strength (Elith and Graham 2009). We evaluated the models and verified that they performed better than random using two evaluation methods. The first was a threshold-dependent binomial test based on omission and predicted area. It uses the extrinsic omission rate (i.e., the fraction of test points that fall into pixels not predicted suitable) and the proportional predicted area (i.e., the fraction of all pixels that are predicted as suitable) (Phillips et al. 2006).

The second evaluation method is a threshold-independent receiver operating characteristic (ROC) analysis that characterizes model performance at all possible thresholds by a single number: the area under the curve (AUC) (Phillips et al. 2006). Traditionally, researchers used the ROC curve to evaluate the accuracy of the model and each variable's predictive power (Hanley and McNeil 1982). The ROC curve represents the relationship between the percentage of presences correctly predicted (sensitivity) and one minus the percentage of the absences correctly predicted (specificity). The AUC measures the ability of the model to classify correctly a species as present or absent. AUC values can be interpreted as the probability that a site with the species present will have a higher predicted value than a site with the species absent when both are drawn at random. We modified the AUC to accommodate presence-only data by using presence versus random, rather than presence versus absence comparisons. Although using AUC test statistics has received criticism in recent years (Lobo et al. 2008), it is still viewed as an important metric for evaluating predictive performance (Elith and Graham 2009). Following Araujo and Guisan (2006), a rough guide for classifying the model accuracy is 0.5–0.6 = insufficient, 0.6–0.7 = poor, 0.7–0.8 = average, 0.8–0.9 = good, and 0.9–1.0 = excellent.

Sample Analysis

We used a sample analysis to identify the environmental limiting factors that collectively define the environmental space or habitat needs of each species. The sample analysis GIS function, located within the Spatial Analyst Tools of ArcMap 9.3, creates a table that shows the values of cells from a raster or set of rasters for defined location points (ESRI 2009). We used sample analysis to compare the environmental space identified from known point locations and current climate and geology with the modeled potential distributions. The sample analysis generated specific values from the predictor variables at each occurrence point. This provided a range of observed values for each predictor variable for each species and defined the species' environmental space. We compared the values and defined environmental space generated from the sample analysis with the response curves and environmental space identified by Maxent.

For the sample analysis, we considered a variable to be a good predictor of environmental space if the current and potential distribution ranges of variables were within 10 millimeters for precipitation, within 1°C for temperature, and limited to one geology layer. We compared these variables with the top five predictor variables from the Maxent distribution model.

(a) Sample Analysis

Species	BIO 1	BIO 2	BIO 3	BIO 4	BIO 5	BIO 6	BIO 7	BIO 8	BIO 9	BIO 10	BIO 11	BIO 12	BIO 13	BIO 14	BIO 15	BIO 16	BIO 17	BIO 18	BIO 19	Geology
Astragalus humillimus		x	x											x						x
Carex specuicola			x										x	x	x	x		x	x	x
Cycladenia humilis var. jonesii		x	x											x	x					
Erigeron maguirei	x		x			x	x	x					x	x	x					
Erigeron rhizomatus			x										x	x	x					
Eriogonum pelinophilum	x		x											x						x
Gilia caespitosa		x	x		x		x		x				x	x	x	x			x	x
Glaucocarpum suffrutescens			x										x	x	x					x
Lesquerella congesta	x		x										x	x		x		x	x	x
Packera franciscana		x	x																	x
Pediocactus bradyi			x										x	x	x		x			
Pediocactus despainii		x	x				x							x	x		x		x	x
Pediocactus peeblesianus var fickeiseniae			x										x	x	x		x			
Pediocactus winkleri		x	x							x	x		x	x			x			x
Physaria obcordata	x	x	x											x						x
Schoenocrambe argillacea														x			x			
Sclerocactus glaucus	x	x	x			x		x	x	x	x		x	x	x		x			x
Sclerocactus mesae-verdae	x	x	x			x								x	x		x		x	x
Sclerocactus wrightiae			x					x						x			x		x	x
Townsendia aprica	x	x	x								x			x	x		x		x	
Totals	7	10	19	0	1	3	3	3	1	2	3	0	10	19	12	3	9	2	7	13

(b) SDM Variable Importance

Species	BIO 1	BIO 2	BIO 3	BIO 4	BIO 5	BIO 6	BIO 7	BIO 8	BIO 9	BIO 10	BIO 11	BIO 12	BIO 13	BIO 14	BIO 15	BIO 16	BIO 17	BIO 18	BIO 19	Geology
Astragalus humillimus							1		2		3			4	5					
Carex specuicola		5	1						4						2					3
Cycladenia humilis var. jonesii			2	1		4			3								5			
Erigeron maguirei			2						1					4	3					5
Erigeron rhizomatus			2		4				5						3					1
Eriogonum pelinophilum			3						1		5				4					2
Gilia caespitosa			3		2				4						1				5	
Glaucocarpum suffrutescens			1				3		2									5		4
Lesquerella congesta							2	3	1										5	4
Packera franciscana			3	4			2							5					1	
Pediocactus bradyi			4						2							5	1			3
Pediocactus despainii			4				5		1			3							2	
Pediocactus peeblesianus var fickeiseniae			4			2			5								3			1
Pediocactus winkleri			3				2		1			5							4	
Physaria obcordata		4					3	1	2											5
Schoenocrambe argillacea			3	1			5		2											4
Sclerocactus glaucus	3		5	2				4	1											
Sclerocactus mesae-verdae			5	1		3			2					4						
Sclerocactus wrightiae							4		2			3			5				1	
Townsendia aprica		3							1						4				2	5
Totals	1	3	15	5	2	3	6	7	18	0	2	3	2	3	8	1	2	1	7	11

FIGURE 8.2. Variable importance for (a) sample analysis and (b) species distribution modeling (SDM). The top five predictor variables with corresponding number of importance are indicated, with 1 = highest importance.

Extinction Risk

We calculated extinction risk for each species. We compared the percentage of habitat (or pixels) lost at each time step to initial suitable habitat. We categorized species' extinction threat using a modified version of the IUCN Red List threat categories (IUCN Species Survival Commission 2004). We calculated the habitat loss with no dispersal using the extent of the Colorado Plateau, and we calculated habitat loss with full dispersal using the western United States extent. We categorized a species as extinct if it had projected habitat loss of 100%, critically endangered if projected habitat loss was more than 80%, endangered if projected habitat loss was more than 50%, vulnerable if projected habitat loss was more than 30%, and stable if habitat loss was less than 30% or there was a gain in potential habitat. We calculated risk totals by adding total species for each scenario. We created ensemble maps by overlaying all species and identifying areas where multiple species find suitable habitat under current and extreme scenarios.

Dispersal Analysis

In addition, we applied two types of dispersal assumptions to the species distribution modeling: no dispersal and full dispersal. The no-dispersal assumption prevents movement between pixels from the present to the future. Therefore, pixels were identified as suitable in the future only if they were identified as suitable habitat in the present-day Colorado Plateau. The full dispersal assumption allowed suitable habitat to be located anywhere in the western United States.

Model Results

Model predictive performance varied between the species and two test metrics. AUC scores varied between species; all were considered very good or excellent (table 8.2). *Sclerocactus wrightiae* had the lowest training and test AUCs of 0.8952 and 0.8555, respectively, which rank as "good," and all other species had AUC scores of 0.9 and above, which rank "excellent." The current potential distribution varies in range size from the smallest, 741 square kilometers (*Erigonum pelinophilum*), to the largest, 67,105 square kilometers (*Erigeron rhizomatus*). The threshold-dependent test metric identified that nearly all model omission rates were less than 1% (see supplementary tables).

Environmental Space and Variable Importance

The sample analysis identified isothermality (mean diurnal temperature range divided by the annual temperature range) and precipitation of the driest month as

TABLE 8.2

Model performance ranked by area under the curve (AUC) values for training and testing each species distribution model and the number of location points used for each species.

	Training Points	Test Points	AUC Training	AUC Test
Astragalus humillimus	13	5	0.993	0.987
Carex specuicola	77	31	0.983	0.988
Cycladenia humilis var. *jonesii*	26	9	0.984	0.975
Erigeron maguirei	21	8	0.9956	0.9987
Erigeron rhizomatus	24	9	0.991	0.921
Eriogonum pelinophilum	38	15	0.9997	0.9995
Gilia caespitosa	24	9	0.9998	0.9998
Glaucocarpum suffrutescens	46	19	0.9994	0.9964
Lesquerella congesta	49	21	0.9997	0.9996
Packera franciscana	10	4	0.9993	1
Pediocactus bradyi	29	12	0.9972	0.9791
Pediocactus despainii	17	7	0.9963	0.9931
Pediocactus peeblesianus var. *fickeiseniae*	40	17	0.9928	0.9866
Pediocactus winkleri	10	4	0.9966	0.9975
Physaria obcordata	35	14	0.994	0.9988
Schoenocrambe argillacea	57	24	0.9999	0.9996
Sclerocactus glaucus	188	80	0.9987	0.9963
Sclerocactus mesae-verdae	54	23	0.9969	0.9935
Schlerocactus wrightiae	81	34	0.8952	0.8555
Townsendia aprica	51	21	0.9964	0.9925

important variables for nineteen of the twenty species (fig. 8.2a). Isothermality is a quantification of how large the day-to-night temperature oscillation is in comparison to the summer-to-winter oscillation. A value of 100 would represent a site where the diurnal temperature range is equal to the annual temperature range. A value of 50 would indicate a location where the diurnal temperature range is half of the annual temperature range. Geology and precipitation seasonality were also considered important for more than half of the species. Maxent provides a variable importance ranking for how environmental variables were used to build each model. This ranking identified mean temperature of the driest quarter as the strongest predictor variable for the greatest number of species (eighteen of twenty; fig. 8.2b). Isothermality and geology were also strong predictors for more than half the plants.

Isothermality and geology are important variables for predicting endemism on the Colorado Plateau according to the comparison of SDMs and sample analyses. The unique geologic patterns of the plateau correspond with the high number of endemics. Isothermality will have important consequences as temperatures rise over the next hundred years.

Habitat Loss or Gain

Under the no-dispersal assumption, all three extreme climate change scenarios (BCCR SRB1, NIES SRA2, and NIES SRA1B) projected that most species will lose a majority of their habitat in the 2010–2039 time step (fig. 8.3). The NIES SRA1B model projected 100% loss for all but one species (*P. obcordata*), which showed a net 22% loss over the next hundred years (fig. 8.3). The full-dispersal assumption resulted in more species having habitat gain for all three models, although 75% of plants still showed a loss. Under the full-dispersal assumption, seven species showed the majority of loss during the 2010–2039 time step (*A. humillimus, C. specuicola, G. caespitosa, P. franciscana, P. despainii, S. mesae-verdae,* and *T. aprica*). The remainder show major losses from 2040–2069, the second time step. For plants that gained suitable habitat, the majority of gains were equal between time steps but varied dramatically between scenarios.

Risk Analysis

The highest extinction risk assuming no dispersal and a warmer, dry future (NIES SRA2) was in 2070–2099, with twelve of twenty species predicted to become extinct (fig. 8.4). The lowest risk of extinction assuming no dispersal was found with scenario BCCR SRB1 in the 2010–2039 time step, with twelve out of twenty plants stable. Comparing results from the three extreme scenarios (fig. 8.4) during the 2070–2099 time step, the warmest scenario (NIES SRA2) has twelve species in the extinct category, five critically endangered, one vulnerable, and two stable. The driest (NIES SRA1B) predicted ten in the extinct category, seven critically endangered, one vulnerable, and two that are stable. The wettest scenario (BCCR SRB1) predicted one extinct, two critically endangered, six endangered, five vulnerable, and six stable species. *Pediocactus bradyi* and *P. obcordata* were found to be stable in the majority of scenarios, suggesting that these species may do well in a warmer climate.

Climate change imparts lower extinction risk when full dispersal is assumed (i.e., when species are given an unlimited ability to move across the landscape in the models). Comparing the three extreme scenarios in 2070–2099 with full dispersal, the warmest (NIES SRA2) predicted six species in the extinct category, five critically endangered, zero endangered, one vulnerable, zero stable, and five that gain habitat (fig. 8.4). The driest (NIES SRA1B) predicted four in the extinct category, six critically endangered, one endangered, one vulnerable, zero stable, and

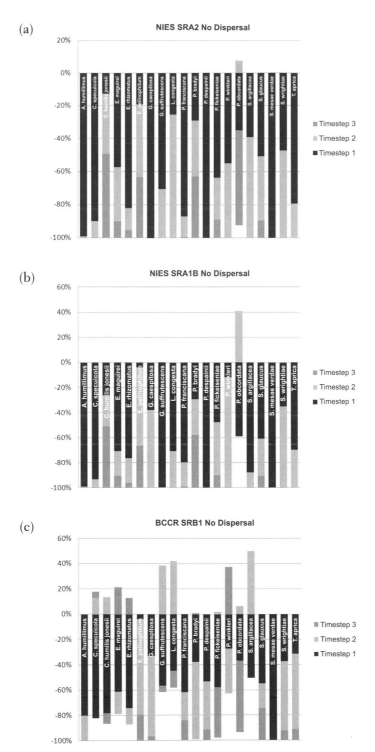

FIGURE 8.3. Percentage of the initial habitat that is lost at each time step for **(a)** the warmest (NIES SRA2), **(b)** driest (NIES SRA1B), and **(c)** wettest (BCCR SRB1) climate change scenarios, where species have no dispersal ability.

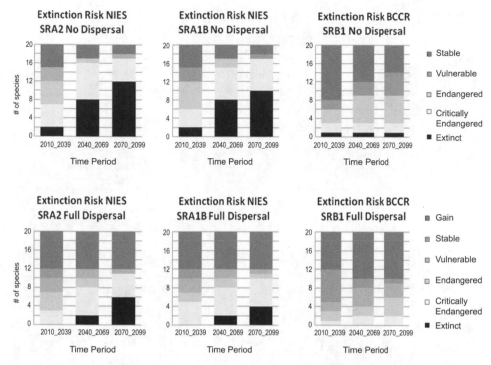

FIGURE 8.4. Extinction risk under no dispersal and full dispersal for the warmest (NIES SRA2), driest (NIES SRA1B), and wettest (BCCR SRB1) climate change scenarios through time.

eight that gain habitat. The wettest (BCCR SRB1) predicted no species in the extinct category, two critically endangered, four endangered, three vulnerable, one stable, and ten that gain habitat. *Glaucocarpum suffrutescens, P. despainii, P. winkerlii,* and *T. aprica* were categorized as extinct for the driest and warmest scenarios.

Areas of Change

Ensemble maps of all species for the three extreme scenarios show a concentration of suitable habitat in the northwestern portion of the plateau under the no-dispersal assumption (NIES SRA2 shown in fig. 8.5). This area has suitable habitat for six species: *C. humilis* var. *jonesii, E. maguirei, P. bradyi, P. winkleri, S. wrightiae,* and *T. aprica*. Land management of this area is through Capital Reef National Park, Mount Ellen–Blue Hills BLM Wilderness Area, and the BLM.

Full dispersal ensemble maps have a concentration of suitable habitat in the area northwest of the plateau and farther north in Utah. The area outside the plateau has a concentration of six species: *E. maguirei, G. suffrutescens, P. wink-*

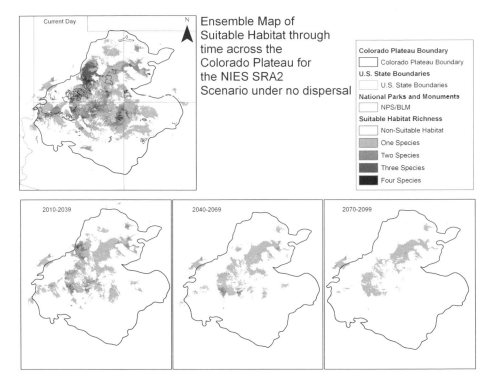

FIGURE 8.5. Ensemble maps of suitable habitat for all plants for the NIES SRA2 climate change scenario under no dispersal in current day, 2010–2039, 2040–2069, and 2070–2099 time steps.

leri, S. argillacea, S. wrightiae, and *T. aprica.* The BLM provides land management in this area.

Discussion

Threatened and endangered plants of the Colorado Plateau are already vulnerable due to habitat loss, fragmentation, and invasive species. Climate change may put these species at even higher extinction risk. By understanding a species' environmental space, we may be able to begin protecting important future habitats. The Colorado Plateau is uniquely located between the Rocky Mountains to the east and the Great Basin desert from the west—two opposing moisture trajectories—which provides for an ecosystem that could be among the most sensitive to past and future drivers of climate change (Schwinning et al. 2008). Climate change projections for the Colorado Plateau vary between models, but a consensus of twenty-two downscaled models predicts annual temperatures to exceed the

1950–1999 range of variability by the 2030s (Garfin et al. 2007). Annual precipitation in the southern region of the Colorado Plateau may decrease by 6% by the end of the century, with spring precipitation declining by 11–45% in the twenty-first century (Garfin et al. 2007). However, increases in summer monsoon rain patterns are predicted across the plateau. This predicted variability in precipitation may be especially important to rare plants on the plateau because winter–spring recharge of soil moisture is important for all plants (Ehleringer et al. 1991). For the Colorado Plateau, soil moisture recharge results primarily from winter precipitation, whereas summer precipitation recharges only the uppermost layers of soil (Fernandez and Caldwell 1975). Related to these changes are higher summer temperatures, which may limit moisture intake and were also significant factors in the models. Lin and colleagues (1996) found evidence of niche differentiation among plants with respect to soil moisture use in arid and semiarid woodland ecosystems.

In addition to geology, the narrow range of isothermality and precipitation of driest month for the majority of species suggest that these variables are controlling factors for the geographic distribution of many of these threatened plants. Isothermality may play an important role with a changing climate because increased nighttime temperatures may accelerate carbon loss through increased respiration rates (Alward et al. 1999) and may decrease temperature sensitivity (Luo et al. 2001; Atkin and Tjoelker 2003). Importance rankings from the sample analysis and models matched some species with geology and isothermality. For the majority of species, minimum temperature was a controlling factor, and maximum temperature was less influential for suitable habitat predictions. A majority of climate change models predict a warmer future. This may be less influential on these Colorado Plateau plants, because maximum temperatures may not be constraining their distributions.

Practical Implications for Traditional Restoration Work

Preparing conservation strategies requires that we can predict when a plant might lose a majority of its suitable habitat. For the three extreme scenarios, the majority of habitat was lost in the first time step (2010–2039). The ensemble maps predict areas where suitable habitat might persist into the future. These areas should have higher conservation value not only for the species that can naturally disperse but also for conservation and relocation efforts. It is important to note that these temporally stable areas identified in the ensemble maps are not currently on protected lands. These findings may provide conservation planners better guidance for identifying places where endangered plants may find suitable habitat in the future and where future protected areas could provide the most protection for the most species.

Areas of Need and Research Opportunities

SDMs provide tools to better understand a species' environmental space and characteristics of the species' niche. Although many models can incorporate dispersal mechanics, competition, and many other biotic factors that control species' distribution, continued research on how best to model these factors is needed and will help predict climate change impacts.

Prospects and Cautions for Appropriate Use of MR

For the species we studied and for many other endangered plants, the extent to which distributions are limited by suitable habitat or by barriers to movement and stochastic processes is not well understood (Wiser et al. 1998). This issue can be seen with some of the plants modeled under the full dispersal assumption; with dispersal limitations and other biotic restraints removed, the models identify suitable habitat outside their current range. Yet it is unlikely that these plants will be able to disperse to new areas without human assistance. Whether these areas can be used in the future for reintroduction remains to be tested.

What makes this study unique among distribution applications is that we tracked habitat through time to identify areas where a species might be able to move with the changing climate. For many of the species studied, the no-dispersal assumption is the closest to reality. These species will not be able to disperse outside a 1-square-kilometer grid cell. The ability to move across the landscape is important for many of these species if they are to persist in the future. For many endangered plants of the plateau, dispersal greater than 1 kilometer is unlikely. A review of the species literature and biology revealed that eight of the twenty plants have no dispersal ability outside a 1-square-kilometer pixel.

The other twelve plants probably have some dispersal ability with assistance from seed dispersers. Areas along riparian corridors for hanging garden species may also allow longer dispersal. However, we examined neither of these dispersal mechanisms in this study. None of the plants have dispersal abilities that resemble the full-dispersal model. A more realistic, intermediate assumption between the extremes of no dispersal and full dispersal may provide a better understanding of how species might be able to disperse to suitable habitat in the future. The full-dispersal model allowed us to identify suitable habitat available for MR. It is clear from our findings that MR may become the only viable option for endangered Colorado Plateau plants to persist in the future. Species distribution modeling can help identify areas of current and future suitable habitat. Although there are few publications on the use of species distribution modeling and its use with MR, we believe that identifying suitable habitat areas outside a species' historical range provides insights on how to manage for species conservation. It is essential for reintroduction projects to identify areas where a species will persist in the future under a changing climate.

Summary

Although the climate models make different predictions, the majority of the models trend toward a drier and warmer future. If these assumptions are correct, this study shows that more than half the endangered and threatened plants of the Colorado Plateau may go extinct in the next hundred years. If future climates are warmer and wetter, the extinction risk is dramatically lower. Although precipitation data from GCMs are still not as reliable as temperature data, scenario comparison provides us with a better understanding of the range of potential change in the future and the source of uncertainty. Our results are comparable to extinction risks predicted by others. For example, Thomas and colleagues (2004) found that as many as 37% of species may be destined for extinction. Midgley and colleagues (2003), working on plant diversity in the Cape Floristic Region of Africa, predicted that four of the twenty-eight Proteaceae species would experience range elimination, seventeen would experience range contractions, and eleven would experience range expansions. As others have found, our results show variability in possible responses under climate change. There is growing evidence of potential shifts in species' ranges under future climate change. Predictions formed from SDMs are considered very useful, but there are still uncertainties. Some key assumptions of SDMs are crucial for understanding model performance: A species is at equilibrium with the environment, the local environment has been sampled so gradients can be identified, and dispersal limitations will constrain movement based on parameters set (Wiens et al. 2009; also see Elith and Leathwick 2009 for review). As in all modeling experiments, model calibration and execution should be well known to best understand predictions.

The Colorado Plateau provides a unique setting with numerous protected areas. Although many endangered plants currently occupy some of these protected areas, it is predicted that the majority of endangered plants will not find suitable habitat in protected areas in the future. We argue that modeling results from numerous future climate scenarios, different dispersal abilities, and GIS sample analysis can facilitate strategic conservation decisions by providing insights about a species' environmental space in current and future timeframes. Only with this information will it be possible to make decisions about reintroductions.

The authors thank Aimee Stewart at the University of Kansas Biodiversity Institute for processing the dispersal models and the University of New Mexico Center for Advanced Research Computing for the use of their high-performance computing center to run the Maxent models. Neil Cobb, Tina Ayers, and Randall Scott provided useful comments and suggestions on a previous version of the manuscript. This research was funded in part by National Science Foundation grant 0753336 and the Gloria Barron Wilderness Society Scholarship for CMK.

TABLE 8.S1

Correlation matrix showing the values of the correlation coefficients that depict the relationship between two environmental variable grids.

Layer	1	2	3	4	5	6	7	8	9	10	11	12	13	14	15	16	17	18	19
1	1	0.15985	0.18925	0.02879	0.94682	0.92927	-0.03998	0.63168	0.64268	0.97279	0.95432	-0.59457	-0.30378	-0.8667	0.48572	-0.38675	-0.83725	-0.4465	-0.4174
2	0.15985	1	0.7272	-0.2161	0.18921	0.00676	0.28321	0.29403	0.01824	0.09591	0.18668	-0.08108	0.2602	-0.23684	0.55809	0.22596	-0.28388	0.22967	-0.20874
3	0.18925	0.7272	1	-0.80644	0.00634	0.28091	-0.44631	0.25	0.31758	0.0024	0.395	0.31153	0.70035	-0.1504	0.74238	0.64268	-0.14325	0.62204	0.24902
4	0.02879	-0.2161	-0.80644	1	0.30463	-0.23577	0.85654	-0.003	-0.3477	0.24371	-0.25533	-0.62085	-0.842	-0.13319	-0.51004	-0.80036	-0.18377	-0.77537	-0.60628
5	0.94682	0.18921	0.00634	0.30463	1	0.79837	0.25568	0.59249	0.48639	0.9874	0.82983	-0.75393	-0.52855	-0.87264	0.32183	-0.59693	-0.85953	-0.64692	-0.59127
6	0.92927	0.00676	0.28091	-0.23577	0.79837	1	-0.37101	0.5362	0.73038	0.85789	0.97557	-0.38834	-0.11362	-0.76518	0.50024	-0.20048	-0.70884	-0.2689	-0.18671
7	-0.03998	0.28321	-0.44631	0.85654	0.25568	-0.37101	1	0.04927	-0.43046	0.13929	-0.29705	-0.54185	-0.63542	-0.11482	-0.31009	-0.6004	-0.18573	-0.56701	-0.61806
8	0.63168	0.29403	0.25	-0.003	0.59249	0.5362	0.04927	1	0.36169	0.6085	0.59841	-0.46393	-0.10986	-0.66802	0.53174	-0.19361	-0.68486	-0.17086	-0.44823
9	0.64268	0.01824	0.31758	-0.3477	0.48639	0.73038	-0.43046	0.36169	1	0.55215	0.72574	-0.13376	0.06577	-0.50334	0.42012	0.00369	-0.4138	-0.08304	0.05636
10	0.97279	0.09591	0.0024	0.24371	0.9874	0.85789	0.13929	0.6085	0.55215	1	0.87446	-0.71639	-0.48418	-0.87423	0.35602	-0.55572	-0.8549	-0.60934	-0.53883
11	0.95432	0.18668	0.395	-0.25533	0.82983	0.97557	-0.29705	0.59841	0.72574	0.87446	1	-0.39964	-0.06783	-0.7972	0.59652	-0.15943	-0.75121	-0.22581	-0.22478
12	-0.59457	-0.08108	0.31153	-0.62085	-0.75393	-0.38834	-0.54185	-0.46393	-0.13376	-0.71639	-0.39964	1	0.82407	0.74742	-0.1405	0.86809	0.80555	0.85824	0.94237
13	-0.30378	0.2602	0.70035	-0.842	-0.52855	-0.11362	-0.63542	-0.10986	0.06577	-0.48418	-0.06783	0.82407	1	0.38961	0.41278	0.98854	0.40902	0.97421	0.71429
14	-0.8667	-0.23684	-0.1504	-0.13319	-0.87264	-0.76518	-0.11482	-0.66802	-0.50334	-0.87423	-0.7972	0.74742	0.38961	1	-0.54643	0.47405	0.97292	0.52251	0.62172
15	0.48572	0.55809	0.74238	-0.51004	0.32183	0.50024	-0.31009	0.53174	0.42012	0.35602	0.59652	-0.1405	0.41278	-0.54643	1	0.33057	-0.58187	0.30049	-0.20183
16	-0.38675	0.22596	0.64268	-0.80036	-0.59693	-0.20048	-0.6004	-0.19361	0.00369	-0.55572	-0.15943	0.86809	0.98854	0.47405	0.33057	1	0.49438	0.98359	0.7469
17	-0.83725	-0.28388	-0.14325	-0.18377	-0.85953	-0.70884	-0.18573	-0.68486	-0.4138	-0.8549	-0.75121	0.80555	0.40902	0.97292	-0.58187	0.49438	1	0.52561	0.71809
18	-0.4465	0.22967	0.62204	-0.77537	-0.64692	-0.2689	-0.56701	-0.17086	-0.08304	-0.60934	-0.22581	0.85824	0.97421	0.52251	0.30049	0.98359	0.52561	1	0.70551
19	-0.4174	-0.20874	0.24902	-0.60628	-0.59127	-0.18671	-0.61806	-0.44823	0.05636	-0.53883	-0.22478	0.94237	0.71429	0.62172	-0.20183	0.7469	0.71809	0.70551	1

The correlation matrix represents the pixel values from one raster layer as they relate to the pixel values of another layer. The correlation between the two layers is a measure of dependence between the layers. It is the ratio of the covariance between the two layers divided by the product of their standard deviations. Because it is a ratio, it is unitless. Correlation ranges from +1 to −1. A positive correlation indicates a direct relationship between two layers; for example, when the cell value of one layer increases, the cell values of another layer are also likely to increase. A negative correlation means that one variable changes inversely to the other. A correlation of zero means that two layers are independent of one another.

TABLE 8.S2

Climate data trends of mean annual temperature and annual precipitation for the eleven model scenarios, the three time periods, and current day data from the Worldclim dataset.

Model Scenario	Time Period	Mean Annual Temperature (°C)	Annual Precipitation (mm)
BCCR 1PTO2X	2010	9.47	346.54
BCCR SRA1B	2010	9.89	340.60
BCCR SRB1	2010	9.94	347.72
CSIRO 1PTO2X	2010	9.60	361.79
CSIRO SRA1B	2010	9.82	344.57
CSIRO SRB1	2010	9.52	347.57
NIES 1PTO2X	2010	9.97	354.90
NIES 1PTO4X	2010	9.97	354.90
NIES SRA1B	2010	11.06	335.70
NIES SRA2	2010	10.97	327.46
NIES SRB1	2010	10.89	319.41
BCCR 1PTO2X	2040	10.47	361.12
BCCR SRA1B	2040	11.30	352.28
BCCR SRB1	2040	10.72	344.87
CSIRO 1PTO2X	2040	10.41	326.80
CSIRO SRA1B	2040	10.60	327.93
CSIRO SRB1	2040	10.07	345.40
NIES 1PTO2X	2040	9.97	380.11
NIES 1PTO4X	2040	11.57	380.11
NIES SRA1B	2040	12.90	283.26
NIES SRA2	2040	12.84	279.07
NIES SRB1	2040	12.14	307.13
BCCR 1PTO2X	2070	11.03	317.59
BCCR SRA1B	2070	12.33	355.14
BCCR SRB1	2070	11.14	388.20
CSIRO SRA1B	2070	10.75	344.18
CSIRO SRB1	2070	11.14	334.35
NIES 1PTO2X	2070	12.81	330.44
NIES 1PTO4X	2070	13.39	348.75
NIES SRA1B	2070	14.61	272.23
NIES SRA2	2070	15.30	281.25
NIES SRB1	2070	13.07	318.29
Worldclim	Present day	8.91	362.07

Note that IPCC did not provide data for CSIRO 1PTO2X 2070–2099, so that scenario is not included in this table.

TABLE 8.S3

Threshold-dependent test results.

Species	Suitable Pixels	Unsuitable Pixels	Total Pixels	Test Points	Incorrect Prediction	Omission Rate	Proportional Predicted Area	Probabilities of the Binomial Test
Astragalus humillimus	38,376	703,421	741,797	5	0	0.00	0.05	0.77
Carex specuicola	46,629	695,168	741,797	31	0	0.00	0.06	0.13
Cycladenia humilis var. jonesii	57,366	684,431	741,797	9	1	0.11	0.08	0.48
Erigeron maguirei	16,237	725,560	741,797	8	0	0.00	0.02	0.84
Erigeron rhizomatus	61,555	680,242	741,797	9	1	0.11	0.08	0.46
Eriogonum pelinophilum	741	741,056	741,797	15	1	0.07	0.00	0.99
Gilia caespitosa	1,493	740,304	741,797	9	0	0.00	0.00	0.98
Glaucocarpum suffrutescens	7,142	734,655	741,797	19	0	0.00	0.01	0.83
Lesquerella congesta	963	740,834	741,797	21	0	0.00	0.00	0.97
Packera franciscana	5,526	736,271	741,797	4	0	0.00	0.01	0.97
Pediocactus bradyi	26,999	714,798	741,797	12	2	0.17	0.04	0.64
Pediocactus despainii	8,949	732,848	741,797	7	0	0.00	0.01	0.92
Pediocactus peeblesianus var. fickeiseniae	34,738	707,059	741,797	17	1	0.06	0.05	0.44
Pediocactus winkleri	8,859	732,938	741,797	4	0	0.00	0.01	0.95
Physaria obcordata	1,798	739,999	741,797	14	2	0.14	0.00	0.97
Schoenocrambe argillacea	1,496	740,301	741,797	24	1	0.04	0.00	0.95
Sclerocactus glaucus	13,382	728,415	741,797	80	3	0.04	0.02	0.23
Sclerocactus mesae-verdae	11,408	730,389	741,797	23	2	0.09	0.02	0.70
Sclerocactus wrightiae	11,914	729,883	741,797	34	1	0.03	0.02	0.58
Townsendia aprica	14,952	726,845	741,797	21	1	0.05	0.02	0.65

For each species we report suitable, unsuitable, and total pixels, test points, incorrect prediction omission rate, proportional predicted area, and the probabilities of the binomial test. The binomial test was based on omission and predicted area using the extrinsic omission rate, which is the fraction of test points that fell into pixels predicted to be unsuitable, and the proportional predicted area, which is the fraction of all pixels that were predicted to be suitable.

Chapter 9

Using Population Viability Analysis to Plan Reintroductions

Tiffany M. Knight

Matrix population models and population viability analysis (PVA) are useful tools for plant conservation, allowing managers to project the trajectory of populations, compare alternative management scenarios, and suggest actions that might promote persistence (Menges 2000; Caswell 2001; Morris and Doak 2002). The same quantitative tools can be used to optimize reintroductions of rare plant species into habitats where they once persisted, into newly mitigated habitats (Falk et al. 1996; Menges 2008), or even as part of a managed relocation program in the face of global climate change (Vitt et al. 2010). Matrix population models and PVA (see also Menges 2008) can answer several questions stemming from rare plant reintroductions. Which sites should be chosen for reintroduction? Should reintroductions be created using seeds? Or would reintroductions of seedlings or larger plants, despite the greater effort involved, be more successful? How many propagules (of seeds, seedlings, or larger plants) are needed to provide reasonable assurance of creating a viable population?

In this chapter, I provide an overview of the utility of PVA for planning and evaluating reintroduction success. First, I review the few published studies that have used PVA and extensive demographic information to plan and evaluate success of reintroductions with endangered plant species. These studies represent the gold standard of the use of PVAs for reintroduction in that they contain several years of data on both natural and reintroduced populations and are able to make quantitative projections (Menges 2008). However, conservation practitioners often lack the resources (time, money, or otherwise) to perform such detailed PVA studies, have extremely limited sample size, or need to take immediate action in order to mitigate habitat disturbance of species particularly vulnerable to extinction. In such cases, it is often not possible or practical to collect demographic data on natural populations in order to make decisions for reintroductions. To address this issue, I next discuss how the quantitative tools can be useful even when extensive data are unavailable. Using guidelines from PVA models on natural plant populations, I ask whether there are any general rules that a practitioner can apply when planning reintroductions. I synthesize matrix population models created for eleven plant species (twenty-three total populations) that were studied in multiple years (allowing temporal environmental variation to be quantified). I use these estimates to assess how many seeds or transplants are necessary to create a viable new population. I present results from a new case study examining the reintroduction success of the forest herb *Trillium grandiflorum* by comparing two fundamental principles that emerge from the synthesis of published PVA studies: Reintroduction success will depend on (1) the number and life stage of propagules used to initiate the population and (2) habitat-specific characteristics influencing a population's growth rate.

Matrix Population Models and PVA of Reintroduced Populations

The ultimate goal of reintroductions is to create new populations that are able to persist on their own with minimal intervention and maintenance. Demographic data (i.e., vital rates including stage-specific survivorship, growth, and fecundity) incorporated into matrix population models can reveal population trajectories that might not be obvious from other sampling techniques (e.g., counts of individuals through time). For example, it might take many years for significant increases in the number of individuals or the population density of a long-lived species to be observed, but demographic models can project whether populations are on a positive growth trajectory and can estimate the time it will take for a target population size to be reached. In addition, detailed demographic monitoring of natural and reintroduced populations can identify the environmental drivers of vital rates and quantify how critical they are to population growth. Therefore, comparing demographic models of reintroduced and natural populations can allow a more comprehensive understanding about the environmental conditions that must be present for reintroductions to succeed (appendix 1, #36).

Matrix population models and PVAs are best used as comparative tools (Silvertown et al. 1993, 1996; Ramula et al. 2008; Buckley et al. 2010). These models are simplifications of nature and will probably not capture enough detail to forecast accurately the number of individuals that will be present in the population at some time in the future. However, models that incorporate relevant demographic aspects (stage structure, environmental factors that influence vital rates) can synthesize this information into comprehensive response variables (e.g., population growth rate) and can be used to evaluate the likely outcomes of different reintroduction strategies. For example, matrix population models can help evaluate which vital rates and environmental drivers are most important influences on population growth rate and can compare different strategies for reintroduction, such as using seeds versus transplants or estimating the number of propagules needed (Menges 2008).

Although many studies have used matrix population models to study the ecology of natural plant populations, few have applied this approach to plant reintroductions; exceptions are studies on *Cirsium pitcheri* (Bell et al. 2003), *Pseudophoenix sargentii* (Maschinski and Duquesnel 2007), and *Centaurea corymbosa* (Colas et al. 2008). Each of these studies asked whether reintroduced populations had similar demographic vital rates as natural populations and thus similar probabilities of persistence. Most reintroductions collect some demographic data (e.g., whether transplants survived) to measure success. However, these three studies measured vital rates at all phases of the life cycle. Thus, they were able to make more comprehensive evaluations of the status of reintroduced populations relative to natural populations and the environmental conditions necessary to create a viable reintroduced population.

In these three studies, vital rates differed significantly between natural and reintroduced populations. This provided valuable insight into the number and stage of propagules needed to restore a population successfully and identified the environmental drivers of the system. For example, in a reintroduced *C. pitcheri* population at Illinois Beach, transplanted individuals had lower fecundity than naturally recruited individuals at this site, and thus the number of transplants needed to establish a viable population was higher than would have been predicted based on the demography of natural populations (Bell et al. 2003). Alternatively, for *C. corymbosa*, reintroduced populations had greater fertility than natural populations, probably because the genetic diversity of seeds used in these reintroductions was high, increasing mate availability for self-incompatible flowering individuals. This result highlights the importance for founder diversity in all future reintroductions of this species (Colas et al. 2008).

Matrix population models are an excellent way to quantify the relative importance of environmental drivers on population growth, because both the effect of the driver on vital rates and the role of the vital rates in determining population growth are simultaneously assessed. In both natural and restored populations of *C. pitcheri*, vital rates were poor and population growth rate was low during years with low precipitation (Bell et al. 2003). This has serious consequences for reintroductions. As a result of high variation in environmental conditions, matrix population models revealed that 1,600 seedling transplants were necessary to create a viable population. For *P. sargentii*, little was known about the environmental drivers of vital rates and population dynamics before the reintroduction because only one wild population existed. However, the comprehensive demographic analyses of the thirteen reintroduced populations revealed that topographic factors had overriding importance to reintroduction success (Maschinski and Duquesnel 2007; Monks et al., this volume).

A Quantitative Synthesis

Although the case studies discussed here illustrate the power and utility of having detailed demographic data for modeling PVAs and making reintroduction recommendations, detailed demographic data such as these cost considerable resources and time. Often, collecting detailed demographic data on natural populations before conducting reintroductions is not possible (e.g., this is the situation in many of the Hawaiian case studies discussed by Kawelo and colleagues, this volume). In cases such as these, it would be useful to have general reintroduction rules (see appendix 1).

Several important generalities have emerged from more than three hundred matrix population models constructed for natural plant populations (reviewed in Silvertown et al. 1993, 1996; Ramula et al. 2008; Buckley et al. 2010). Here, I reexamine these studies in the context of population reintroduction to discern whether there are any general guidelines for creating viable reintroduced populations. Specifically, I use demographic data on natural populations to project population size into the future, considering an initial population of only seeds or juveniles, as would be the case in reintroductions. I ask whether there are any rules regarding the habitats into which populations should be reintroduced and whether population performance, plant life history, and related traits can inform which life stages (e.g., seeds or juveniles) and how many individuals are needed to maximize the probability of reintroduced population persistence.

From the database of demographic studies presented by Ramula and colleagues (2008) and Buckley and colleagues (2010) (established from keyword searches in the Web of Science [ISI] electronic database for 1975–2006), I selected studies that (1) measured plant species native to the ecosystem (i.e., exotic plants were excluded), (2) had data for at least 4 consecutive years (so that stochasticity in demographic vital rates could be considered), and (3) had an average population growth rate (λ) greater than or equal to 1. I excluded populations with $\lambda < 1$ because the population will go extinct no matter what the initial population size. Although such populations might be viable if the environment is altered or if rare years with sporadic recruitment are possible, without quantitative data on these possibilities, it is not useful to consider these populations in this review. I also eliminated populations located in disturbance-maintained ecosystems (e.g., fire-adapted plants) because incorporating periodic disturbances requires a different modeling approach than incorporating stochastic environmental variability. Finally, I considered only populations that have at least one juvenile stage class (i.e., perennials). This allowed me to compare seeds with juvenile plants for founding viable populations (Bell et al. 2003; Menges 2008; Albrecht and Maschinski, this volume). This reduced the available sample size to N = 23 populations (table 9.1). Although this is much lower than the more than three hundred original plant studies, these populations are the most appropriate ones to answer the questions posed, and this sample size is adequate for quantitative synthesis.

For the purposes of this chapter, I had four a priori predictions relevant to creating viable reintroduced populations: (1) Populations with higher mean population growth rates (λ) will need fewer individuals for successful reintroduction, (2) species that live in more variable environments will need more individuals for successful reintroduction, (3) long-lived species will need more seeds than short-lived species, and (4) planting stage classes larger than seeds will be more beneficial for longer-lived species than for shorter-lived species.

To examine these predictions, I used the data available from the matrix population models (table 9.1). All analyses considered natural rather than reintroduced populations. I assumed that the general results would apply broadly to plant reintroductions, especially if the natural and reintroduced populations had similar vital rates. For each population, the matrix of vital rates for each year represented a possible state for the environment. I assumed that each state had an equal probability of occurring in the future and that environmental conditions in one year were independent of conditions in the previous years (i.e., environments did not cycle in a predictable manner). To determine the number of introduced individuals needed to create a viable population, I started the population with ten individuals (all seeds or all juveniles) and calculated the ending population size and extinction probability for the population after one hundred years. I used $n_{t+1} = A^*n_t$ to project population size from t to $t+1$, over one hundred successive time intervals, using a matrix of vital rates drawn at random each time interval. A population was considered extinct if population size dropped below ten individuals at any time step (quasi-extinction threshold). For each starting population size (and structure), I replicated this simulation one thousand times and calculated the proportion of those thousand populations that dropped below ten individuals (extinction probability). I continued to increase the starting population size by one individual until an extinction probability of less than .05 was reached and the minimum number of individuals needed to create a viable population was determined. For all these analyses, I modified MATLAB code provided in Morris and Doak (2002).

TABLE 9.1

Natural populations studied for at least 4 consecutive years (three matrices) used in a quantitative synthesis on the use of matrix population models for plant reintroductions.

Species	Family	Age 1st Repro (yr)	Mean λ	Variance in λ	# Seeds	# Juveniles	Effect Size	Pop. ID	Years Studied	Location	Citation
Agrimonia eupatoria	Rosaceae	7.96	1.05	0.01	130	12	2.38		1994–1998	Oregon, USA	Kaye and Pyke 2003
Astragalus scaphoides	Fabaceae	7.56	1.17	0.09	49	23	0.76	Haynes Creek	1987–1991	Montana, USA	Lesica 1995
Astragalus scaphoides	Fabaceae	7.56	1.66	1	160	82	0.67	Sheep Corral Gulch	1987–1991	Montana, USA	Lesica 1995
Astragalus tyghensis	Fabaceae	2.55	1.01	0.03	400	40	2.30	10	1991–1999	Oregon, USA	Kaye and Pyke 2003
Astragalus tyghensis	Fabaceae	2.55	1.02	0.09	800	110	1.98	13	1991–1999	Oregon, USA	Kaye and Pyke 2003
Astragalus tyghensis	Fabaceae	2.55	1.03	0.15	8,200	4,410	0.62	25	1991–1999	Oregon, USA	Kaye and Pyke 2003
Calathea ovandensis	Marantaceae	3.99	1.05	0.02	800	320	0.92	25	1982–1985	Laguna Encantada, Mexico	Horvitz and Schemske 1995
Cimicifuga elata	Ranunculaceae	6.05	1.18	0.13	54	14	1.35	Eugrass	1992–1995	Oregon, USA	Kaye and Pyke 2003
Haplopappus radiatus	Asteraceae	5.44	1.03	0.05	2,200	130	2.83		1991–1999	Oregon, USA	Kaye and Pyke 2003
Lathyrus vernus	Fabaceae	11.69	1.19	0.01	180	11	2.80	K	1988–1990	Tullgarnsnäset, Sweden	Ehrlén 1995
Lathyrus vernus	Fabaceae	11.69	1.05	0	500	11	3.82	L	1988–1990	Tullgarnsnäset, Sweden	Ehrlén 1995
Lathyrus vernus	Fabaceae	11.69	1.08	0.01	600	12	3.91	F	1988–1990	Tullgarnsnäset, Sweden	Ehrlén 1995
Lathyrus vernus	Fabaceae	11.69	1	0	630	12	3.96	A	1988–1990	Tullgarnsnäset, Sweden	Ehrlén 1995

Species	Family	Age 1st repro	mean λ	variance	# seeds	# juveniles	Pop. ID	effect size	years studied	location	reference
Lomatium cookii	Apiaceae	8.65	1.1	0.03	73	13		1.73	1994–1998	Oregon, USA	Kaye and Pyke 2003
Lupinus tidestromii	Fabaceae	5.00	1.14	0.04	1,790	18	Abbotts Lagoon	4.60	2005–2009	California, USA	Dangremond et al. 2010
Primula farinosa	Primulaceae	6.48	1.05	0.01	36	12		1.10	1996–1998	Southeast Sweden	Lindborg and Ehrlén 2002
Primula farinosa	Primulaceae	6.48	1.04	0.01	74	18		1.41	1996–1998	Southeast Sweden	Lindborg and Ehrlén 2002
Primula veris	Primulaceae	13.49	1.04	0.02	3,100	11		5.64	1995–1997	Southern Sweden	García and Ehrlén 2002
Trillium grandiflorum	Melanthiaceae	16.07	1.02	0	1,100	10	DR	4.70	1999–2001	Pennsylvania, USA	Knight et al. 2009
Trillium grandiflorum	Melanthiaceae	16.07	1.01	0	1,300	10	DH	4.87	1999–2001	Pennsylvania, USA	Knight et al. 2009
Trillium grandiflorum	Melanthiaceae	16.07	1.09	0	2,900	10	WC	5.67	1999–2001	Pennsylvania, USA	Knight et al. 2009
Trillium grandiflorum	Melanthiaceae	16.07	1.03	0	6,700	10	RH	6.51	1998–2001	Pennsylvania, USA	Knight et al. 2009
Trillium grandiflorum	Melanthiaceae	16.07	1.01	0	13,000	10	DC	7.17	1998–2001	Pennsylvania, USA	Knight et al. 2009

For each population, this table shows the time it takes for seeds to reach adulthood (Age 1st repro = age at first reproduction), mean population growth rate (mean λ), variance across years in population growth rate (variance λ), number of seeds needed to create a viable population (# seeds), number of juveniles needed to create a viable population (# juveniles), log response ratio of the number of seeds versus juveniles needed to create a viable population (effect size), the author's name for the population (Pop. ID), years studied, the location of the population, and the reference.

Effects of Mean λ and Variation in λ on the Number of Seeds Needed to Restore a Viable Population

If a restored environment allowed plants to have high mean vital rates and low variation in vital rates from year to year, I expected that few seeds would be needed to create a viable reintroduced population. If two populations have the same arithmetic mean growth rate, the population with the higher variation in growth rate from year to year will have lower growth in the long term (lower geometric mean growth rate) and will have a higher probability of extinction (Morris and Doak 2002). If we extend these results to reintroductions, it leads to the prediction that populations occurring in variable environments will need more propagules to create a viable population than those in more constant environments.

For the twenty-three populations considered here, I conducted a multiple regression analysis considering three independent variables (mean λ, variance in λ, and age of first reproduction) and one dependent variable (minimum number of seeds needed to create a viable reintroduced population). As expected, the number of seeds needed to create a viable reintroduced population decreased with increasing mean λ ($p = 0.008$) and increased with increasing variance in λ ($p = 0.02$). The importance of variance in λ can be illustrated across populations of *Astragalus tyghensis* (table 9.1); populations #25 and #10 have similar mean λ, but population #25 has much greater variation in λ. As a result, it would take 8,200 seeds to reintroduce a viable population at a site with demographic vital rates similar to population #25 and only 400 seeds to reintroduce a viable population at a site with demographic vital rates similar to population #10.

Given the need to find recipient sites that would maximize mean λ and minimize variation in λ from year to year, it is useful to know the relationship between environmental factors and vital rates for candidate reintroduction species. Sometimes the factors that influence λ of a species are straightforward and can be easily quantified for a reintroduced population. For example, if a species has higher vital rates and higher mean λ in wet than in dry conditions, then wet sites should be targeted for the reintroduction. However, for some species the factors that influence λ are subtle or have changed over time due to anthropogenic factors. For example, for three of the species in this synthesis, *Lupinus tidestromii* (Dangremond et al. 2010), *Lathyrus vernus* (Ehrlén 1995), and *T. grandiflorum* (Knight et al. 2009), herbivory negatively influenced vital rates and λ; for *L. tidestromii* and *T. grandiflorum*, herbivory has increased in recent years. Therefore, identifying recipient sites where these plants would have low levels of herbivory (e.g., inside fenced exclosures) would be key to creating a new viable population successfully (appendix 1, #22, #23). Additionally, this information can be used to identify critical ecosystem components needed to achieve a viable reintroduced population. For example, species growing in fire-maintained ecosystems might need a regular fire frequency for persistence. Studies of natural populations can quantify fire frequencies that maximize λ (e.g., Menges and Quintana-Ascencio 2004). This information can be used to restore the necessary fire regime to sites where rare plants will be reintroduced.

PVA is a useful tool that allows environmental factors to be linked with vital rates and persistence. However, there are also several quicker and easier methods that can help identify sites that will allow a viable reintroduced population. Much can be learned about the conditions a population needs for persistence by simply measuring abiotic and biotic factors at sites and correlating these with species presence or absence (see Maschinski et al. [chap. 7], this volume). For example, before *Cirsium pitcheri* was reintroduced at Illinois Beach (Bowles et al. 1993), a habitat study in natural populations at Indiana Dunes indicated that this species occurred in sites that had 70% or greater bare sand (McEachern 1992).

Simple presence of a species at any site does not necessarily indicate that a population is thriving there (appendix 1, #23). Increased herbivory in populations of *Lupinus tidestromii* (Dangremond et al. 2010) and *Trillium grandiflorum* (Knight et al. 2009) led to PVA models projecting extinction of some populations, but individuals are still present at these sites. Even where populations are declining, it is possible to measure relationships between biotic factors and indicators of population persistence. For example, populations of *T. grandiflorum* with high levels of herbivory are dominated by plants in nonreproductive stage classes (Knight et al. 2009). Such indicators of population persistence can provide a rapid assessment of the environmental conditions that are desirable for a reintroduction site. Maschinski and colleagues (chap. 7, this volume) recommend a recipient site assessment (table 7.1) and experimental approaches for determining reintroduction sites.

Effects of Plant Longevity and Founder Stage on Reintroduction Success

The contribution of a stage class to λ depends in part on the reproductive value of the stage for that population (Caswell 2001). Seeds will have high reproductive value if they have a high probability of surviving to adulthood, if they mature in a short period of time, or if they produce numerous offspring once they are adults. In many long-lived species, such as trees, seeds have a low probability of surviving to adulthood and a long maturation time; both of these result in a low reproductive value of seeds compared with individuals in larger size classes (e.g., saplings). For long-lived plants, changes in vital rates associated with survivorship of larger size classes have the greatest influence on λ, whereas for short-lived plants changes in growth and seed production have the greatest influence on λ (reviewed in Silvertown et al. 1993, 1996; Ramula et al. 2008; Buckley et al. 2010). Extending these results to the consideration of restoring viable populations, as the longevity of the plant increases, the reproductive value of seeds should decrease, and thus the number of seeds needed to restore a viable population should increase (appendix 1, #16).

For the multiple regression considering three independent variables (mean λ, variance in λ, and age of first reproduction) and one dependent variable (minimum number of seeds needed to create a viable reintroduced population), I calculated age of first reproduction using the mean matrix for the species across all populations and years and a formula provided by Cochran and Ellner (1992). As expected, as age of first reproduction increased, the number of seeds needed to create a viable population also increased ($p = 0.03$).

As the longevity of the plant increases, because of the lower expected reproductive value of seeds, the difference between the number of seeds and the number of plants needed to create a viable population should become increasingly disparate. That is, if practitioners introduced juvenile or adult plants rather than seeds (Guerrant 1996a), the probability of establishment should be greatly improved for long-lived species. I determined the minimum number of juvenile plants that would be needed to create a viable population using the same technique as described earlier but initiated the starting population with individuals in the largest possible juvenile size class for each population. Then, to control for the variation observed in λ between years across different populations, I compared the difference between the number of seeds needed with the number of juveniles needed to create a viable population. I predicted that this difference would increase with increasing plant longevity (age of first reproduction). To test this, I calculated the log response ratio of minimum number of seeds versus minimum number of juveniles necessary to create a viable population: ln(min seeds) – ln(min juv). I regressed this founder size against the age of first reproduction of the plant and found a strong negative relationship (fig. 9.1). This suggests that restoring juveniles will have a higher probability of success than restoring seeds with long-lived species. This result concurs with the meta-analysis of Albrecht and Maschinski (this volume) on actual reintroductions.

A cautionary note: Sometimes these analyses show that a very small number of juvenile plants is necessary to restore a viable rare species population (table 9.1). For example, about forty individuals are predicted to be necessary to restore a population of the rare *Astragalus tyghensis* (Kaye and Pyke 2003), leading one to conclude that the restoration of this threatened species should be straightforward. However, it is important to point out that the studies used for this meta-analysis were restricted to natural populations for which $\lambda \geq 1$, and thus analyses considered habitats that were suitable for the long-term persistence of these populations. In fact, several populations of *A. tyghensis* studied by Kaye and Pyke (2003) were excluded from this analysis because mean $\lambda < 1$. Therefore, restoration will actually be much more difficult for this and other rare plant species, because appropriate sites for the reintroduction may be difficult to find or to create.

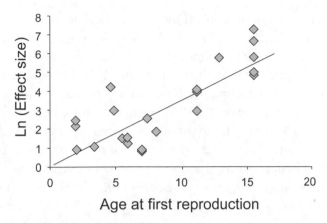

FIGURE 9.1. Relationship between founder size (ln[effect size]) and the age in years at first reproduction, shown for twenty-three natural plant populations. Founder size compares the log response ratio of minimum number of seeds with the minimum number of juveniles necessary to create a viable population using the equation ln(min seeds) – ln(min juv).

Some Practical Advice to Practitioners

Although this review points to critical features of populations that will influence the likely success of reintroductions, the question remains as to what information the practitioner needs and what actions should be taken to achieve successful restorations. As discussed earlier, the two critical features of a successful restoration are (1) finding a habitat where a population's λ is greater than 1 and (2) introducing enough individuals (seeds or juveniles) to be able to break through demographic and environmental stochasticity of low populations to achieve a viable population (appendix 1, #22–25).

Of course, practitioners will not know the potential λ of a species in a habitat where it does not currently persist. Therefore, it will be most useful to have information on extant populations and, in particular, the primary environmental drivers that influence their λ, including both abiotic conditions (e.g., soil, precipitation, temperature) and biotic conditions (e.g., consumers, mutualists, invasive species). If sites can be found that have environmental conditions likely to allow $\lambda > 1$, reintroductions to those sites are more likely to be successful, provided enough individuals are introduced. However, if those environments are rare or nonexistent, additional restoration activities, beyond simply reintroducing propagules, will be necessary. This can include controlling overabundant (Ruhren and Handel 2003) or introduced herbivores (Cordell et al. 2008) or introduced (invasive) competitors (Thomson 2005), mitigating effects of declining mutualists (Pemberton and Liu 2008c), or restoring historical disturbance regimes, such as fire (Menges and Quintana-Ascencio 2004). My results suggest that habitat selection for reintroductions is one of the most critical aspects of reintroductions, overshadowing issues of propagule number and stage class. However, a recent meta-analysis indicated that reintroduction sites are typically selected based on coarse habitat measurements, such as community type (Dalrymple et al., this volume), whereas fine-scale factors are important to consider for population persistence (Maschinski et al. [chap. 7], this volume). I hope this book encourages more careful site selection in future reintroductions (appendix 1, #22–24).

Provided that a suitable habitat exists or is restored, the number of individuals that will need to be reintroduced as founders depends on the realized λ in that habitat and the variation in λ across years; greater λ will require fewer propagules, and greater variation will require more propagules. The practitioner might not know the variation in λ expected to occur at the recipient site, but knowledge of the important environmental drivers that influence λ of the species will provide relevant information. For example, if precipitation is an important driver in natural populations of the species, and precipitation is known to vary significantly from year to year in the region, then it is likely that there will be high variation in λ in the restored site and that a large number of propagules will be necessary to create a viable reintroduced population.

Finally, the relative success of introducing seeds versus larger life stages (e.g., juveniles) depends on both the longevity of the species and the relative influence on λ of the stage. Longer-lived species will be more successfully reestablished with juvenile transplants relative to seeds, whereas the extra costs associated with growing and transplanting juveniles relative to planting seeds may not be worthwhile for shorter-lived species. In addition, some reintroduction projects have very specific goals, such as establishing a population that contains at least fifty reproductive plants. Achieving this goal could take a long time for a perennial plant population founded from seed. Demographic tools can be used to quantify the amount of time it will take the population to reach stable stage distribution and to reach a target population size.

A Reintroduction Case Study: Relative Importance of Environmental Conditions versus Propagule Number in Establishing Populations of Trillium grandiflorum

Large, white-flowered reproductive individuals of *T. grandiflorum* are an iconic symbol of the spring flora of northeastern North America. Though historically common and widespread, this species has been declining throughout its range, primarily as a result of dramatic increases in the abundance of native white-tailed deer (*Odocoileus virginianus*), which devour *T. grandiflorum* when they bloom in the spring (Augustine and Frelich 1998; Knight 2004; Knight et al. 2009).

From 1998 to 2002, as part of a larger study examining the population-level effects of pollen limitation and herbivory on the dynamics of *T. grandiflorum*, I measured vital rates (e.g., seed germination, survivorship, fecundity) and calculated population growth and associated demographic parameters of twelve *T. grandiflorum* populations. Several important parameters related to λ varied across populations. I found that herbivory by deer, which selectively ate old reproductive individuals, appeared to be the primary driver causing variation in λ between populations; higher rates of deer herbivory led to lower λ and faster rates of predicted decline toward local extinction for many populations (Knight et al. 2009).

In the spring of 2001, I transplanted reproductive individuals into experimental populations located directly adjacent to natural populations in my demographic study. Thus, the experimental populations probably experienced similar abiotic and biotic environmental conditions as the natural populations. In each experimental population, I created twelve plots that were 4 × 4 meters in size and separated by at least 50 meters. I transplanted adult *T. grandiflorum* into those plots at different densities (ranging from 1 to 117 adults per plot). This experiment was initiated to examine whether pollination and reproductive success in *T. grandiflorum* were density dependent (see Knight 2003b for more detail). However, these plots can also be considered as experimental reintroductions to sites with different environmental conditions that used a range of founding propagule sizes. From the matrix population models created for each natural population, I projected the number of plants in each size class that was expected in 2010 (9 years into the future) in each plot using a simulation technique identical to that described earlier for projections 100 years into the future.

Nine years after transplantation, in the spring of 2010, I revisited these experimental populations and counted the number of small (one-leaf and small three-leaf plants) and large (large three-leaf and reproductive adults) plants in each plot. For the purposes of this comparison, I chose two locations that differed dramatically in white-tailed deer herbivory and mean λ and thus were projected to have very different population growth trajectories (WC and WH; Knight et al. 2009). In particular, I expected adults transplanted at WC to have a low incidence of herbivory and the total number of individuals in the plots to increase (mean $\lambda = 1.08$ for the natural population), whereas I expected adults transplanted at WH to have high herbivory and the total number of individuals in the plots to decline toward extinction ($\lambda = 0.97$).

In both sites, even the plots with the fewest founding individuals (one individual transplanted) were generally still present because of the long lifespan of this species (individuals can live for decades; Hanzawa and Kalisz 1993). However, the two populations differed greatly in size by 2010 (table 9.2). In the reintroduced population adjacent to the site with high levels of herbivory (WH), although many transplanted individuals were still present, there was very little recruitment into the smaller size classes. Many of the observed juveniles were almost certainly individuals that had been transplanted as adults but had regressed in size, probably due to deer herbivory (Knight 2003a). Most of the plots at WH had fewer individuals in 2010 than were planted in 2001. The matrix population model predicted low recruitment, and in general none of these populations, even those with the highest densities of adult plants introduced, was expected to be viable. Alternatively, the other reintroduced population (WC), which was less than 15 kilometers away and sustained a lower level of herbivory (Knight et al. 2009), had higher recruitment of new individuals. For all but one plot, the number of individuals present in 2010 was greater than the number planted in 2001. Furthermore, in many plots the number of individuals was similar to or exceeded the population size predicted from the matrix population model, except for the highest initial densities, which may have experienced some density dependence (table 9.2).

This experimental reintroduction with T. grandiflorum revealed a fundamental generality. Establishing a viable reintroduced population will be difficult, if not impossible, in habitats where the environmental conditions lead to $\lambda < 1$, regardless of the number of propagules or the stage of propagules used to initiate the population. That is, the species niche that determines λ is of utmost importance for determining the habitats where viable reintroductions can be achieved. Importantly, it is useful to understand the underlying environmental drivers of λ. Because the edaphic conditions among the sites in this case study were similar, deer herbivory was clearly the environmental driver influencing λ. Thus, reintroduction planning must focus on the environmental driver first to create conditions where $\lambda \geq 1$, such as deer fencing or deer population control, before considering details of the reintroduction actions, such as the number of propagules or life stages needed to initiate the population.

Summary

Through a variety of direct and indirect actions, humans have greatly altered the population growth and distribution of numerous species. Although conservation strategies remain essential for minimizing these effects in situ, the restoration of species into habitats where they do not currently occur is increasingly becoming an essential tool for the preservation of biodiversity on the planet. Just as PVA modeling has provided an essential tool for conservation ecology, the same tools can provide important information for reintroducing populations into habitats from which the species has gone extinct or for mitigating the effects of extinctions by introducing a species to new habitats. Several generalities in PVA modeling for restoring populations emerged. First, successful reintroduction requires establish-

TABLE 9.2

Experimental reintroductions of Trillium grandiflorum.

Population	Adults Planted (2001)	Large Plants Expected (2010)	Large Plants Observed (2010)	Small Plants Expected (2010)	Small Plants Observed (2010)
WC	1	1.33 (1.03–1.61)	1	4.51 (2.28–7.30)	0
WC	1	1.33 (1.03–1.61)	1	4.51 (2.28–7.30)	7
WC	1	1.33 (1.03–1.61)	2	4.51 (2.28–7.30)	10
WC	1	1.33 (1.03–1.61)	3	4.51 (2.28–7.30)	14
WC	2	2.66 (2.07–3.25)	6	8.99 (4.67–14.8)	0
WC	2	2.66 (2.07–3.25)	43	8.99 (4.67–14.8)	36
WC	4	5.33 (4.14–6.51)	5	18.0 (9.3–29.6)	11
WC	8	10.6 (8.2–12.9)	8	35.7 (18.9–58.7)	20
WC	23	30.6 (23.7–37.5)	28	104 (51.5–165)	101
WC	46	61.0 (47.7–75.1)	44	205 (105–328)	33
WC	77	101.6 (77.2–125.1)	70	341 (178–556)	64
WC	117	155 (120.3–190.4)	108	521 (260–840)	201
WH	1	0.71 (0.41–1.09)	0	0.83 (0.44–1.41)	0
WH	1	0.71 (0.41–1.09)	0	0.83 (0.44–1.41)	1
WH	1	0.71 (0.41–1.09)	0	0.83 (0.44–1.41)	1
WH	1	0.71 (0.41–1.09)	0	0.83 (0.44–1.41)	0
WH	2	1.39 (0.86–2.09)	1	1.69 (0.87–2.77)	2
WH	2	1.39 (0.86–2.09)	0	1.69 (0.87–2.77)	2
WH	4	2.81 (1.65–4.25)	2	3.36 (1.75–5.56)	2
WH	8	5.55 (3.32–8.45)	1	6.64 (3.48–11.1)	2
WH	23	16.0 (9.56–23.9)	8	19.1 (9.65–31.4)	3
WH	46	32.2 (18.4–49.8)	0	38.5 (20.2–65.9)	2
WH	77	52.4 (31.9–79.8)	44	63.5 (33.4–105.8)	3
WH	117	81.2 (49.5–125)	27	97.4 (52.2–162)	11

In 2001, adult individuals were planted in 4×4-meter plots at densities that ranged from 1 individual to 117 individuals per plot at two sites, WC and WH. These reintroduced sites were adjacent to natural populations for which demographic data were collected and matrix population models were constructed from 1999 to 2002 (three matrices for each site). The WC site had a mean $\lambda = 1.08$ and a lowest $\lambda = 1.04$. The WH site had a mean $\lambda = 0.97$ and a lowest $\lambda = 0.90$. I used the matrices from the natural populations to simulate how many large (flowering + large nonreproductive plants) and small (small nonreproductive + one-leaf plants) plants were expected in each plot in 2010. I revisited each plot in May 2010 to determine the number of plants actually present.

ing propagules into a population where the expected population growth rates are positive (i.e., $\lambda \geq 1$). Second, the number of propagules necessary to establish a viable population depends on the expected λ and its variance; lower λ or higher variance in λ will require more propagules. Third, for long-lived species, reintroductions of large stage classes (e.g., greenhouse-grown juveniles) will need many fewer propagules than those using seeds and may afford a greater probability of restoration success with less proportional effort and cost.

Chapter 10

Influence of Founder Population Size, Propagule Stages, and Life History on the Survival of Reintroduced Plant Populations

MATTHEW A. ALBRECHT AND JOYCE MASCHINSKI

The reintroduction of rare and endangered species is now widely practiced as a conservation tool to reestablish species within their historic range (Guerrant and Kaye 2007; Seddon et al. 2007; Menges 2008). The fundamental goal of a plant reintroduction is to create a self-sustaining population with evolutionary potential that can resist ecological perturbations (Maunder 1992; Guerrant 1996a). Reintroduction practitioners face many challenges and complex choices, many of which have been formulated into generic guidelines for reintroducing a species (Akeroyd and Wyse-Jackson 1995; Falk et al. 1996; Kaye 2008). After choosing reintroduction sites based on biological, logistical, and historical attributes (Fiedler and Laven 1996; Maschinski et al. [chap. 7], this volume), reintroduction practitioners must decide how many individuals to introduce, which propagule stage to introduce, and what follow-up management treatments and aftercare are necessary to ensure that a population survives and reaches critical demographic benchmarks, including sexual reproduction and recruitment of the next generation (Pavlik 1996; Menges 2008). To complicate matters, reintroduction practitioners must balance tradeoffs between the availability of propagules, which are often limited with rare and endangered plants, and population dynamic theory, which predicts that the initial size and composition of the founding population are important determinants in the survival of reintroduced populations (Guerrant 1996a; Kirchner et al. 2006; Maschinski 2006; Armstrong and Seddon 2007). Consequently, a major question in the field of reintroduction biology is whether the persistence of a reintroduced population is affected by the size of the founding population and by the developmental stage of the propagules (Guerrant 1996a; Maschinski 2006; Armstrong and Seddon 2007).

In this chapter we explore how demographic factors and life histories influence plant reintroductions. In comparison to Knight (this volume), who analyzed how demography of natural populations could inform decisions about selecting size and number of propagules for reintroductions, we draw on a rapidly expanding body of plant reintroduction data made available in the literature and an international database. Here we synthesize results from published and unpublished studies and quantitatively analyze the effects of founder population size and founder propagule stage on the persistence of reintroduced plant populations. Specifically, we address two fundamental questions in plant reintroduction biology: (1) How does initial population size influence the survival of reintroduced plant populations? and (2) What are the effects of using different propagule stages on long-term persistence? We explore these questions in relation to plant life history with the goal of understanding broader generalizations that can be applied to future reintroductions to improve their success. Finally, we consider the implications of our results for reintroducing plant populations in a rapidly changing climate.

Relevant Theory: An Overview

The short-term demographic goals of any plant reintroduction are to avoid local extinction and to maximize the initial rate of population growth (Guerrant 1996a). The small population paradigm (sensu Caughley 1994) provides the theoretical basis for plant reintroduction programs because it is based on the extinction risk and rarity of small populations (table 10.1). For populations consisting of few individuals (less than fifty), demographic stochasticity, or random births and deaths of individuals, threatens long-term survival; this is especially relevant to endangered plant reintroductions because founder sizes and propagule stages are constrained by the limited availability of propagules and the biology of the species. Newly founded populations can also suffer from inbreeding depression and genetic drift, which can erode genetic variation and reduce the potential for reintroduced populations to adapt to local conditions over the long term (see Neale, this volume). Thus, stochastic genetic and demographic processes can interactively influence the extinction risk of reintroduced plant populations and must be considered when new populations are designed (Guerrant 1996a). Other stochastic factors affecting the persistence of reintroduced populations include environmental stochasticity and random catastrophes. In addition to these stochastic forces, deterministic mechanisms, such as positive density dependence (i.e., Allee effects) in which low conspecific density or size can reduce fitness and per capita growth rate, can also drive small populations to extinction. Allee effects may be

TABLE 10.1

Key principles of the small population paradigm (Caughley 1994) that are directly applicable to rare plant reintroductions.

Small Population Paradigm Principle

Small populations are susceptible to demographic stochasticity, which results from random differences in survival rates between individuals within a population.

Environmental stochasticity causes population growth rates to vary between years as external factors (e.g., weather, predation, competition) cause vital rates to vary. Moderate fluctuations in the environment can dramatically increase extinction risk.

Chance fluctuations in allele frequencies lead to genetic drift and result in loss of genetic diversity in small populations.

Mating between close relatives is more common in small populations, leading to inbreeding depression.

The size below which a population is at imminent risk of extinction is the minimum viable population (MVP).

The number of individuals that will contribute equally to the gene pool in the next generation is the effective population size.

particularly prevalent in rare self-incompatible or animal-pollinated plants with small, isolated populations (Groom 1998; Hackney and McGraw 2001).

Guerrant (1996a) used the theoretical principles of the small population paradigm to develop methods for reintroducing plant populations. He explored the effects of different propagule stages on extinction risk and population growth rate using simulations with demographic data from natural populations of species with contrasting life histories. These simulations demonstrated that populations founded with seed or seedlings experienced a greater risk of extinction and slower population growth than populations founded with plants in larger size classes. An important outcome from these simulations was the realization that achieving rapid population growth was more challenging than simply reducing extinction risk, even when the largest propagule stages were planted. Based on these results, Guerrant (1996a) hypothesized that many reintroduction projects would be characterized by small, persistent populations that do not grow rapidly, because of the inherent limitations (e.g., costs and resources) of using larger size classes.

Similarly, Van Groenendael and colleagues' (1998) guidelines for plant reintroductions considered the demographic risk associated with sowing seed rather than transplants, but they offered more specific rules for founding new populations. They proposed a strategy of transplanting at least fifty adults for founding new populations for two reasons. First, because establishment rates of seeds are typically lower than those of transplants, populations founded with seed are expected to be small and slow growing, thereby making them more susceptible to stochastic effects (see table 10.1). Second, using adult transplants that can begin producing seed eliminates the need to collect large numbers of seed from local wild sources, which might threaten their viability (Menges et al. 2004). Of course, this method cannot be applied universally, because of the inherent challenges and associated costs of propagating and maintaining certain plant taxa ex situ, especially for long-lived perennials that might take years or even decades before they reach reproductive size classes (e.g., *Pseudophoenix sargentii*; Maschinski and Duquesnel 2007).

Building on this theoretical framework, more recent studies applied population viability analysis in a restoration context to determine the initial size and propagule stage needed to found populations with an extinction probability of less than 5% over a 100-year period. Bell and colleagues (2003) used population viability analysis (PVA) on restored populations of the endangered pitcher's thistle (*Cirsium pitcheri*). They found that 400 1-year-old rosettes, or 1,600 seedlings, or 250,000 seeds would be needed to found populations with high probability of long-term survival. In contrast, Kirchner and colleagues (2006) discovered that just a thousand seeds distributed over a few sites was enough to create viable populations of *Centaurea corymbosa*, a rare self-incompatible species endemic to rock outcrops. Similarly, Münzbergová and colleagues (2005) simulated reintroduction of the perennial herb *Succisa pratensis* with scenarios that varied propagule availability and the spatial distribution and size of potential reintroduction sites. They learned that when seed availability was low (less than a thousand seeds available), population size after 100 years was greatest when seeds were distributed over the largest habitat patches, whereas when seed availability was not limiting (more than 100,000 seeds available), population size was maximized by distributing seeds over as many suitable sites as possible.

Selection of Studies for Review and Analysis

To determine the role of founder population size and propagule stage on the outcome of reintroduction projects, we compiled a data set derived from published studies (peer-reviewed and gray literature) on plant reintroductions and projects in the *CPC International Reintroduction Registry* (Center for Plant Conservation 2009; appendix 2). All studies included in the data set focused exclusively on rare plant reintroductions rather than on dominant species used in community-level restoration projects. Additionally, we used only studies with detailed information for founding a new population, either to a site where the species was not known to occur (introduction) or to a site where the species formerly occurred (reintroduction). Therefore, we use the term *reintroduction* broadly to define any attempt to introduce propagules to an unoccupied patch. Although it is by no means an exhaustive review of all known rare plant reintroductions, the data set consists of a wide range of taxa, life histories, and outcomes (success and failures). For purposes of this analysis and for two reasons, we excluded augmentation studies, where propagules were added to extant populations. First, in some instances it was often impossible to distinguish between founding individuals or their offspring and those in the recipient population. Including these studies would have precluded analyses aimed at understanding founder size and propagule stage effects on population persistence. Second, the influence of founder size is confounded by the size and structure of the recipient population, which may not always be known or documented at the time of the reintroduction.

In this chapter, we defined a reintroduction attempt as the introduction of propagules to one specific location. If propagules were planted multiple times at the same location, we considered this as one reintroduction attempt. We focused specifically on reintroductions at the population level, because this is the most commonly reported scale and most relevant management unit. For each reintroduction attempt, we recorded the following variables: (1) propagule stage, whether seeds, seedlings, or plants were outplanted; (2) founder size, the number of individuals outplanted for each propagule stage; (3) life history, whether the taxon was an annual, a biennial, a herbaceous perennial, or a woody perennial; (4) reintroduction year, the first year that propagules were outplanted at a site; and (5) monitoring period, measured as the number of years between study onset and the last census. Seedlings were defined as individuals less than 1 year old that were propagated from seed, and plants included whole plants more than 1 year old, plant parts (e.g., cuttings, rhizome and root fragments), or plants derived from in vitro micropropagation. If different types of propagule stages were planted, we analyzed them separately when analyzing the effects of propagule stages.

For statistical analyses we scored each reintroduction attempt as a binary outcome, based on whether the population was persistent at the last known monitoring period (success = 1) or failed to establish a persistent population according to the authors or when population size fell below ten individuals (failed = 0), a typical extinction threshold set in population viability analyses (Morris and Doak 2002). We used generalized linear mixed models with binomial error structure (SAS Institute 2001) to test whether population survival varied across propagule stages (seed, seedlings, and plants) independent of life history. We then tested whether propagule stage and life history (perennial herbs vs. woody plants) interacted to influence population survival. We excluded annuals from the analysis because sample sizes with transplants (seedlings and plants) were too low. Using separate logistic regressions for each propagule stage, we examined whether larger initial population sizes increased the probability of survival of reintroduced populations independent of life histories. We used a two-way logistic regression to explore the interactive effect of initial population size and life history (perennial herbs vs. woody) on population survival. For this analysis, we summed seedlings and plants together to increase the sample size. To test whether populations founded with more than fifty individuals were more likely to survive than those founded with less than fifty individuals, we used separate log-likelihood goodness of fit (G) tests for seedling and plant founders.

How Does Propagule Stage Influence the Survival of Reintroduced Plant Populations?

In the reviewed studies, we found that seeds are typically the least favored choice in founding new populations of rare plants (also see Dalrymple et al., Guerrant, this volume). Of the 174 reintroduction attempts included in the analysis, only 22% ($n = 39$) used seeds or a combination of seeds and whole plants. However, founder propagule stages were not randomly distributed with respect to life histories. Only 13% of attempts with perennials used seed, or a combination of seed and whole plants as founders, whereas 92% of annuals were founded with seed. Clearly the most popular choice in founding new perennial populations was to use seedlings ($n = 79$) and plants ($n = 56$), and in most instances these were propagated ex situ.

As expected, establishment rates were consistently greater with transplants than seed founders when attempts were pooled across life histories (fig. 10.1). Low germination rate with direct seeding often resulted in low proportions of seedlings in plots relative to transplants within the first 1.5 years of outplanting. When pooling data across all reintroduction attempts, we found differences in the survival of reintroduced populations founded with different propagule stages ($F = 3.04$, $p = 0.05$; fig. 10.2a). Populations initiated with whole plants were more likely to survive to the last monitoring period than populations initiated with seeds ($p = 0.03$) or seedlings ($p = 0.05$); survival of populations using seeds or seedlings did not differ. These results are consistent with demographic simulations that show that the long-term extinction risk of reintroduced populations decreases when the largest propagule stages are used as founders (Guerrant 1996a; Knight, this volume). Furthermore, there was a greater incidence of using horticultural management (e.g., water and fertilization) for transplants (about 25%) than for seed (less than 3%) founders, which may have also increased the likelihood of transplants persisting over the long term.

Experimental trials with different propagule stages confirm that using transplants as founders is often superior to using seed as founders (table 10.2). Of the twelve studies that experimentally examined the effects of using different founder propagule stages in plant reintroductions, 75% of them found that using trans-

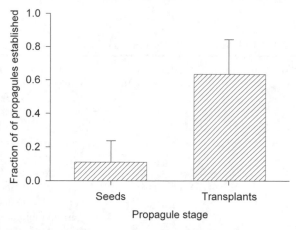

FIGURE 10.1. Mean proportion of individuals that established from seeds and transplants. Data pooled from studies in the *CPC International Reintroduction Registry* that reported founder sizes and the number of individuals established for at least 0.5 years but no longer than 1.5 years. Establishment proportions are significantly greater for transplants ($n = 21$) than seeds ($n = 12$) based on t test ($p < 0.01$). Error bars are 95% confidence intervals.

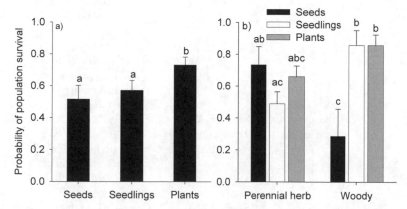

FIGURE 10.2. Effects of (a) propagule stage and (b) life history on the probability of survival (mean ± 1 standard error) of reintroduced plant populations. Different letters indicate significant differences based on post hoc least square means. Population survival was scored as a binary event based on data pooled from literature review and the *CPC International Reintroduction Registry*.

plants raised ex situ was favored over using seed sown in situ. Transplants typically survived better, grew faster, and attained reproductive maturity sooner than plants from seed founders. Another advantage of using transplants as founders is that they can be outplanted directly into microsites that match the regeneration niche of the species (Grubb 1977) and can be planted when climate or seasonal

TABLE 10.2

Summary of studies that experimentally examined different propagule stages on reintroduction success with rare and endangered plant species.

Species	Life History	Propagule Stage (founder size)	Outcome	Source[a]
Acacia aprica	Woody perennial	Seeds (1,500) Seedlings (1,102)	Easier to establish a persistent population with transplants. After 7 yr, only one out of 1,500 seeds germinated, whereas ~80% of seedling transplants survived.	1
Acacia cochlocarpa ssp. *cochlocarpa*	Woody perennial	Seeds (1,500) Seedlings (370)	Seedling transplants favored over directly sowing seeds due to the low germination rates (5.4%) in the field.	1
Arabis koehleri var. *koehleri*	Herbaceous perennial	Seeds (567) Seedlings (190)	Easier to establish population with seedlings. After 2 yr, no plants from seed survived, whereas ~10% of transplants survived.	1
Asclepias meadii	Herbaceous perennial	Seeds (96–243) Juveniles (29–148)	Seedling transplants favored over directly sowing seeds. Survival rates among plants from seed and transplants were similar but had low field seed germination rates and long pre-reproductive period for plants grown from seed.	2
Brachycome muelleri	Herbaceous annual	Seeds (unknown) Transplants (60)	Easier to establish a persistent population with transplants. Survival and growth greater in plants from seedlings than in plants from seed.	4
Holocarpa macradenia	Herbaceous annual	Seeds (1,080) Seedlings (750)	Germination rates too low to establish a population from sowing seed in situ and recruitment of next generation too low to establish a population with seedling transplants.	5
Ipomopsis sancti-spiritus	Herbaceous perennial	Seeds (1,800) Seedlings (169–212)	Seedling transplants favored over directly sowing seed. After 5 yr, only one plant from seed sexually matured, whereas sites with seedling trans-	1

TABLE 10.2

Continued

Species	Life History	Propagule Stage (founder size)	Outcome	Source[a]
			plants all contained persistent populations with reproductive plants.	
Jacquemontia reclinata	Herbaceous perennial	Seeds (264) Plants (18–924)	Transplants favored over sown seed even though the former is expensive and time consuming. Germination rates of field sown seed <3% but only when watered for 3 mo.	7
Lilium occidentale	Herbaceous perennial	Seeds (640) Bulbs (120)	Plants have persisted for >12 yr from both propagule types, but none have reproduced. Plants from bulbs emerged and grew at higher rates than plants from seed.	1
Lupinus sulphureus ssp. *kincaidii*	Herbaceous perennial	Seeds (3,400) Seedlings (109)	Plants have persisted for >5 yr from both propagule types. Germination rates 6–22%, and seedling transplant survival rate 2–10%. Seedlings from in situ sown seed survived between 1% and 10%.	3
Pediocactus knowltonii	Semiwoody perennial	Seeds (288–2,250) Transplants (102–150)	Both propagule types considered suitable for founding populations, but only 5% of plants from seed became established adults. Recruitment of next generation slow using either seeds or transplants.	1
Ziziphus celata	Woody perennial	Seeds (1,728) Transplants (144)	Transplants favored over directly sowing seed. After 4.5 yr, establishment rates <2% for seed versus 72–77% for transplants.	6

[a]*Sources:* [1]CPCIRR (2009); [2]Bowles et al. (2001); [3]Thorp et al. (2008b); [4]Jusaitis et al. (2004); [5]Holl and Hayes (2006); [6]Weekley and Menges (2008); [7]Machinski and Wright (2006).

Founder size represents the number (or range of numbers) of individuals planted at each site.

conditions are most appropriate for establishment (Jusaitis et al. 2004; Albrecht and McCue 2010), thereby maximizing the probability of persistence. On the other hand, microsites can change suitability, as seeds remain dormant for prolonged periods before they germinate (Schupp 1995).

In most studies that explored the effects of using seeds and transplants, the latter was the preferred founder stage because they reduced the demographic cost of reintroduction (Guerrant et al. 2004a). Germination rates are almost always greater when seeds are propagated in a controlled setting than when they are directly seeded in the field. Thus, the demographic benefits of using transplants rather than direct seeding often outweigh the greater financial costs of growing plants ex situ. For example, Bowles and colleagues (2001) found that even though survivorship of Mead's milkweed (*Asclepias meadii*) seedlings from seed sown in situ was greater than that of transplanted juveniles, transplants were the preferred propagule stage for reintroduction because germination rates in the greenhouse were 74%, compared with only 33% in the field. Furthermore, they found that plants from seed took much longer to reach reproductive stage than juvenile transplants, slowing population growth and increasing its susceptibility to demographic stochasticity.

Kaye and Cramer (2003) calculated the economic costs of using seeds and transplants grown ex situ for reintroducing the threatened prairie plant Kincaid's lupine (*Lupinus sulphureus* ssp. *kincaidii*). They found that when seed was scarce, raising transplants ex situ was an efficient use of limited seed. However, when seed source populations produced abundant seed, direct seeding was the most cost-effective method.

There are clearly circumstances in which using seed as founders is the only practical or possible method for preventing the imminent loss of a rare plant population (appendix 1, #16). For example, Smith (1999) moved layers of topsoil, which presumably contained dormant seed of the rare winter annual *Geocarpon minimum*, from one rock outcrop in imminent danger of destruction from a highway development project to an adjacent rock outcrop site. The new population successfully established in the short term, although its long-term viability remains uncertain.

Propagule Stage and Life History

The survival probabilities of reintroduced populations varied across different propagule stages for perennial herbs and woody plants (propagule stage × life history interaction, $F = 4.51$, $p = 0.01$). In comparison to woody plants, perennial herb populations were more likely to survive when founded with seed or whole plants than when founded from seedlings, whereas woody plant populations had best survival if seedlings or whole plants were used as founders (fig. 10.2b).

Most data on using seed as founders for woody perennials are available from xeric systems. Establishment rates of woody perennials with seed founders are consistently low, often less than 5% (table 10.2). Low germination rates resulted in very small populations that had a low probability of survival (fig. 10.1), as predicted from population dynamic theory (table 10.1). However, in 75% of these attempts, founder sizes were greater than a thousand seeds, suggesting that large numbers of seed would be needed to overcome demographic bottlenecks associated with severe seedling recruitment limitations. Annual populations that failed to survive exhibited similar recruitment limitations; too few second-generation plants recruited and sexually reproduced to maintain persistent populations (Helenurm 1998; McGlaughlin et al. 2002; Holl and Hayes 2006). In cases with woody perennials where seed founders did form persistent populations, those populations exhibited slower growth, fewer reproductive individuals, or much longer pre-reproductive periods than populations founded with transplants of the same species (see table 10.2 and Jusaitis 2005).

Why is it so challenging to found populations of woody plants in xeric habitats with seed? In arid systems, seedling recruitment of woody perennials often increases with soil moisture, which can sometimes be enhanced by the presence of neighboring plants (i.e., nurse effects) or inhibited by competitive interactions (Padilla and Pugnaire 2006). In reintroductions with *Purshia subintegra* on limestone outcrops, for example, seedling recruitment within herbivore-excluding cages was restricted largely to moist microsites and was greater in the presence of shrubs than in open spaces (Maschinski et al. 2004a). On the other hand, in experimental seeding trials with *Acacia whibleyana* in southern Australia, populations persisted in weed-free plots but not in plots with weeds present (Jusaitis 2005). In contrast to nurse effects, removing vegetation noticeably increased moisture availability for young A. *whibleyana* seedlings and presumably facilitated their survival and growth. Thus, variation in moisture availability across habitats can alter interactions with vegetation, changing them from facilitative to inhibitive. Patches of the cactus *Pediocactus knowltonii* founded with seed remained stable (recruitment approximately equaled mortality) over a 14-year period in plots with and without native vegetation, although seedling recruitment declined during drought years (Sivinski 2008; CPC 2009). Perhaps as more data become available, patterns in the success rates using seed founders with woody perennials can be evaluated across a wider range of habitats.

Most surviving perennial herb populations founded with seed were located in open habitats, such as grasslands and rock outcrops. This finding counters previous research suggesting that establishing populations of rare perennial herbs from seed would be challenging (Drayton and Primack 2000; Lofflin and Kephart 2005). In a large-scale reintroduction experiment where several species of forest herbs were seeded into an isolated forest preserve, few populations established and fewer populations recruited second-generation plants, even when microsites were experimentally created by manual disturbance (Drayton and Primack 2000). However, the founding population sizes in a majority of the attempts were fewer than one hundred seeds, sometimes even between fifteen and thirty. This is smaller than the minimum founder sizes needed to create self-sustaining populations in many perennial herbs (see Knight, this volume).

How Does Founder Population Size Influence the Survival of Reintroduced Plant Populations?

In our review, we found that although seeds were the least commonly used propagule stage to found new populations, seeds were typically sown in much larger quantities than nonseed transplants. When using seed as founders, 45% of attempts used more than a thousand seeds, whereas only 8% of attempts used less than a hundred seeds. In contrast, 44% of attempts with seedlings and 64% of attempts with plants had founder sizes less than a hundred, suggesting that approximately half of all reintroductions using transplants start with small population sizes. Nevertheless, founder size varied widely among nonseed transplants, ranging from 3 to more than 3,500 per attempt.

When we pooled data across all life histories, we found that introducing more seed founders did not increase the chance of population survival (table 10.3). These results are not surprising given the overall low success rates of using seed as founders, the wide variability in the number of seeds introduced, and the low frequency of studies reporting using seeds in plant reintroductions. However, despite the lack of broader trends among the studies reviewed, there are several examples where increasing seed founders clearly decreased the extinction risk of introduced populations. For example, as an illustration of the feedback between population size and genetic processes, introducing more seed to multiple sites increased the probability of mate availability and lowered the long-term extinction risk of *Centaurea corymbosa*, a self-incompatible perennial herb (Kirchner et al. 2006). Similarly, Münzbergová and colleagues (2005) found that increasing the number of seeds introduced to unoccupied grassland sites decreased the extinction risk of *Succisa pratensis*, a perennial herb of temperate grassland. In contrast to the observed trends with seeds, there was a general positive relationship between founder size and population survival when nonseed transplants were used, although the relationship was statistically significant only with plants (table 10.3).

TABLE 10.3

Effects of initial population size on the survival of reintroduced populations across life histories and for transplants in perennial species (herbs vs. woody).

Across Life Histories	Wald's Z	p
Seeds	−0.28	0.78
Seedlings	1.80	0.07
Plants	2.24	0.03

Perennials (herbs vs. woody)	F	p
Initial population size (log)	2.61	0.11
Life history	5.97	0.01
Life history × initial population size (log)	1.21	0.27

Separate logistic regressions were conducted for seeds, seedlings, and plants. For perennials, a two-way logistic regression tested whether initial population survival interacted with life history to affect the survival of reintroduced populations; annuals were not included in the analysis due to small sample size.

Although increasing the number of transplants (seedlings and plants) often lowered the extinction risk, populations founded with few individuals were not all doomed to failure. One of the more spectacular examples began in 1970 when fourteen individuals of the globally endangered herb *Cochlearia polonica* were translocated to a site along the Centuria River in southern Poland. Thirty years later, the introduced population consisted of more than 30,000 individuals spanning an area of 2,000 square meters and represented the only demographically stable population in the wild (Cieslak et al. 2007). Other examples of small founder populations persisting, but for shorter time periods (5–10 years), include *Jacquemontia reclinata* (n = eighteen seedling founders) in a coastal strand in Florida (Maschinski and Wright 2006), and *Prostanthera eurybioides* (n = ten seedling founders) in rock outcrops of south Australia (Jusaitis 2005). The long-term fate of these small populations remains unknown, and it will be important to monitor how they respond to environmental perturbation over timescales longer than 5 years. In a study that surveyed rare plant populations over the course of a decade, Matthies and colleagues (2004) showed that although a majority of small populations are prone to extinction, small rare plant populations are sometimes capable of developing into larger ones if the habitat quality remains suitable.

Variation within Species

The most informative examples of variation in the effects of initial population size on persistence arise when equal numbers of the same species are introduced to different sites. Comparisons within species control for the potential confounding effects of differences in species traits that might potentially influence success rates and can help disentangle the relative influence of founder population sizes and environmental stochasticity on population survival (see Forsyth and Duncan 2001 regarding bird introductions on islands). For example, Bottin and colleagues (2007) introduced 450 adult *Arenaria grandiflora* plants, derived from in vitro culture, to each of three rock outcrop sites in France. After 7 years, one of the introduced populations tripled in size, whereas introduced populations at the other two sites declined dramatically (fewer than fifty individuals), presumably due to rabbit predation and localized competition with grasses. Helenurm (1998) introduced equal numbers of seed of the winter annual *Lupinus guadalupensis* to three sites and found that desiccation and/or herbivory decreased population survival at two sites but not the third. These studies highlight the importance of site-specific effects in determining reintroduction outcomes (Knight, this volume).

Context-specific factors in the local environment play a key role in determining whether reintroduced populations survive or die. Using simulation models, Rout and colleagues (2007) demonstrated that site quality rather than the number and timing of reintroductions determined long-term population persistence. In most failed reintroductions we reviewed, the underlying drivers were unclear (also see Dalrymple et al., this volume), probably because of subtle and changing habitat characteristics not readily observed (Primack 1996). For example, one hundred lakeside daisy (*Tetraneuris herbacea*) plants were introduced to each of three dry gravel prairie sites in Illinois, and within 5 years two populations went extinct, whereas the other persisted for 15 years. The reasons for differential survival were uncertain, as no major differences in disturbance events were noted between sites (McClain and Ebinger 2008). Similarly, Severns (2003) experimentally introduced equal numbers of scarified and unscarified Kincaid's lupine (*Lupinus sulphureus* ssp. *kincaidii*) seeds at two different upland prairie sites; after 3 years, a population established at one site but not the other, despite both sites supporting seemingly suitable upland prairie habitat.

Founder Population Size and Life History

The effects of initial population size were largely consistent across life histories. Although reintroduced populations of woody plants were more likely to survive than perennial herbs, there was no significant interactive effect between life history and initial population size (table 10.3). The number of attempts with seeds was too low to test for differences between life histories. When all attempts for perennials were pooled together, reintroduction attempts with fewer than fifty seedlings and plants were significantly more likely to fail than ones started with more than fifty seedlings and plants (fig. 10.3).

In the reviewed studies, there were several examples of rare annual and perennial herbs of grasslands or shrublands going extinct even with large initial population sizes (more than five hundred propagules). Attempts to reintroduce seeds and

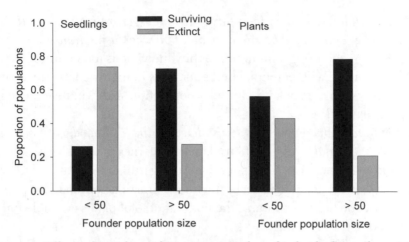

Figure 10.3. Effects of initial population size (number of individuals) on the survival of reintroduced populations founded with **(a)** seedlings and **(b)** plants. Reintroduction attempts were summed across perennials. *G* tests indicated that reintroduced populations founded with less than fifty individuals were more likely to go extinct than those founded with more than fifty individuals for both seedlings ($n = 59$ attempts, $G = 11.11$, $p = 0.0009$) and plants ($n = 75$ attempts, $G = 3.67$, $p = 0.05$).

seedlings of *Holocarpha macradenia*, an annual California grassland herb, to multiple sites failed to establish self-sustaining populations (Holl and Hayes 2006). Similarly, less than half of the perennial herbs reintroduced into small, fragmented western Australia grasslands as seedling transplants failed to persist for more than 5 years (Morgan 1999). Despite being introduced with large founder sizes (more than 1,500 seedlings), populations of two threatened species that persisted for more than 5 years were small (fewer than sixty individuals) and suffered severe recruitment limitation (Morgan 1999). A population of the annual herb *Stephanomeria malheurensis*, reintroduced to its type locality in 1987, has been maintained only by repeated seedling outplantings over many years; otherwise the population crashes (Guerrant and Kaye 2007). In all these cases, reintroduced population persistence was low because of lack of next-generation seedling recruitment, caused by either poor seed production and viability or lack of appropriate microsites that facilitate seed germination and seedling survival (see Maschinski et al. [chap. 7], this volume).

Implications for Traditional Reintroductions

Our examination of the effects of founder population size and founder propagule stage on the persistence of reintroduced plant populations yielded several broad generalizations that may improve success rates of future rare plant reintroductions:

- *Using founder sizes larger than fifty individuals generally increased the probability of establishing a persistent population when using transplants.* Thus, we recommend using a minimum of fifty individuals for a reintroduction. When working with perennial herbs and sites in highly competitive environments such as grasslands, founder population sizes will need to be larger (appendix 1, #25).
- *Using transplants rather than seeds increased the likelihood of establishing a persistent population.* Using transplants overcomes the problems associated with losing germplasm from poor field seed germination. Particularly when few seeds are available, we recommend maximizing germination under nursery conditions to optimize the number of total plants available for a reintroduction (appendix 1, #16).
- *Regardless of the total seeds used, seed reintroductions had low establishment rates.* Because there was no relationship between the number of seed founders and population survival, we recommend developing a demographic model for the species to determine the optimum founder size (see Knight, this volume). When seeds are the only option, as is usually the case for annuals, we recommend using an experimental protocol that involves watering with drip irrigation in the field until seeds germinate and become established, a practice often used with long-lived perennials (appendix 1, #26).
- *Recruitment limitation presents a severe challenge for creating persistent populations, especially for seeds of annuals and woody perennials.* Reintroduction practitioners may need to create or seek conditions that will increase the probability of next-generation establishment, which would require knowing something about wild seedling establishment patterns (appendix 1, #26, #31).
- *Large founding size alone does not guarantee population persistence.* Factors such as genetic composition (Neale, this volume), habitat conditions (Maschinski et al. [chap. 7], this volume), horticultural treatment (Haskins and Pence, this volume), and competitive or disturbance regimes may overshadow any benefits of large founding population and therefore should be taken into consideration in the design and execution of any reintroduction.

Research Needs

- *How do some populations grow and survive with small founder sizes?* Understanding mechanisms whereby small founding populations persist will help us hone our reintroduction practice. Studies done with equal numbers of propagules placed into multiple sites (McClain and Ebinger 2008) or conditions can help us refine factors that can optimize reintroduction persistence.

- *Do certain traits or characteristics of species influence reintroduced population persistence?* Some studies have found that certain traits facilitate the invasion of species introduced into new habitats, making it possible to screen species for invasive potential before they are moved outside their range (Reichard et al., this volume). Similar studies are needed for rare species that are candidates for reintroduction to determine whether certain traits may make them more or less likely to establish persistent populations. However, addressing this question will be possible only with a larger sample size to document reintroductions properly (Dalrymple et al., this volume).

- *What are the appropriate conditions for reintroducing seeds?* There are large gaps in our understanding of the field conditions needed to promote seed germination and long-term establishment, especially for annuals. Multiple factors such as sowing techniques (e.g., burial depth, mulching, watering), protection from herbivores (caging) or competitors (weeding), and microsite conditions (e.g., disturbance regime, presence of nurse plants, canopy cover) influence germination and persistence of reintroduced species. As yet there are too few studies to draw generalizations about specific recommendations; therefore, this is a critical area for future research. Reintroductions can provide an experimental arena for testing basic seed ecology and population persistence (Young et al. 2005).

Implications for Managed Relocation

With climate change, increasing the total number of rare plant populations will become more urgent. Because of small sample sizes, our analysis did not separately examine reintroductions outside the known range and those done within the known range; therefore, we have little empirical evidence to support or refute managed relocation (MR) efficacy based on founding population size or propagule type alone. Because population demography varies with site (Pulliam 2000), as has been illustrated by within-range introductions, careful selection of appropriate habitat is key for any MR. We suggest that the practices advised for reintroductions within range mentioned earlier should be followed for any attempts outside the known range. The probability of successful colonization is likely to be higher if introductions are done with whole plants of long-lived perennials than with seeds or annuals. As previously recommended by Falk and colleagues (1996) for traditional reintroductions, we recommend that any MRs be done as experiments. When transplants are used to found new populations, a minimum of fifty transplants for any given treatment will improve the likelihood of establishment and our ability to infer generalizations between treatments and across species. Tracking demography of populations introduced outside the known range and comparing with extant populations is advised.

Summary

We reviewed 174 studies from the literature and the *CPC International Reintroduction Registry* to examine whether founding population size and the type of propagule planted influenced the persistence of reintroduced populations. We found that the type of propagule planted was skewed by life history; the majority of reintroductions done with annuals used seeds, whereas reintroductions of perennials allowed comparisons of propagule type influence on population persistence. When all life histories were combined, reintroductions using whole plants had the greatest persistence. The effect of propagule type on perennial population persistence differed between herbaceous and woody perennials. A significantly greater proportion of woody plant populations were more likely to survive when founded with seedlings and plants than when founded with seeds. For perennial herbs, differences were not significant; however, there was a nonsignificant trend for greater population persistence when the population was founded with plants versus seedlings or seeds.

Similarly, examining the influence of founding population size on reintroduction success was also skewed by life history. In comparison with the quantities of seedlings or plants used in reintroductions, exponentially greater numbers of seeds have been sown in reintroductions. When seeds were used, there was no difference in population survival across a tremendous range of founding population sizes. However, for seedlings and plants in perennial species, there was a trend for increased proportions of populations persisting with increased founder size. The exception was herbaceous perennials reintroduced to highly competitive habitats.

We recommend using founding populations of more than fifty whole plants for perennial reintroductions. For annual species that may require using seed, experimental reintroductions will be needed on a case-by-case basis to understand biotic (e.g., competitors) and abiotic (e.g., climate) factors that influence population establishment and population survival.

We are deeply indebted to Rick Luhman, Ed Guerrant, and Kathryn Kennedy for their time and role in making the *CPC International Reintroduction Registry* possible. We thank Ed Guerrant for thoughtful discussion on designing plant populations for restoration and two anonymous reviewers whose comments improved the manuscript. We also thank Juan Carlos Penagos, Julia Maritz, and Adriana Cantillo for help in organizing data for the analysis of plant reintroduction projects.

Chapter 11

Determining Success Criteria for Reintroductions of Threatened Long-Lived Plants

LEONIE MONKS, DAVID COATES, TIMOTHY BELL, AND MARLIN BOWLES

Increasingly, species reintroductions are being attempted to counteract the loss of biodiversity worldwide. Plant reintroductions aim to create or maintain self-sustaining populations capable of surviving in both the short and long term (Vallee et al. 2004). They have been increasingly used to recover threatened species and to mitigate against habitat loss. Some species persist in the wild only through reintroduced populations (e.g., *Sophora toromiro*, Maunder et al. 2000; *Allocasuarina portuensis*, Vallee et al. 2004). It is difficult to ascertain the number of threatened taxa worldwide that are currently in reintroduction programs, because many projects are unpublished or in difficult-to-access reports. But it is estimated that there are between 94,000 and 144,000 threatened plant species worldwide (Pitman and Jorgensen 2002). Target 8 of the 2002 *Global Strategy for Plant Conservation* (IUCN 2002) recommends that "60% of threatened plant species" should be "in accessible *ex-situ* collections, preferably in the country of origin and 10% of them" should be "included in recovery and restoration programs" by 2010. From these figures, by the end of 2010 an estimated 5,640 to 8,640 species worldwide should be in reintroduction programs. Given the resource-intensive nature of reintroductions, this is a huge task for reintroduction practitioners.

To date only a few reintroductions have been cited in the literature as successful. After 4 years and at three sites, *Filago gallica*, a rare annual, was considered successfully reestablished based on annual population counts (Rich et al. 1999). Maschinski and Duquesnel (2007) cautiously considered the reintroduction of the long-lived regionally endangered *Pseudophoenix sargentii* a success despite the lack of reproduction, as determined by population viability analysis (PVA). Of 249 reintroductions reviewed in New Zealand, 53 (21%) were considered successful based on 50% or higher survival (Coumbe and Dopson 1999; also see Dalrymple et al., Guerrant, this volume).

Even scarcer than documented examples of successful reintroductions are examples in which reintroduced populations have contributed to the downlisting or delisting of a threatened species. *Potentilla robbinsiana*, a long-lived perennial alpine herb endemic to New Hampshire, was removed from the US Federal List of Endangered and Threatened Plants after known populations were considered adequately protected and two reintroduced populations reached or surpassed the minimum viable population size of fifty plants (Lynch and Weihrauch 2002).

With reintroductions increasingly being used as part of species recovery programs, it is important to determine adequate, meaningful, and measurable criteria for success. Species assessment bodies, such as the International Union for Conservation of Nature (IUCN), need appropriate reintroduction success criteria to enable them to assess changes in a species listing as a result of reintroductions. Also, practitioners need to be able to learn from and evaluate their work to determine success and failure. Clearly, as more reintroductions occur, more resources are needed, and we need some way of understanding when success has been achieved in order to allocate and prioritize financial resources appropriately.

In this chapter, we review traditional and alternative methods for measuring success of plant reintroductions. Using detailed examples, we highlight ways of assessing reintroduction success for long-lived plants and emphasize the challenges practitioners face.

Reintroduction Success

Reintroduction success has been measured in many ways. Pavlik (1996) set four goals (abundance, extent, resilience, and persistence) for assessing reintroduction success. Practitioners commonly assess reintroduction success using Pavlik's measures of abundance (establishment, vegetative growth, fecundity, and population size), resilience (genetic variation,) and persistence (self-sustainability), but many of the suggested criteria—extent (dispersal, number of populations, distribution of populations), resilience (recovery from perturbation, dormancy), and persistence (microhabitat variation, community membership)—are not commonly reported in the literature. Clearly no one measure of reintroduction success will provide sufficient certainty to declare a population viable and self-sustaining or provide the foundation for recommending the downlisting or delisting of a threatened species. A combination of measures will provide the best foundation to predict success confidently.

By far the most commonly reported assessment of reintroduction success is survival or first-generation establishment. *Ranunculus prasinus* achieved its reintroduction goals when a thousand individuals occurred at the site (Gilfedder et al. 1997), and Olwell and colleagues (1990) considered 83% survival of *Pediocactus knowltonii* cuttings 3 years after planting to be encouraging. Vegetative growth, reproductive output, and recruitment are often considered in conjunction with survival to indicate whether a new location is suitable.

Reproductive success determined through flower and viable seed production and recruitment of subsequent generations is a critical measure of long-term population persistence. Factors such as pollination and the mating system significantly influence reproductive success. Changes in pollinator behavior and effectiveness due to fragmentation of plant populations can influence gene flow patterns, genetic quality of pollen received by plants, and resultant progeny fitness due to inbreeding effects (Krauss et al. 2007; Yates et al. 2007b). Patterns of mating and pollination are therefore likely to be good indicators of population persistence and reintroduction success, particularly where comparisons can be made with natural populations.

Genetic variation can also be used to assess reintroduction success. Cost-effective molecular markers can allow comparison of genetic variation between reintroduced and wild populations to determine short-term success where inbreeding may be a factor and long-term success where resilience and adaptation to changing conditions may become important (Neale, this volume). Significant change in genetic variation will be particularly important to record for managed relocations (MRs) or introductions done outside the known range of a species, a notion now given serious consideration in the context of climate change predictions (Broadhurst et al. 2008).

To evaluate reintroduction success, comparing wild reference populations with introduced populations is advised (Pavlik 1996; appendix 1, #36). Reproductive success, pollination, mating system, and genetic variation are part of a broad range of measures that can be used for comparison. Vital rates, such as survival, growth, and fecundity, and even more specific measures of plant health, such as transpiration and photosynthetic efficiency, could be benchmarked or referenced against similar-aged plants in reference populations. Without comparison there is no way to ascertain whether the reintroduction is headed along a self-sustaining or failing trajectory. "While higher vital rates may be preferable to lower vital rates, placing results into the context of wild populations may provide some solace," particularly when the vital rates are low (Menges 2008, p. 193). This was the case for the limestone endemic *Purshia subintegra*. Although survival of reintroduced seedlings after 5 years was low (7% for the best site), when compared with survival of wild seedlings over the same time period (0.3%), this was seen as encouraging (Maschinski et al. 2004a). Comparative information not only can be used to evaluate success but can also help define project goals (White and Walker

1997) and should be an integral part of any reintroduction program (Menges 2008).

Finally, most information used to determine reintroduction success is aimed at a specific step or process in the reintroduction program and does not necessarily facilitate prediction of long-term success. PVA is one method to use a range of demographic data to model the trajectory and persistence of a population. Although widely used to predict long-term trends in natural plant populations (e.g., Menges 2000), PVA has only recently been used for reintroduced populations (e.g., Bell et al. 2003; Knight, this volume).

Long-Lived Species

Long-lived plant species provide significant challenges for reintroduction because of the timeframes over which they may reproduce, recruit, and disperse. First, flowering and fruiting may take years, even decades for some species. For example, a 26-year-old specimen of *Magnolia sinica* has still not flowered (see box 4.2), and the endangered palm *Pseudophoenix sargentii* may take more than 30 years to become reproductively mature (Maschinski and Duquesnel 2007). Many of the methods suggested to assess reintroduction success depend on species moving through life stages to indicate acclimation to the new site. With funding cycles often significantly shorter than the decades it may take to demonstrate success, reintroduction of long-lived plants poses significant challenges to practitioners. Additionally, recruitment may be linked to infrequent and sporadic disturbance events, such as fire for *Acacia aprica* (Yates and Broadhurst 2002) or fire, flood, or storm for *Eucalyptus salmonophloia* (Yates et al. 1994), making a potentially long wait until natural recruitment occurs.

Reproductive Output, Recruitment of Subsequent Generations, and Response to Disturbance

A primary measure of reintroduction success is completion of the life cycle at the new site. Reproduction and next-generation recruitment are clear indicators of this. For example, good *Cordylanthus maritimus* ssp. *maritimus* reproduction and population growth suggested a successful reintroduction (Parsons and Zedler 1997). Australian grassland daisy (*Rutidosis leptorrhynchoides*) showed no evidence of lower fitness in reintroduced populations compared with natural populations after 5 to 10 years (Morgan 2000). Seed set per inflorescence, seed germination, and second-generation seedling growth for five small reintroduced populations were equal to or greater than two large natural populations, indicating that reintroduced populations had the potential to be self-sustaining.

In long-lived perennials, recruitment is episodic and seedling establishment is a rare event; thousands of seeds may produce only few individuals (Sutter 1996). Therefore, observing recruitment in reintroduced populations provides tangible evidence that conditions are suitable for species persistence. Mistretta and White (2000) recorded hundreds of progeny of the perennial species *Eriogonum ovalifolium* var. *vineum* and *Erigeron parishii* 6 to 7 years after planting at the reintroduction site. Transplanting the clonal *Rhus michauxii* was considered a viable conservation option after aboveground shoots increased by 37% and 219% at two reintroduction sites (Braham et al. 2006).

Species from environments prone to disturbance (e.g., fire, flood, tornados) must be able to survive and recover from such a disturbance when it occurs. Long-lived species will probably be exposed to disturbance in their lifetime and therefore are often adapted to the particular disturbance type and regime that occurs in their habitat. Some recruit only, or substantially, after such a disturbance. For example, in many ecosystems fire plays a key role in plant establishment (Abbott and Burrows 2003). Yates and Broadhurst (2002) showed that the threatened woody shrub *Acacia cochlocarpa* ssp. *cochlocarpa* had greater germination in burnt plots than in unburnt plots, with no seedlings emerging in unburnt plots. Two reintroductions of this species commenced in 1998. To date, despite substantial viable seed production, no natural seed germination has occurred (L. Monks, personal observation). It is unlikely that substantial seedling recruitment will occur without fire, and this was taken into account when practitioners developed success criteria for the reintroduction (see box 11.1).

Population Viability Analysis

PVA uses demographic data to project the future status of a population (Morris and Doak 2002) by determining population characteristics such as growth rates, probabilities of extinction, and critical life stages of study species. Although PVA is often described as a quantitative conservation tool for assessing the fates of rare and endangered species (Schemske et al. 1994; Brook et al. 2000; Brigham and Thomson 2003), it is rarely applied for this purpose (Morris et al. 2002), especially for reintroduced populations (Menges 2008) or long-lived plants (Schwartz 2003). Few published reintroduction PVAs have compared demographic characteristics of reintroduced and natural populations (Bell et al. 2003; Maschinski and Duquesnel 2007; Colas et al. 2008; Knight, this volume) or have evaluated introduction strategies (Bell et al. 2003; Satterthwaite et al. 2007). The scarcity of reintroduction PVAs may be a consequence of few available data sets and small initial population sizes, differences in vital rates between reintroduced and naturally recruited plants, and lack of representation of all stages. These issues are often most problematic for reintroductions of long-lived species.

In reintroduced populations, vital rates of outplanted and naturally recruited

Contributed by: Leonie Monks
Species Name: *Acacia aprica* Maslin & A.R. Chapman (Fabaceae)
Common Name(s): Blunt wattle
Family: Mimosaceae

We reintroduced the critically endangered *Acacia aprica* in 1998 to a reserve north of Perth in southwest Western Australia. For each reintroduced individual, we followed seedling survival, growth, flowering, and fruiting. Concurrently we monitored these same attributes in four natural *A. aprica* populations. Comparing flowering and fruiting attributes between the reintroduced and several natural populations showed the 3-year-old reintroduced population had a level of reproductive output similar to that of the natural populations (table 11.1). Two natural populations (populations 1 and 3) produced more seeds per branchlet than the reintroduced population, but reintroduced and natural populations produced similar seeds per legume. Natural variation may account for some of the differences seen, as would the younger age of the reintroduced plants compared with the plants in the natural populations. More detailed monitoring of reproductive output is needed to confirm whether the reintroduced population continues to be comparable to the natural populations. However, the results are seen as encouraging and suggest that the criteria for short-term reintroduction success for reproductive capability have been met (Monks 2002; Monks and Coates 2002).

TABLE 11.1

Mean reproductive attributes of a 40-cm branchlet for the reintroduced and four natural populations of Acacia aprica in 2001.

Attribute Measured	Reintroduced Population	Population 1	Population 3	Population 5	Population 7
Mean no. inflorescences/branchlet	18	88	111	59	44
Mean no. legumes/branchlet	3	14	34	3	4
Mean no. seeds/legume	7	7	9	8	5
Mean no. seeds/branchlet	22	92	317	23	18
Inflorescence to legume ratio	6:1	6:1	3:1	20:1	11:1

individuals can differ within a single population, and demographic matrix modeling requires separate stage classes for these individuals (Bell et al. 2003). For example, survival, growth, and fecundity of transplants of the short-lived mono-carpic perennial *Cirsium pitcheri* were lower than for naturally recruited plants, even years after the transplant event (Bell et al. 2003). Similar contrasts may occur between transplants and naturally recruited plants in long-lived species, but few data are available. For the long-lived Mead's milkweed (*Asclepias meadii*), plants established from sown seeds had not yet reached the initial size of transplanted individuals even after 13 years (Bell et al. 2003). Therefore, inferences about population viability based solely on vital rates of transplants may lead to improper management conclusions.

All transition stages are less likely to be present in reintroduced than in natural populations, especially in the early years after the reintroduction. As a result, important variables, such as fecundity, may be difficult to obtain or may differ from those of natural populations. This problem is exacerbated by small population sizes, where the number of individuals in size classes or stages is often too small for accurate estimation of vital rates and will require pooling demographic data over years (Schwartz 2003). Thus, whereas a natural population may require only 2 years of demographic monitoring to develop a transition matrix if all stages are present in sufficient numbers each year, many more monitoring years are necessary for reintroductions, where transplants must complete their life cycle before naturally occurring cohorts can be established and develop all stages. In addition, because transplant transition frequencies may differ from those of naturally recruited individuals, reliable transition matrices cannot be constructed for a reintroduction until all stages are present among naturally recruited individuals. Thus, published reintroduction PVAs are limited to one (Maschinski and Duquesnel 2007; Colas et al. 2008) or three (Bell et al. 2003) transition matrices for reintroduced populations, despite 10 (Bell et al. 2003; Colas et al. 2008) or up to 14 (Maschinski and Duquesnel 2007) years of demographic data from natural populations. For the short-lived *Cirsium pitcheri*, all stages were present in the reintroduction 5 years after transplanting began, but a transition matrix for naturally recruited individuals from that reintroduction could be constructed only 8 years after the reintroduction took place.

The problem of inaccurate estimation of vital rates is most acute for slow-growing, long-lived plants, because transitions between stages occur infrequently (Schwartz 2003). Population viability analyses for reintroductions of long-lived plants, such as *Asclepias meadii* (see box 11.2) and *Pseudophoenix sargentii*, are further hampered by slow growth and lack of data on reproductive stage classes (Bell et al. 2003; Maschinski and Duquesnel 2007). For example, a stage-based matrix for introduced *P. sargentii* plants consisted of pooled demographic data

The difficulty of using population viability analysis (PVA) to evaluate the success of a long-lived plant reintroduction is illustrated by *Asclepias meadii* reintroductions, which began in Indiana and Illinois in 1994 with a combination of seeds and greenhouse-grown 1-year-old juveniles (Bowles et al. 2001). Little progress has been made in developing a PVA for *A. meadii* reintroductions since the original analysis, despite an additional 9 years of demographic data. It was originally projected that seedlings would take 12 years or more to reach reproductive maturity, but that estimate has been revised to 25–30 years because of suppression of seedling growth by competition. Even with transplanted *A. meadii* juveniles, seed production has occurred only when plants have reached a flowering threshold size as measured by a leaf area index of 80 (fig. 11.1). These results illustrate the need to estimate vital rates from multiple cohorts that span the entire life history of the species in order to develop a PVA for reintroductions of long-lived plants (Bell et al. 2003).

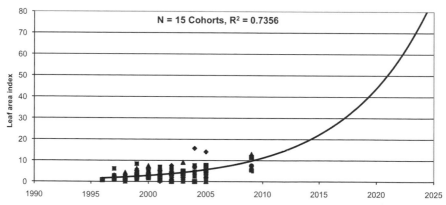

FIGURE 11.1. Exponential growth projections of *Asclepias meadii* juveniles based on demographic monitoring over 15 years. Symbols represent different cohorts. Flowering occurs when plants reach a threshold leaf area index of 80. Modified from Bell and colleagues (2003).

from introduced and wild plants over all sites because introduced plants were not represented in all stages and none of the introduced plants reproduced, even after 14 years (Maschinski and Duquesnel 2007).

The differences between reintroduced and natural populations illustrate that traditional PVA approaches must be customized for reintroduced populations. The difficulty in developing complete matrix models for reintroductions of long-lived plants, because of missing or poorly represented stages, highlights the need to introduce multiple stage classes, especially mature stage classes, to accelerate completion of plant life cycles. Establishing multiple stage classes would have the potential benefit of allowing investigators to assess reintroduction success through PVA in a timely manner.

Comparisons with natural populations will be important for developing and interpreting matrix models in reintroductions of long-lived plants. The viability of a reintroduction consisting mostly of transplanted individuals could be incorrectly interpreted unless compared with natural populations.

Mating Systems

A key factor influencing reproductive output and ultimate population persistence is the mating system, which controls the way in which gametes combine within a population. This major influence on patterns of genetic variation within a species is influenced by floral attributes, such as flowering phenology, compatibility, flower structure, and ecological factors including mode of pollination, population size, and population density (Eckert et al. 2010). There is increasing interest in studying mating systems to understand contemporary patterns of gene flow within populations, pollen pool size, and related factors such as pollinator availability and behavior (Sampson et al. 1996; Young et al. 2000; Yates et al. 2007a). The mating system regulates within-population levels of inbreeding, which in turn significantly influences a wide range of fitness components, including reproductive output, seed weight, germination, seedling establishment, and growth (Fenster and Dudash 1994).

The majority of plant species are hermaphroditic, and more than 50% are self-compatible (Igic and Kohn 2006). Therefore, assessing mating system variation in reintroduced populations and benchmarking against natural populations can improve our ability to assess population resilience and persistence. Variations in the mating system such as levels of outcrossing and the degree to which seed from the same plant has the same father can be readily assessed using genetic marker analysis of seed progeny arrays (Ritland 2002).

In comparison to measuring recruitment directly, comparative mating system studies combined with pollination biology can be accomplished over short timeframes (one or two flowering seasons). For long-lived species they can therefore give vital clues to potential recruitment and reproductive success in subsequent generations. Recent reviews have shown that changes in the mating system can be useful indicators of population processes and can give valuable insight for developing strategies for plant species persistence after anthropogenic disturbance and landscape fragmentation (Coates et al. 2007; Eckert et al. 2010). In eight animal-pollinated, mixed-mating, long-lived species, Coates and colleagues (2007) found that although mating system varies with pollination biology and life history, as populations get smaller and habitat disturbance increases, there is a trend toward increased inbreeding, smaller effective sizes of paternal pollen pools, and greater variation in outcrossing among plants. For example, in the common, bird-pollinated, long-lived *Calothamnus quadrifidus*, as populations got smaller there was significantly greater variation in outcrossing among plants, and the number of fathers contributing to seed production per plant declined. The potential for inbreeding was significantly raised in these populations, and as population size decreased, levels of seed abortion increased dramatically (Yates et al. 2007b).

The potential for elevated levels of inbreeding and increased seed abortion in small populations raises important issues for the conservation and reintroduction of animal-pollinated species with mixed mating systems in fragmented landscapes. For example, five rare and endangered animal-pollinated species investigated by Coates and colleagues (2007) are typical of many rare and threatened plant species in southwest Australia in that they survive in populations of a few hundred plants or less and are often in highly degraded habitats. The long-term persistence of these species would appear problematic without significant intervention involving reintroduction, but the persistence of any reintroduced populations will also depend on adequate population sizes and the presence of appropriate numbers and types of pollinators.

Apart from providing critical data on mating and inbreeding, mating system studies also allow predictions about pollinator behavior (see box 11.3). From a reintroduction perspective such data will be particularly useful for benchmarking mating system variation in a reintroduced population and provide valuable guidance for reintroduction planting design (appendix 1, #29).

Genetic Variation

The availability of high-quality propagule sources with suitable levels of genetic variation of rare species can be problematic. Many rare species exist as small, isolated remnant populations that may contain low levels of genetic variation due to genetic drift. To prevent overharvesting, founding populations may have been collected from a restricted sample. As a result, reintroduced populations may be go-

---

BOX 11.3. COMPARING MATING SYSTEMS IN NATURAL AND REINTRODUCED POPULATIONS OF *LAMBERTIA ORBIFOLIA*

Contributed by: David Coates
Species Name: *Lambertia orbifolia* C.A. Gardner
Common Name(s): Round-leaf honeysuckle
Family: Proteaceae

A mating system study of the natural populations of the threatened, bird-pollinated *Lambertia orbifolia* before reintroduction showed that plant density strongly influenced variation in outcrossing (or selfing) among plants (Coates and Hamley 1999). Measures of biparental inbreeding indicated little crossing between related plants and probably minimal genetic structure. In low-density populations, when pollinators occasionally move between plants, they affect significant outcrossing, but they often stay on individual plants, leading to higher levels of geitonogamous self-pollination. With this information in mind, we reintroduced *Lambertia orbifolia* ssp. *orbifolia* in a design with varying plant densities (Monks 2009). We conducted another mating system study comparing natural and reintroduced populations once the reintroduced population had significant reproduction. Preliminary results based on initial seed production suggest that significantly fewer fathers are contributing to seed production in the reintroduced population than in the natural populations, and this may be due to differences in pollinator behavior (Coates et al., unpublished data). Further studies are under way to establish whether this is a transient effect.

---

ing through genetic bottlenecks, resulting in negative demographic and genetic consequences. This may be particularly evident if the source populations are fewer than 200 reproductive plants (Young et al. 2000; Lowe et al. 2005; Yates et al. 2007a). Genetic variation of the founder population could be a useful measure of population resilience, and it is likely to influence both short- and long-term reintroduction successes. It is potentially a valuable predictive measure of reintroduction success, particularly in long-lived species where more traditional measures may take many years to evaluate. Assessing genetic variation therefore is more frequently being suggested as a tool for monitoring reintroduced populations (Krauss et al. 2002; Ramp et al. 2006; Neale, this volume).

Although a number of recent studies demonstrate the use of molecular markers in assessing reintroduction success (Neale, this volume, see box 11.4), there

---

BOX 11.4. USING GENETIC STUDIES TO PROMOTE REINTRODUCED POPULATION HEALTH

Contributed by: Leonie Monks
Species Name: *Grevillea scapigera* A.S. George
Common Name(s): Corrigin grevillea
Family: Proteaceae

The critically endangered *Grevillea scapigera* was established at two new locations in 1996, 1997, and 1998 using tissue-cultured plants. Ten plants chosen from forty-seven known wild plants were selected as genetically representative founders. Krauss and colleagues (2002) used molecular markers (amplified fragment length polymorphisms) to genotype plants in the reintroduced populations to assess the maintenance of genetic variation after establishment. Genotyping revealed only eight founding clones. Four of these clones were producing 85% of the seed, and the second generation was 22% more inbred and 20% less heterozygous than the founding population. To prevent further genetic decline, they suggested a range of strategies, including equalizing founder numbers, adding new genotypes when discovered, and changing planting and maintenance regimes to promote multiple paternity and reduce inbreeding.

---

are few studies where other genetic measures such as chromosome variation have been used. Following an analysis of allelic richness and chromosomal variation in five reintroduced and two natural populations of the Australian grassland daisy *Rutidosis leptorrhynchoides*, Young and Murray (2000) found lower allelic richness in the reintroduced populations and greater frequency of chromosomal abnormalities. They suggested that high numbers of chromosome abnormalities arose from disjunctional errors caused by inbreeding and interpopulation gene flow, leading to hybridization between diploid and tetraploid cytotypes. These findings indicate that the reintroduced populations may have lower reproductive capabilities and may not persist in the long term. In contrast, Morgan (2000) found no reduction in reproductive fitness, as measured by seed set, seed germination, and second-generation seedling growth. Young and Murray (2000) recommended that future reintroduction attempts take into account chromosome variation in source populations and potential consequences of pollen flow from natural populations and strive for larger populations to maintain allelic richness.

Establishment and Survival

Possibly the most widely used measure of reintroduction success is population establishment and survival. This is usually defined as a specified proportion of the propagules surviving over a specified time period. The new population must be a minimum size to be self-sustaining in the long term so that genetic consequences of small population size can be avoided. Using nine life history characteristics (longevity, breeding system, growth form, fecundity, ramet production, survivorship, seed duration, environmental variation, and succession status), Pavlik (1996) estimated that 50 to 2,500 individuals would be necessary to found a population (but see Knight, this volume). Guerrant (1996a, p. 194) suggested that the "founding population should be as large as possible," with the "ceiling set primarily by practical and other strategic considerations." A recent review of seed sourcing for restoration showed that populations of less than two hundred are likely to be poor sources of seed (Broadhurst et al. 2008). Theory suggests that, at least for obligate outcrossing species, reintroduced populations of fewer than two hundred plants are likely to be unsuccessful due to inbreeding difficulties (see Neale, Albrecht and Maschinski, this volume).

Survival can vary vastly depending on the timeframes of measurement. "The longer an experimental population is monitored, the more stochastic factors will affect the population persistence" (Wendelberger et al. 2008, p. 550). There is often an initial decline in plant numbers in the first year after planting. For reintroduced plants, initial mortality may be caused by shock (Bell et al. 2003), or for reintroductions started from seed, high seedling mortality may occur. For example, the proportion of *Purshia subintegra* seedlings surviving 1 year after planting declined to between 13% and 48%, depending on habitat and treatment (Maschinski et al. 2004a). Most deaths after experimental reintroductions of *Prostanthera eurybioides* (up to 85%), *Acacia cretacea* (up to 95%), and *A. whibleyana* (up to 98%) occurred within the first year, and then plant numbers stabilized (Jusaitis 2005).

The timeframes over which survival is measured for long-lived species need to take into account the longevity of the plants. Monitoring long-term survival (more than 10 years) is essential (Guerrant and Pavlik 1998), especially given the lifespan and time needed to reach reproductive maturity of many long-lived species (appendix 1, #38).

Population trends may become apparent only many years after reintroduction. After seeding, the reintroduction of the annual *Cordylanthus maritimus* ssp. *maritimus* appeared successful when the population increased from approximately 5,000 plants in year 1 to 13,997 plants in year 4 (Parsons and Zedler 1997). However, because opportunities for seed germination and seedling establishment were limited, it was unclear whether the population would be sustainable in the long term and whether additional monitoring was needed. For the long-lived perennials *Eriogonum ovalifolium* var. *vineum* and *Erigeron parishii*, survival was 86% and 75%, respectively, 1 year after reintroduction (Mistretta and White 2000). Monitoring after 6–7 years showed only a slight decline in survival of the initial plantings (to 77% and 66%, respectively). Thus, survival and progeny counts of 325 for *E. ovalifolium* var. *vineum* and 452 for *E. parishii* indicated that the population had successfully established and was increasing. For twenty-four perennials, Mottl and colleagues (2006) found that a promising 91% mean survival after 1 year declined to a mean of 57% after 7 years. However, the authors still considered the reintroduction a success, in part because the mean percentage survival had stabilized between 5 and 7 years after planting. At a grassland restoration site in southeastern Australia thirty-three species of eighty-five restoration species persisted after 15 years (McDougall and Morgan 2005). Just three species established sufficiently well to form the dominant community components, as they would in natural sites, whereas thirty species remained as minor vegetation components. After 10 years Guerrant and Kaye (2007, p. 369) thought that "it was still too early to determine whether populations" of *Lilium occidentale* "are self-sustaining, and to what degree they are resilient to environmental perturbation." The danger in considering a reintroduced long-lived species stable and successful after the first couple of years is that it might decline before significant reproduction occurs (see box 11.5). If this occurs, the population may undergo a significant genetic bottlenecking event or simply become extinct.

Vegetative Growth

Measuring growth parameters can provide evidence that conditions are suitable for a species at a new site. Changes in height, volume, stem diameter, and number of leaves are often used along with monitoring plant survival to indicate reintroduction success. For example, Mistretta and White (2000) considered that an indicator of successful *Eriogonum ovalifolium* var. *vineum* and *Erigeron parishii* reintroduction was the growth of reintroduced plants by a mean of 4.32 and 0.96 centimeters, respectively, over the first 5 years. A mean stem diameter growth of 5 millimeters for reintroduced *Pediocactus knowltonii* over 3 years was considered appreciable, given that the plants usually average only 13.1 millimeters in diameter (Olwell et al. 1990). Plant volume of the reintroduced *Amorpha herbacea* var. *crenulata* increased in the months after planting but varied in subsequent years depending on season of measurement (Wendelberger et al. 2008) and provided one indication that the reintroduced population was performing well. Increased plant height, stem diameter, and number of leaves were used to indicate how well

BOX 11.5. CHANGING SURVIVAL RATES AFTER 1, 4, AND 10 YEARS

Contributed by: Leonie Monks
Species Name: *Grevillea calliantha* Makinson & Olde
Common Name(s): Foote's grevillea
Family: Proteaceae

Foote's grevillea is a federally endangered long-lived woody perennial from the southwest of Western Australia. A total of 466 propagules were planted in 1998, 1999, 2001, 2002, and 2005 in a single site north of Perth in Western Australia. Monitoring initially gave an optimistic outlook for population survival. Initial establishment of each planting cohort was good (55–95% plants survived the first year; table 11.2). Four years after the first planting, in 2002, 68% of plants survived, suggesting the population had established well. However, monitoring thereafter revealed a less positive outlook, with just 18% survival after 10 years (table 11.2). Herbivory of unfenced seedlings planted in 2002 resulted in the death of all plants in that cohort (other cohorts were entirely or partially fenced), and drier-than-average years may have contributed to declines of all cohorts between 2002 and 2008. Age to first reproduction was 2–4 years, with substantial reproduction occurring after 4–6 years. The reintroduced population declined substantially before any significant reproduction occurred. Additional planting at the reintroduction site will need to take place in order to avoid inbreeding in subsequent generations (Monks and Coates, unpublished data).

TABLE 11.2

Number (%) of outplanted Grevillea calliantha *plants surviving after 1 year and then in the medium (4 years) and long term (10 years).*

Cohort Number	Year Planted	No. Planted	No. (%) Surviving After 1 Year (Initial Establishment)	No. (%) Surviving in 2002 (4 Years After Initial Planting; Medium Term)	No. (%) Surviving in 2008 (10 Years After Initial Planting; Long Term)
1	Planted 1998	106	73 (69)	45 (42)	20 (19)
2	Planted 1999	115	109 (95)	88 (76)	50 (43)
3	Planted 2001	114	63 (55)	63 (55)	2 (2)
4	Planted 2002	106	N/A	106 (100)	0 (0)
5	Planted 2005	25	15 (60)	N/A	13 (52)
All	All years	466	N/A	302 (68)	85 (18)

four native tree species established in restoration sites in a Mexican cloud forest (Alvarez-Aquino et al. 2004). Plants generally survived better in degraded remnants of cloud forest but grew much quicker in cleared areas adjacent to cloud forest remnants, indicating that restoration of Mexican cloud forest can be successfully undertaken using the four native species.

Although measurements of growth can be indicators of reintroduction success, they can also be used to gauge the usefulness of various reintroduction techniques or suitability of sites (see box 11.6). Maschinski and Duquesnel (2007) measured height, trunk diameter, number of leaves, and presence or absence of a leaf spike to gauge the performance of slow-growing *Pseudophoenix sargentii*. This allowed comparisons between microsites, with the tops of coastal berms found to be the microsite where the fastest growth occurred. It also enabled the authors to assign plants to stage classes to undertake PVA. Measurements of the height and length of the longest leaf were used to compare the success of two planting techniques and two sites for the reintroduction of *Rhus michauxii* (Braham et al. 2006). Although no trends were found for height, mean length of leaf differed between planting years, suggesting that allowing plants to grow in the nursery for 1 year was a better technique than direct transfer of plants between donor and recipient sites. In another study Jusaitis and colleagues (2004) found that the number of leaves per plant and plant diameter were useful indicators of weed impacts on reintroduction success for *Brachycome muelleri*.

Practical Implications for Traditional Reintroduction Work

We have indicated that assessing reintroduction success in long-lived species can be challenging using more traditional approaches that rely on reproductive output and recruitment given the timeframes over which such species may reproduce, recruit, and disperse. Although traditional approaches are important and needed, we suggest that other approaches can allow more timely assessments of reintroduction success and should be given consideration. PVA is a powerful tool that can help predict long-term reintroduction success. However, despite its potential there can be limitations with PVA, particularly in estimating vital rates, because transitions between stages for long-lived plants are infrequent. Practitioners undertaking reintroductions of these species may need to reconsider the frequency of monitoring because measurements taken for the development of PVA models will need to be at a rate that accurately charts movement of an individual from one life stage to another (appendix 1, #35, #37).

Assessing mating system variation and levels of genetic variation in the reintroduced populations are also approaches that can allow some indication of long-term reintroduction success. Both can be readily assessed using molecular genetic

---

Box 11.6. Using Growth Measurements to Judge Suitable Reintroduction Sites

Contributed by: Leonie Monks
Species Name: *Prostanthera eurybioides* F. Muell
Common Name(s): Monarto mint
Family: Lamiaceae

The federally endangered perennial Monarto mint is known from only a small number of populations in South Australia. Jusaitis (2005) used small-scale experimental reintroductions to investigate the suitability of three sites in preparation for large-scale reintroduction. In conjunction with survival, the change in growth index over 8 years gauged which site was the most suitable for the species. Plants at site 1 thrived with 80% survival and increased size by eight and a half times (table 11.3). Site 2 had a lower survival of 20% and just over four times increased growth. However, plants at site 3 fared poorly, with growth declining over the first year before all plants died. The combination of survival and growth measurements provided good evidence on which to base site selection for larger-scale reintroductions (Jusaitis 2005).

TABLE 11.3

Initial and final growth index (GI) of Prostanthera eurybioides *plants translocated in June 1996 to three different microsites at Monarto, South Australia (Jusaitis 2005).*

Site	Initial GI	GI at Final Monitoring (8 Years after Planting)
1	80	680
2	100	420
3	100	0

Growth index (GI) is calculated as (Height + Crown width 1 + Crown width 2)/3.

---

markers (Neale, this volume), but there are constraints with such approaches. We emphasize that these approaches do not necessarily act as direct surrogates for more traditional measures of establishment, survival, vegetative growth, reproductive output, and recruitment. For example, frequently used and readily available neutral molecular markers, such as microsatellites, may provide very good measures of genetic variation and small population size effects, but it is not entirely clear how critical neutral variation might be for adaptation and persistence. Mating system analysis will require collecting seed from individual plants, so at least an initial stage of reproduction is needed before progeny can be genotyped and mating patterns determined. Despite these caveats we believe that these approaches and PVA can be valuable adjuncts to the more traditional methods for assessing reintroduction success, particularly in long-lived species.

Areas of Need and Research Opportunities

Using PVA, mating system variation, and genetic variation to measure reintroduction success efficiently and effectively raises a number of key research questions. For all three approaches, comparisons with natural populations are important for benchmarking and interpretation. As previously mentioned, few studies have used PVA to compare reintroduced and natural populations, so there is a need for more comparisons and explorations of customized PVA approaches for long-lived reintroduced populations.

Benchmarking is particularly critical in considering mating system and genetic variation. Although changes in parameters that estimate these variables can be monitored within the reintroduced population, determining key thresholds at which failure is likely will be extremely difficult. A predetermined level based on an assessment of natural populations is therefore essential. For example, at what level does genetic variation become critically low due to reduced population size, and when is augmentation needed? At what point does population reproductive output become severely compromised by reduced outcrossing and increased inbreeding? Is this effect caused by a change in plant–pollinator interactions?

Where reintroductions involve threatened species that are critically endangered, benchmarking may not be feasible because the remaining natural populations are already too small, there is no recruitment, and they are found only in highly disturbed habitats. In these cases, it would be worth exploring the feasibility of comparisons with sister taxa or other species with similar functional traits.

Prospects and Cautions for Appropriate Use of MR

Introducing a plant species outside its natural range in response to the threat of climate change has been raised recently as a key tool in plant conservation (Vitt et al. 2010; Haskins and Keel, this volume). Assessing the success of such introductions to novel geographic areas would not necessarily involve different approaches to those outlined here. The challenge with long-lived plants lies in our ability to assess reproductive output and recruitment given the timeframes over which these processes are likely to occur. In cases of MR, a useful criterion for predicting success would be the level of genetic variation in the introduced population. It is still extremely difficult to predict where a species might be best located under current climate change modeling regimes. Yet maintaining high levels of genetic variation and maximizing evolutionary potential in introduced populations would be an important target for achieving success (see Broadhurst et al. 2008).

Managed relocation may lead to unexpected interactions between the introduced species and other species in the recipient site. Such interactions may be negative (e.g., Riccardi and Simberloff 2009a), and perhaps success criteria might need to be broadened to cover not only the factors that would be predicted to indicate persistence of the introduced species but also factors that might indicate a negative impact on the recipient community (see Reichard et al., this volume). Becoming invasive and negative genetic interactions such as outbreeding depression and genetic swamping, following interbreeding between native and introduced populations, are potential undesirable outcomes. In these situations the MR might be viewed as a failure despite the persistence of the introduced species.

Summary

Defining reintroduction success for long-lived plants, where reproduction and recruitment of second and subsequent generations may take place long after the funding for the reintroduction has ceased, is a major challenge for conservation agencies and conservation practitioners. We need to determine when a reintroduction is viable and capable of persisting. If multiple species reintroductions are needed in a single region, we need to understand when a satisfactory level of success has been achieved and reallocate resources. Traditional measures of success consider vital rates such as survival, growth, and reproduction, preferably over multiple generations before the likelihood of reintroduction success is evaluated. Although these measures can still be valuable for long-lived species, assessing them over multiple generations is problematic when many years or decades may pass before the plants reach sexual maturity. Their value can be greatly increased by benchmarking against natural populations. We concur with Menges (2008) that comparative information should play a key role in assessing success in any reintroduction program.

Equally, we suggest that there are a range of other measures and approaches such as PVA, mating system variation, and genetic variation that are likely to be especially effective for predicting reintroduction success. These not only will be specifically relevant to long-lived species but could be more broadly considered for plant reintroductions. We emphasize that the development of effective means of predicting long-term success for reintroductions of plant species is becoming increasingly critical where resources for reintroductions are limited and where the strategic reallocation of resources is needed to deal with the dramatically escalating number of threatened plant species worldwide.

Chapter 12

Unique Reintroduction Considerations in Hawaii: Case Studies from a Decade of Rare Plant Restoration at the Oahu Army Natural Resource Rare Plant Program

H. KAPUA KAWELO, SUSAN CHING HARBIN, STEPHANIE M. JOE, MATTHEW J. KEIR, AND LAUREN WEISENBERGER

The extreme isolation of the Hawaiian Islands coupled with their sequential volcanic origin and gradual erosion from 4,000 meters above sea level down to sea mounts created a large variety of habitats from rainforest to desert within a small geographic area. This mix of ecological settings resulted in a highly endemic fauna and flora often lacking the typical assemblage of herbivores and carnivores found in continental situations (Carlquist 1970; Stone and Scott 1984). After Cook's discovery of the islands, large numbers of nonindigenous species were introduced, and some of them have proven to be directly or indirectly detrimental to the native species. Most plants face multiple threats from predators, insects, feral ungulates, weeds, and loss of pollinators and dispersers. Hawaii's unfortunate reputation as the home of 37% of the nation's federally listed plants (US Fish and Wildlife Service [USFWS] 2009) reveals a conservation crisis where restoration efforts are of critical importance. Indeed, conservation efforts have been under way since the early 1900s (Mehrhoff 1996), but rare plant conservation in Hawaii faces numerous challenges, many of them overlapping. Hawaii's experiences with rare plant reintroduction elucidate the challenges managers confront in both island situations and continental environments.

The US Army Garrison Hawaii (Army) is responsible for stabilizing fifty-one endangered plant taxa on Oahu to offset potential impacts from Army training. Twenty-four of the fifty-one taxa have fewer than three hundred individuals remaining in the wild, and of these, fifteen taxa have fewer than fifty individuals. Therefore, reintroduction plays a large role in achieving stability goals for many of these rare plants. However, the efforts reported in this chapter are unprecedented in scale, both with respect to the number of species involved and in the detail of the planning effort.

In 1994 the USFWS Hawai'i and Pacific Plant Recovery Coordinating Committee proposed a standard of stability for endangered plants in Hawaii. A plant is considered stable when it has three populations with a minimum of either 25 mature and reproducing individuals of long-lived perennials (>10 year life span), 50 mature and reproducing individuals of short-lived perennials (<10 year life span), or 100 mature and reproducing individuals of annual taxa per season (<1 year life span). In addition to numerical criteria, genetic storage must be in effect for the taxon and all major threats must be controlled.

The Army has been involved in rare plant reintroduction on the island of Oahu since 1998. The Oahu Army Natural Resource Program (OANRP) began this work in 1995 with four employees and has a current staff of fifty-one. The Army's responsibility was outlined in Endangered Species Act, Section 7, "Consultations between the USFWS and the Army." The Army was tasked to prepare documents detailing stabilization steps defining population and management units and a milestone schedule of actions. The stabilization plans must often be adapted to address myriad problems, particularly in five major areas where Hawaii has unique challenges: genetic considerations, pollination biology and breeding systems, obtaining propagules, threat control strategies, and site selection limitations. We provide case studies that best illustrate the five major problems from our species recovery efforts.

Problem Area 1: Genetic Considerations

Reintroducing taxa with small populations entails potential risks of inbreeding and outbreeding depression (Neale, this volume). We sought to reduce those risks by maximizing founder representation in outplantings without losing population-specific alleles. To develop stabilization strategies we consider factors such as species biology, morphological variation, genetic diversity, breeding and mating systems, historic distributions and life cycles, and threats from invasive species. Unfortunately, these details are poorly understood for most endangered Hawaiian plant taxa. For many species, extinction appears imminent and stabilization actions need to proceed immediately, in conjunction with research. However, continually adapting genetic management strategies to incorporate research results is complicated. It can be difficult to differentiate between risks due to poor fitness (which can be addressed through reintroduction design strategies) and invasive species threats (which can be addressed via threat control). For example, slugs may severely limit seedling recruitment, overshadowing genetic considerations (Joe and Daehler 2008). Deleterious effects of inbreeding may be delayed and difficult to assess quickly enough to guide genetic management decisions (Husband

and Schemske 1996). Molecular variation may not be strongly correlated with the ability to evolve (Frankham et al. 2002). Indeed, some small populations may have historically low genetic variability and may or may not benefit from increased gene flow (Ellstrand and Elam 1993). In some species such considerations are moot: *Cyanea superba*, now extinct in the wild, was reestablished from two individuals.

In order to meet stabilization goals, outplantings are necessary. Outplantings use nursery-grown plants with known maternal founders that are individually tagged. Plans for augmentation, reintroduction, and introduction consider the genetic implications of each type of outplanting. Augmentation plans consider the impact of introducing new individuals into an existing breeding group. Reintroduction plans consider mixing sources and the potential impacts on plants near the proposed site and the relatedness of the source populations. Plans for introductions into previously unoccupied habitat consider the potential impacts on undiscovered plants and on related species. All outplantings are established over a span of a few years until all available founders are equally represented at each site. Ex situ genetic storage collections are essential for supplying the appropriate founders for each outplanting. These collections consist of seed collected from in situ founders or first-generation offspring from either intrapopulation crosses or selfed pollinations of plants growing in ex situ living collections. These genetic storage collections represent "pure" seed stocks, which maintain original genetic variability for individuals and populations. If certain founder combinations are discovered to be an important variable in outplanting success, these collections will be available for future efforts. The degree to which founders are mixed in an outplanting depends on considerations of historic records, genetic research, and breeding systems and is discussed in this chapter for each species.

Populations of taxa that are probably facultative selfers are potentially more genetically isolated and locally adapted and may have purged their strongly deleterious alleles (Charlesworth and Charlesworth 1987). A strategy to increase gene flow between populations may not be necessary. If founders from different sites are mixed in an outplanting, there may be a risk of reducing potential local adaptations and causing outbreeding depression (Ellstrand and Elam 1993).

Case Study: Delissea waianaeensis

Delissea waianaeensis Lammers (Campanulaceae), endemic to the Waianae Range on Oahu, was reduced to a total of twenty-seven plants from seven different sites in 2005. Known threats include predation and habitat degradation by feral goats and pigs, seedling predation by slugs, and competition with invasive plants.

It is presumed to be a facultative outcrosser because of floral morphology and protandry. The production of seeds by isolated plants despite the loss of the putative Hawaiian drepanid pollinators demonstrates that it is capable of autogamy.

We considered several factors in the development of the stabilization plan for D. waianaeensis by reviewing the historic and current distribution to determine how long populations have been isolated or occurring in different habitats, incorporating observations of single plants producing viable seed, and documenting morphological variation in leaf characteristics. The reintroduction strategy is to maximize the number of founders within certain designated areas but not to mix sites that were morphologically or genetically distinct unless inbreeding depression is shown to be a limiting factor. Although twenty-seven founders are available, they will be mixed in reintroductions only with plants from nearby groups.

Before conducting large-scale reintroductions, small outplantings are established to represent each group. Collections from these small outplantings are used to supplement the ex situ stored seed collection of "pure" stock from each site. This reduces the potential for loss of any local adaptations or unique alleles (outbreeding depression) of a particular population site that may occur if all the founders are mixed. Seeds collected from the larger mixed outplantings are stored separately. The benefits these collections may have from the influx of additional genotypes can then be applied to future outplantings.

Our caution in mixing founders from management sites was validated when two additional taxa (*Delissea takeuchii* and *Delissea sinuata*) were described within the previous taxonomic treatment of D. waianaeensis (Lammers 2005). In addition, when new plants of unknown provenance were discovered near an outplanting site, molecular studies showed that plants at each site were genetically distinct (mean expected level of heterozygosity $H_t = 0.27$) (James 2009). The strategy for D. waianaeensis strives to preserve genetic and morphological variation in genetic storage while experimenting with increasing gene flow in outplantings and awaiting genetic research and pollinator observations.

Case Study: Schiedea obovata

Schiedea obovata Sherff (Caryophyllaceae), endemic to the Waianae Range on Oahu, is hermaphroditic but facultatively autogamous. It is extirpated from most of its recorded locations, including all sites in the southern Waianae Range (Wagner et al. 2005). There are only twenty-four mature plants remaining in three sites within 3.5 kilometers of each other. The observed threats to the remaining sites and the presumed reasons for decline include predation and habitat degradation by feral goats and pigs, competition with alien plants, and seedling predation by slugs (US Army Garrison Hawaii 2003; Joe and Daehler 2008). The existing sites differ in elevation and habitat. The variation in leaf morphology observed between the sites is maintained when plants are grown together.

The stabilization goal for this species is to establish and maintain three sites, with one hundred mature reproducing plants in each. Two of the sites will be reintroductions. The third site will be an introduction into historically unoccupied habitat. In an effort to preserve the potential variation in remaining sites, the stabilization plan directs that augmentations use founders from that site only. Maintaining the genetic integrity of every site by keeping all outplantings separate defines the strategy for the facultatively autogamous *S. obovata*. In order to maximize founder representation in outplantings, seed collections were used from extant and previously sampled extirpated founders. Outplantings at two of the three sites used seeds collected from in situ plants or collections from ex situ sources of extirpated sites.

All options are being considered for the introduction at the third site. Parentage of the seeds in genetic storage and morphological and habitat differences between the sites will be considered when designing the strategy for the introductions in the third site. As part of the effort to select the source or mix of sources for this introduction, research is under way to quantify genetic variability using microsatellite data (see Neale, this volume). We will investigate the presence of inbreeding depression within the remaining sites and the potential for outbreeding depression when sources from different sites are mixed. The fitness of offspring produced by selfing and interpopulation and intrapopulation hand-pollinated crosses will be measured. In addition, we will compare the biotic and abiotic elements of the candidate introduction sites with the in situ habitats to guide the strategy for the introduction at the third site.

Problem Area 2: Pollination Biology and Breeding Systems

Hawaii's extinction crisis has affected both plants and the insects and birds that serve as pollinators. Nectivorous birds are important pollinators in Hawaii (Pratt 2005), but many native birds are now largely restricted to elevations of more than 1,220 meters (above the highest elevation on Oahu) because of avian disease (Scott et al. 1986; van Riper and Scott 2001). The Hawaiian flora has the highest level of dioecy of any flora surveyed (Thomson and Barret 1981; Sakai et al. 1995a, 1995b). It has been suggested that various outbreeding mechanisms evolved in some groups after the arrival of their ancestral Hawaiian founders, thereby overcoming founder effects such as low genetic diversity and inbreeding depression (Thomson and Barret 1981; Charlesworth and Charlesworth 1987; Sakai et al. 1995a, 1995b). With the decline of the Hawaiian avifauna (Scott et al. 1986; Pratt 1994) and the loss of many native insect pollinators (Cox and Elmqvist

2000), dependent plant taxa will continue to decline (Kearns et al. 1998). Thus, in order to conserve these plants it will be necessary to assist pollination among individuals.

Case Study: Hesperomannia arbuscula

Hesperomannia arbuscula A. Gray (Asteraceae, Tribe Vernonieae) is a small tree/shrub present in the Waianae Range. It is presumably bird pollinated, has large, conspicuous flower heads, and produces copious amounts of nectar. In 1999 there were approximately sixty-five individuals in seven populations; in 2011, just eleven individuals in five populations remained. The steep decline may have been caused by several factors, including drought, competition from nonnative plants, and ungulates. In 2003 OANRP, the Oahu Plant Extinction Prevention Program, and The Nature Conservancy of Hawaii focused on clonal propagation (air layering) and seed collection of wild plants. Between 2001 and 2004 a few viable seeds were wild collected. However, in 2005–2006 no viable seed was observed or collected at any of the populations. Mature plants occur in only three of the five populations, and some individuals are too distant for generalist pollinators to be effective.

Observations of flowers both in situ and ex situ revealed that although mature pollen is pushed out of the corolla by the elongating style, the stigmas are not receptive until 2 to 3 days later. Protandry and the lack of viable seed production suggest obligate outcrossing. A large effort to cross-pollinate individuals by hand was initiated in 2007. With five extant populations, only two or three of which may have flowering individuals in a given year, pollen donors and crosses have been chosen opportunistically. Every effort was made to include as many crosses as possible, and any combination of individuals both within and between populations was considered appropriate; flowering ex situ individuals were also used as pollen sources. Pollen in the Asteraceae is trinucleate, which typically is difficult to store for long periods of time (Brewbaker 1967). Therefore, fresh pollen was collected and refrigerated for up to 4 weeks.

Since the initiation of hand pollinations in 2007, more than three hundred plants have been produced (fig. 12.1). Preliminary results are encouraging for four reasons. First, no significant differences in vigor (e.g., growth rate, survivorship) between the offspring of various crosses have been observed, although some populations exhibit distinct morphological features (e.g., reddish venation). Second, ex situ plants are able to produce viable seed after cross-pollination. Third, most crosses did produce viable seed. Fourth, some seedlings from the crosses have not survived, although mortality has not been linked to any particular cross.

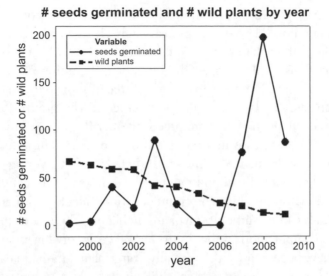

FIGURE 12.1. Number of wild *Hesperomannia arbuscula* plants and viable seeds collected since 1999. Hand pollination efforts began in 2007.

Approximately 50% of the seedlings (169/361) have survived to produce young plants, 128 of which were part of in situ reintroductions in 2009 and 2010, with 96% survivorship. Reintroduction efforts are continuing, and there are plans to quantify fitness as the plants mature. These results indicate that although hand pollination is necessary for the recovery of this species at this time, we hope that the augmented populations will be able to sustain large enough numbers of mature flowering individuals where naturally occurring generalist pollinators will support regeneration.

Case Study: Labordia cyrtandrae

Labordia cyrtandrae St. John (Loganiaceae) is a shrub found in mesic to wet forests on Oahu. Motley and Carr (1998) observed that all species in this endemic genus are functionally dioecious. This species is assumed to be bird pollinated due to the production of nectar and the lack of scent (TJ Motley, personal communication). However, because native birds in Hawaii are uncommon below 1,220 meters, native pollinators for this species are expected to be limited or absent. Nonindigenous species may be fulfilling this role at present.

Early collections for this species are all from the Koolau Range, where currently only one individual is known. All other known plants (about seventy-four mature individuals) are found in the northern end of the Waianae Range. Motley and Carr (1998) reported that the flowering periods of some congeneric populations are asynchronous. There are approximately eighteen groups of this species. The term *group*, rather than *population*, is used by OANRP because none of the eighteen collections of plants meet the functional definition of population (see Glossary, this volume). Each group has only three or four individuals, which are sometimes separated by large geographic distances. Even groups that have both male and female plants may not produce fruit after flowering. This can imply a few scenarios: (1) Both sexes in a group did not flower synchronously, (2) a specific pollinator may also be in decline or extirpated, or (3) flowering plants are not of a sufficient density to attract pollinators. The dehiscent fruit capsules take several months to mature, making the timing of fruit collection difficult. Access to some individuals can be difficult, involving helicopter flights or rappelling. The result of these complications is that in the last 10 years very few mature fruit with viable seeds have been collected, especially from isolated individuals.

In 2009, twelve crosses using ten wild plants were conducted in situ and yielded approximately 13% fruit set, compared with almost none without hand pollination. Two major lessons were learned from this project. The first is that pollen of this species does not store under ambient conditions for more than a few weeks or months. The second is that greenhouse plants did not produce as many successful crosses as in situ plants. Research on this species is ongoing, and techniques for crossing will be refined over the next few seasons. From these results it appears that this species also needs hand pollination initially in order to meet stabilization and recovery goals.

Problem Area 3: Obtaining Propagules

The type and source of rare plant propagules have been two areas of focus for many studies investigating reintroduction success (Guerrant and Kaye 2007; Menges 2008; Neale, this volume). Many of these studies have looked at whether seeds or plants should be used to establish a reintroduction and the number of populations from which propagules should be collected (see Albrecht and Maschinski, this volume, for more on propagule selection). In Hawaii, the target species often do not produce enough seed naturally, or complete representation of extant plants is difficult to achieve due to inaccessibility or a sporadic and uncertain production of seeds.

There are four major obstacles to acquiring in situ collections. The first major obstacle is that access to sites is limited by helicopter flight constraints and helicopter habitat disturbance. Second, many taxa do not reproduce during a predictable time window but flower sporadically year-round. Infrequent flowering at low levels necessitates many repeat visits to obtain seeds. The third obstacle is that some taxa do not produce seeds or produce a very small amount of viable seeds. These plants may be dioecious or self-incompatible, have poor reproductive vigor, reproduce sporadically, be inbred, or lack pollinators to complete fertilization. Genetic storage goals and reintroduction needs may be higher than the amount of seed wild populations can produce. In addition, an adequate amount of seed should remain in situ so as not to deplete the potential for natural regeneration. And finally, some propagules are simply too difficult to collect from wild plants. In this category are plants with fruits that violently dehisce immediately upon maturation (e.g., *Viola chamissoniana* Ging. ssp. *chamissoniana*, Violaceae) or that require climbing structurally unsafe canopy trees (e.g., *Flueggea neowawraea* Hayden, Euphorbiaceae).

For some species, ex situ living collections are integral parts of our reintroduction program. Living collections provide a way to overcome the difficulty of obtaining propagules from the wild. Most species currently propagated by OANRP do not produce ample amounts of seed but are easily cloned. In a nursery environment, fruits are more easily monitored, harvested at maturity, and protected from predators. Fruits that dehisce violently can be bagged with less damage to the plants. Seeds can be harvested much more frequently than in the natural setting. Compared with field pollination attempts, ex situ plants are more easily hand pollinated, a larger number of crosses can be conducted, and more seeds can be collected. All seeds can be collected from ex situ plants without concern about overcollecting. Additionally, using clonal collections reduces the number of generations that nursery-source seeds are removed from the parent plant.

However, there are reasons for limiting this type of propagation. Nursery space is a severe limitation. Propagation is restricted to species with seeds that are difficult to collect. Additionally, there are downsides to managed breeding, where artificial selection may influence the genotypes that are ultimately reintroduced or stored in the seedbank. Most of this selection is avoided by using only clones to produce seeds in the nursery. However, hand pollination has additional levels of artificial selection. Pollen is often stored before pollination, where selection can occur for grains that best survive desiccation. From fertilization to seed set, artificial selection may continue to occur due to the nursery environment. Ultimately, without managed breeding in the nursery, reintroductions would not occur because propagules would not be available. The goal remains to limit selection as much as possible but still produce plants to outplant and seeds to store.

Case Study: Flueggea neowawraea

Currently, thirty-six individuals of *F.neowawraea*, a dioecious tree, exist on Oahu, a substantial reduction from a much wider distribution indicated by the number of snags throughout the Waianae Range. Based on genetic analyses, the longevity of this taxon, and its historically uniform distribution, all *F. neowawraea* in the Waianae Range are managed as a single population. Present threats include *Xylosandrus compactus* Eichhoff (coffee twig borer), weeds, rats, ungulates, and the low number of remaining individuals (Hayden 1987). Pollination in this dioecious species is severely limited by the paucity of plants.

Obtaining propagules from wild plants for reintroductions has been difficult. There is highly reduced living plant tissue on extant *F. neowawraea* due to *X. compactus* damage. Most living material is located on the highest, most difficult to reach branches. Viable seed has been collected from only three wild plants over the past 12 years. Ex situ clonal collections of cuttings and air layers (artificially rooted stems produced in situ) are currently representing twenty-one of the thirty-six wild plants, 70% of the known females and 75% of the known males. Because the clonal material is often from mature branches, the living collection plants can flower within a year after rooting. In situ plants flower only once per year, whereas the ex situ plants often flower twice a year.

The reintroduction goal for this taxon is to represent the thirty-six known founders equally in each of four sites. With such a small number of plants and a continuing decline, reintroductions must be initiated quickly. Every female–male combination possible will be made. A balance of all possible combinations would be ideal at each outplanting, but pollinating only for this goal is not realistic for the following reasons: Fifteen founders are still unrepresented ex situ, older pollen collections are prioritized regardless of the combination of crosses that are underrepresented, and some plants in the living collection are overrepresented because they flower more frequently.

The following protocols and priorities have been established for crossing and outplanting as the sources of males and females become available:

1. Pollen donor within an in situ population site (founders are geographically close). We are uncertain of the extent of gene flow except where individuals are found growing next to or near each other.
2. Pollen donor from an old collection (to test storage longevity and not lose any collection).
3. Pollen donor not yet used (in storage but has yet to produce seed for storage).
4. Pollen donor that is novel for a particular female (a new combination).

During winter 2008–2009, OANRP initiated two reintroductions (twenty-five plants total), using two different female–male combinations. These outplantings were supplemented in the winter of 2010–2011 with an additional twenty-five plants representing five new female–male combinations. Survivorship by fall 2011 was 94%. Approximately fifty plants are available for the 2011–2012 outplanting season, some of which will represent three new male founders. OANRP has also begun grafting stock from wild plants in extremely poor condition onto saplings, as other forms of vegetative propagation have failed numerous times for plants in this condition. OANRP will continue to pursue the additional fifteen founders and follow the OANRP prioritization for acquiring new genetic crosses for reintroduction.

Case Study: Tetramolopium filiforme

Helicopter flights and rappelling are needed to access some populations of *Tetramolopium filiforme* Sherff (Asteraceae). This species is found along dry to mesic ridges in the northern Waianae Range (Wagner et al. 1999). Plants produce a few flower heads throughout the year, and mature seeds are difficult to secure because achenes are wind dispersed immediately after maturing. In situ collections have low seed set, but the cause is unknown.

Cuttings of this species can be made at any time of the year and clones maintained ex situ. The wild population does not have to be visited repeatedly, decreasing the risk of overcollecting and lessening human impact to the habitat. As generations come and go in situ, more wild plants are represented clonally in the living collection than exist at smaller in situ sites.

Nursery-grown cuttings become much larger than the wild plants and produce more fruit with higher seed set. Seeds collected from ex situ clones are stored and will be used along with fresh seeds from additional founder lines for propagating plants and seed-sowing for reintroductions. This combination of stored and fresh seeds allows for a larger number of wild plants represented at the reintroductions than at any in situ sites at any given time. This increased founder representation may provide more genetic variation and ultimately population stability (Ellstrand and Elam 1993; Frankham et al. 2002).

Problem Area 4: Threat Control Strategies

The key threats to native Hawaiian plants are feral animals (disturbance and predation), loss of pollinators, fire, weedy nonindigenous plants, and human activity. The impact of these threats varies for each species. Some threat control strategies, particularly feral animal control, have been used for decades and are well refined. More recently identified threats do not have well-established control methodologies. Extensive research and development are needed to develop effective tools.

The following case studies exemplify rare plant management being conducted by OANRP.

Established Methods

Pritchardia kaalae Rock (Arecaceae) is a fan palm found in the northern Waianae Mountains. The major threats to this taxon are ungulates and rats. Threat control techniques in Hawaii are well established for excluding and eliminating feral ungulates (Stone and Scott 1984; Hawaii Conservation Alliance 2005) and rats (Tobin et al. 1997; Nelson et al. 2002; Witmer and Eisemann 2007). Remaining *P. kaalae* plants are located in steep forest patches on eroded cliffs. Most plants are probably more than 100 years old, based on their trunk size and anecdotal evidence from *Pritchardia* cultivation. In 1995, there were no seedlings or mature fruit of *P. kaalae* in a group of approximately seventy-four trees at Makua Military Reservation (MMR), and rodent damage on fruit and fruit caches were observed. Control of these threats in perpetuity is fundamental in reaching the stability goals of this taxon.

Ungulate Control

The control of feral ungulates is necessary for most restoration efforts in Hawaii. Goat control at MMR was initiated in 1995 by constructing a perimeter fence to isolate the goats from populations in neighboring valleys. The 13.5-kilometer perimeter fence was completed in 2001, enclosing 1,700 hectares. By 2003, goats were browsing new *P. kaalae* seedlings, and an additional fence was constructed immediately around the largest concentration of reproducing *P. kaalae*. Goats were removed by ground- and helicopter-based hunts and by trapping. By 2004, Makua Valley was goat free after 1,747 had been removed. The overall cost of fence construction was $1.2 million, and goat removal cost approximately $600,000. Feral pigs are also a threat at MMR, and in 2003, 70% of the *P. kaalae* outplanted at a reintroduction site were damaged before another 0.5-hectare exclosure could be built. Subsequent plantings have had 100% survivorship.

Rat Control

Rats are ecologically destructive, particularly in island systems, where floras are not adapted to such pressures (Drake and Hunt 2009). Rats consume the fruit of native Hawaiian palms, destroying the embryo (Perez et al. 2008). The uncontrolled presence of rats will eventually lead to extinction of Hawaiian palms (Towns et al. 2006). This is a particular problem for palms because these fruits

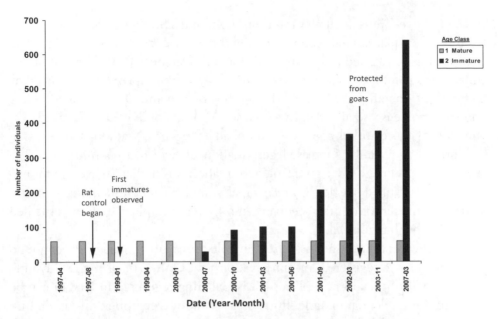

Figure 12.2. *Pritchardia kaalae* population trend of mature and immature plants in the Makua Military Reservation site since 1997. Dates of management activities are highlighted.

have characteristics that are attractive to rats. On Easter Island, the native palm *Jubaea disperta* was driven to extinction by rat predation (Hunt and Lipo 2007). On the island of Oahu there is evidence that the extirpation of the *Pritchardia* palm forest of the Ewa Plain can be attributed to rats (Athens et al. 2002; Athens 2009).

Rat control via snap traps and poison baits began in 1997, costing approximately $60,000. Ripe fruit and thirteen seedlings were first observed below parent plants in 1999. The number of seedlings has increased since rat control began and goat exclosure was established (fig. 12.2). In 2007, 685 immature *P. kaalae* were counted, and now the estimate is more than 1,000 individuals. In contrast to the 1995 situation, there are currently hundreds of ripe fruits, with carpets of immature *P. kaalae* dominating the understory.

Methods under Development

The threat of terrestrial slugs (*Stylommatophora*, Limacidae and *Systellommatophora*, Veronicellidae) illustrates the paucity of control options available to resource managers. Hawaii has no native slugs; the twelve slugs currently established are accidental introductions (Cowie 1997). Slugs are widespread in mesic and wet native forests at all elevations on the islands. Slugs are identified in US-FWS recovery plans as a threat or potential threat for 22% of Hawaiian plant species listed as threatened or endangered (Joe and Daehler 2008). Survival of two endangered plant species (*Schiedea obovata* and *Cyanea superba*) was reduced by half over a 6-month period due to slug herbivory (Joe and Daehler 2008). Seedling emergence is also affected. In a seed-sowing trial conducted in 2008, the application of the certified organic molluscicide Sluggo® (Neudorff Co., Fresno, California) significantly increased germination of *Schiedea obovata* (fig. 12.3) over 6 weeks. Though not significant, germination was higher on average for two additional rare plant species, *Cyanea superba* and *Cyrtandra dentata*. Unfortunately, germination from the seedbank, mostly invasive plant species, remained unaffected by slug removal.

No pesticides against mollusks are available to resource managers because no molluscicides are registered for conservation use. Only four of the twenty-nine pesticides with Special Local Needs (SLN) labeling registered in Hawaii can be used outside residential or agricultural areas (Hawaii Department of Agriculture [HDOA] 2009a, 2009b). Copper barriers and beer traps are alternatives to molluscicides but have drawbacks. Generally used by backyard gardeners, both methods are most appropriately used on a small scale (less than 1 acre) and are labor intensive. Copper barriers cannot practically be used to make exclosures larger than about 1 square meter and run the risk of trapping slugs at the time of construction. During the wet season in Hawaii (November–March), when slugs are abundant, the soil beneath barriers erodes, causing gaps that slugs can penetrate. Thus, to be

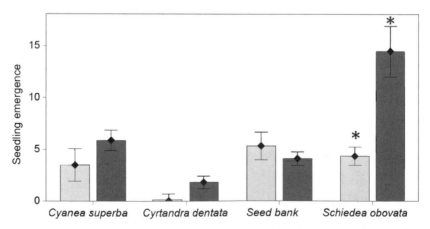

FIGURE 12.3. Mean seedling emergence in slug-treated (dark gray) and control (light gray) plots initially sown with 200 seeds for three species compared with seedlings that emerged from the seedbank ($N = 15$). Asterisks indicate a significant ($p < 0.05$) difference between groups (Tukey's HSD test). Bars are ±1 standard error.

effective, barriers need weekly maintenance. Because slugs cannot detect airborne odors from a distance of more than 10 centimeters (South 1992), beer traps must be placed at an impossibly high density to be effective in the field. In light of these drawbacks, OANRP began testing the safety and efficacy of molluscicides in 2006. OANRP is now in the final stages of SLN label registration for the use of Sluggo in natural areas. Unlike metaldehyde, the active ingredient in popular baits such as Corry's Slug & Snail Death and Deadline products, Sluggo contains iron phosphate. Iron phosphate is not a contact poison, is not toxic to vertebrates, and occurs naturally in the environment (Reilly 1997).

Problem Area 5: Site Selection Limitations

Site selection and availability are a major consideration and limitation for rare plant reintroduction programs in Hawaii. Two case studies illustrate different site selection scenarios. The first covers a taxon with a very limited range. The second demonstrates the challenges presented in managing a taxon in highly degraded habitats.

Case Study: Stenogyne kanehoana

Stenogyne kanehoana Sherff (Lamiaceae) is a scandent vine endemic to the Waianae Range. Only one individual survives in the wild in Haleauau. The species has also been collected from one additional site, Kaluaa, 6 kilometers away. The native habitat is extremely limited, having been destroyed by feral cattle and pigs and overrun by weeds, and much of it is within the safety zone of an artillery range. The total habitat available in the historic range is approximately 27 hectares.

Limiting management of this taxon to these two fenced units would leave *S. kanehoana* extremely susceptible to stochastic events (e.g., hurricanes). Our stabilization strategy does not limit outplanting to documented "historic range" but allows reintroduction into other "likely suitable habitat." This flexibility allows an outplanting to be planned far enough away that a single event is less likely to affect all three sites.

Both the Haleauau and Kaluaa plants have been cloned and used to augment the Kaluaa site. Stock from both sites will be used to establish a new separate population on the leeward side of the mountain range in order to protect from catastrophic loss of all reintroduction work. Recent modeling work suggests that the projected range for *S. kanehoana* encompasses the whole Waianae Range (Price et al. 2007). The establishment of another outlying population will provide a more secure future for this range-restricted, critically rare mint.

Case Study: Hibiscus brackenridgei *ssp.* mokuleianus

Hibiscus brackenridgei ssp. *mokuleianus* A. Gray (Malvaceae) is endemic to Kauai and Oahu. Its low-elevation habitat has been severely altered by fire, development, and invasive species, particularly goat and cattle browsing and competition with invasive grasses (Gon et al. 2006). Nearly all the known habitat has been burned and is still susceptible to wildfire.

The largest population of plants occurs on privately owned land near Waialua, where hundreds of juvenile individuals were observed during surveys in 2004. The site is overrun with *Urochloa maximum*, a fire-adapted grass that provides abundant fuel. *H. brackenridgei* remains on the margins of the grass infestations, on rocky ledges where soil is too shallow for *U. maximum*. In 2007, a catastrophic fire destroyed 95% of the individuals at the Waialua site (fig. 12.4). Immediately after the fire, *H. brackenridgei* seedlings were observed, and the number of individual plants eventually increased. This may be due to the temporary release from competition with *U. maximum*.

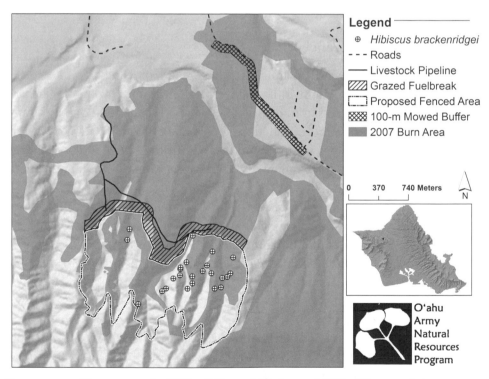

FIGURE 12.4. The extent of a 2007 wildfire, which burned 95% of the remaining occurrences of *Hibiscus brackenridgei* ssp. *mokuleianus*. Also shown is the revised wildfire management regime proposed for the site to protect the lowland plant environment and the upper elevation forest.

After the 2007 fire, the Army reconsulted with the USFWS to analyze the effect on taxon status and management plans. Because of the substantial reduction in numbers of mature plants, the USFWS outlined an elaborate fuel control and grazing plan (USFWS 2008) (see fig. 12.4). The costs associated with this fuel control plan are beyond what was allocated in the stabilization plan. Significant additional funding will be needed for the stabilization efforts for *H. brackenridgei* in this severely degraded habitat, and there is a low probability of success.

The 2007 Mokuleia fire was caused by arson. Therefore, an active outreach component is critical in fire prevention. The Honolulu Fire Department has a very active pre- and post–fire season outreach campaign aimed at educating communities about the consequences of wildfire in Hawaii. The impacts to natural resources are central in this campaign.

Managed Relocation Related to Hawaiian Plant Reintroduction

In an effort to ascertain the implications of climate change for Hawaiian plant restoration, OANRP consulted with local biologists and reviewed the literature. The OANRP management plans do not currently address the potential impacts of climate change but allow flexibility for changing situations. The reasons often cited for considering managed relocation include rarity and increased risk of extinction in current locations (Thomas et al. 2004). Conservation programs in Hawaii are familiar with the elevated extinction risks of small population size. Global patterns predict increased disturbance associated with more frequent or severe storms, wildfires, and ocean level changes, but Timm and Diaz (2009) predicted decreased rainfall in the wet season and a slight increase of rainfall in the dry season for the Hawaiian Islands. Thus, there is no clear prediction regarding the effects of climate change in Hawaii. At worst, these changes will increase the threats already faced by most rare plants in Hawaii today.

Because of Hawaii's geographic location and high rate of endemism, the effects of climate change may have unique impacts on the biota. For example, many management plans for continental species recommend moving to sites furthest north in a taxon's range (United Nations Environment Program 2009). In Hawaii this would amount to moving a taxon to a higher elevation or, in some cases, another island to gain any substantial latitude. The risk to this strategy is in creating unnatural hybrids for many rare Hawaiian plants that have evolved into very narrow distribution ranges with many closely related species. In addition, remaining upland habitat types already support numerous endangered plant taxa. Moving one taxon from a lower elevation into a higher site will compound space limitations in the destination habitat. Another consideration is that little is known about historic distribution, biology, and ecosystem roles of rare Hawaiian plant taxa. Therefore, it is difficult to predict how they will be affected by climate change. New research modeling the geographic range of Hawaiian plants can be used to inform restoration efforts and adapt management plans to global climate change.

A few general themes surfaced in the climate change presentations from the 2009 Hawaii Conservation Conference and in discussions with local biologists. First, habitats maintained close to intact conditions will be better able to withstand a subtle increase in temperature (Kaufman and Mehrhoff 2009). Second, climate change is one in a cadre of threats currently affecting rare plants in Hawaii. Although building climate change into species management plans is something managers should do, these taxa are already affected by too many severe threats having immediate tolls. Third, genetic collections should be secured and used to experiment with a range of plants. If collections are robust, there will be stock to support site selection experimentation and buffers for a changing climate. Fourth, large landscapes should be protected to allow for the possible shift of habitat upward in elevation (Kaufman and Mehrhoff 2009). And finally, an emphasis on keeping wildfire out of managed habitat is needed (Fried et al. 2008).

Summary

The Hawaiian Islands contain a unique island flora with many rare plant species at risk of extinction. We examined five major conservation problems found on Hawaii: low genetic diversity, pollinator loss and unknown breeding systems, inadequate propagule supply, uncontrolled threats, and limited reintroduction sites. Many of these problems are tied to small population size. Whereas mainland rare species may face one of these problems, Hawaiian species often face all of them. The Oahu Army Natural Resource Program has been working for the past 16 years to combat these growing problems and in the process has developed novel protocols and policies to aid in the protection of Hawaii's rare flora. The case studies presented here serve several purposes. They exemplify creative solutions to complex conservation problems. Furthermore, they illustrate the plight of many rare species and promote awareness of conservation needs.

We thank Jane Beachy, Matthew Burt, Joel Lau, Stephen Mosher, Joby Rohrer, Daniel Sailer, Krista Winger (all with OANRP), Michelle Mansker (Department of Defense), Marie Brueggman (USFWS), Sam Gon (The Nature Conservancy in Hawaii), Clyde Imada (Bishop Museum), and Clifford Smith (Pacific Cooperative Studies Unit/University of Hawaii).

PART III

Managed Relocation

"To cherish what remains of the Earth and to foster its renewal is our only legitimate hope of survival."

—*Wendell Berry*

Given the current predictions of global climatic change and threats to biodiversity worldwide, there lurks an uneasy feeling and the burning question, "What else can be done to prevent extinction of rare plant species?" Traditional conservation measures such as habitat protection, ex situ conservation, and reintroduction will continue to play an important role in rare plant conservation. But nontraditional measures have also been debated. Human-assisted plant movement, or managed relocation (MR; also known as assisted migration and assisted colonization), has been put forth as a strategy to conserve species threatened by the changing climate. The premise of MR is that threatened species can be moved to locations where the future climate is predicted to be more favorable for their persistence (Hoegh-Guldberg et al. 2008). Many in the conservation community believe this strategy poses more questions than solutions.

Part III addresses the history, promise, and perils of MR. Climate change may hugely limit our options to conserve some of the rarest plant species within their habitat. For some species MR may be one of the only options available for keeping the species in nature. The history of managed relocation is dissected in chapter 13 from its inception in 2002 to current day. The evolution of the concept in the literature sheds light on the diverse reactions of the academic community to this idea. But how do conservation practitioners and those responsible for overseeing endangered species on public lands feel about the possibility of MR for species under their care? Haskins and Keel (chap. 13) present discussions that arose at the Center for Plant Conservation international symposium "Evaluating Plant Re-introductions as a Plant Conservation Strategy: Two Decades of Evidence" about perceptions and fears related to MR. These demonstrate the range of concerns of plant conservation practitioners and government regulators. In an effort to help earnest land managers faced with species at risk by climate change, Haskins and Keel provide a process for evaluating MR proposals.

One of the major criticisms of MR is the risk that the relocated species may become invasive in its new community. Reichard and colleagues (chap. 14) take an in-depth look at the risk of biological invasion (Ricciardi and Simberloff 2009a). They describe invasion theory and present a test of rare species that have been introduced through horticulture trade using previously established weed risk assessments. This is an important step toward examining evidence needed to allay fears.

Progress in any discipline requires discourse, argument, and counterargument. Only through healthy exchange, suggestions for new directions, and hypothesis testing will we advance the science of reintroduction. Clearly, lessons learned from traditional reintroductions will be invaluable for any MR proposals. And only careful experimentation will provide the evidence we need to examine the efficacy of this strategy (Shirey and Lamberti 2011).

Chapter 13

Managed Relocation: Panacea or Pandemonium?

KRISTIN E. HASKINS AND BRIAN G. KEEL

Human-induced land use change and global climate change are likely to exceed the resilience of many ecosystems in the twenty-first century (IPCC 2007), such that rapid species-level extinctions are predicted (Thomas et al. 2004). Although pollen and fossil records have documented that plant species can move (Delcourt 2002), what is unclear is whether plants will have the ability to move on their own in today's rapidly changing climate. As in the past, some species will be able to colonize new habitats successfully or adapt to changing conditions within their current range without human assistance. However, the current global situation begs the question, "What can be done for species that cannot move on their own or adapt to the new climate?" Answering this question is the challenge at hand. Managed relocation (MR), which implies human-assisted plant movement outside the species' native documented range to counteract negative effects of climate change (Hellman et al. 2008), has been suggested by some as a cure-all for rare species, whereas others view it as a dangerous strategy fraught with uncertainty. This chapter explores the history of MR from its inception to current concepts; examines feelings, fears, and counterarguments related to MR that are commonly held by conservation practitioners; and offers guidance for future MR proposals and projects.

A Brief History

Managed relocation has experienced a long and bumpy road, and the fact that some assisted movements have led to ecological nightmares (e.g., kudzu, *Pueraria lobata*, and Brazilian pepper, *Schinus terebinthifolius*) is enough to make us all stop and think about intervening in the movement of rare species. Perhaps we need to ask ourselves, "Is moving rare species to save biodiversity truly appropriate?" If we deem biodiversity worthy, then "What fears do we have about plant movement?" And finally we ask, "What can we learn from the long history of intentional human movement of plants that will improve our reintroduction practice?"

As the climate changes, plants can exhibit several responses. They may go extinct, but plant conservation practitioners reject this as an option. The second option is migration. Historic plant species' migrations have provided insight into how plants can colonize suitable habitats in response to climate change (Delcourt 2002). But human-modified landscapes have disrupted and fragmented habitats such that few naturally occurring dispersal corridors exist (Safford et al. 2009). The third option is adaptation to a changing climate through genetic evolution. The crux of the problem is that some plants may not be capable of adapting quickly enough to match the pace or magnitude of climate change (Jump and Peñuelas 2005), or they may not have the genetic diversity or capacity to adapt or to persist at their present locations. Plants with high genetic diversity may be able to express phenotypic plasticity in response to climate variations and persist in current locations (Stockwell et al. 2003). But merely having the ability to be plastic is not the same as adapting. Without adaptation, migration will be the only option for survival.

Since 1989 and perhaps before this time, scientists warned that the geographic distributions of individual plants were expected to shift north of shrinking southern boundaries (Davis 1989; Jump and Peñuelas 2005). Models predict that alpine species will move to higher elevations and be displaced by plants from lower elevations (Bartlein et al. 1997; Lesica and McCune 2004; Jump and Peñuelas 2005). Primack and Miao (1992) warned that rare plant species were particularly vulnerable to changing weather patterns, habitat destruction, acid rain, and the extinction of seed dispersers. These findings indicate a need for conservation efforts that focus on plant movement under climate change.

"Assisted migration is the intentional establishment of populations or metapopulations beyond the boundary of a species' historic range for the purpose of tracking suitable habitats through a period of changing climate. This might involve migration between islands, up mountain slopes, and between mountain tops" (Keel 2005, p. 36). The term *assisted migration* was first defined in 2002, and the general concept was introduced to the scientific community in the *Oxford Dictionary of Ecology* shortly thereafter. Keel (2007) constructed a theoretical framework of assisted migration as the concept expanded and gained the interest of conservation practitioners. Since 2007, numerous scientific journal articles, media articles, letters to the editor, and articles in popular magazines for and against the intentional movement of plants have surfaced (table 13.1). Debates

TABLE 13.1

The changing face of assisted migration.

1992	Primack and Miao	Suggested moving plants in response to global climate change
2000	Barlow	Predicted intentional movement of plants as a future conservation tactic
2002	(Keel, personal communication)	Keel first coined the term *assisted migration*
2005	Keel	Defined assisted migration as a concept
2007	Keel	Constructed theoretical framework of assisted migration
2007	Hunter	Introduced the term *assisted colonization* to avoid confusion with the term *migration*
2008	Hoegh-Guldberg et al.	Presented a decision-making framework for undertaking assisted colonization
2008	Managed Relocation Working Group	Introduced the term *managed relocation* to better capture conservation actions
2009a	Ricciardi and Simberloff	Proposed arguments against using assisted colonization
2009	Schlaepfer et al.	Proposed counterarguments to Ricciardi and Simberloff
2009	Richardson et al.	Presented a multidimensional decision-making framework for undertaking assisted colonization that incorporates socioeconomic axes
2010	Minteer and Collins	Proposed addressing ethical issues associated with managed relocation and shifting attention to developing criteria for managed relocation
2010	Seddon	Suggested thinking beyond moving single species to moving and constructing new communities
2011	Shirey and Lamberti	Called for more regulation of plant sales, restricting purchases of endangered species hybrids, and controlled pilot managed relocation by the US Fish and Wildlife Service

over terminology ensued, leading to many synonyms. Because *migration* is often associated with animal movement, Hunter (2007) suggested *assisted colonization*, and Hellman and colleagues (2008) proposed *managed relocation*. It is important to remember that no matter which term is being used, they all refer to human-assisted movement of plants outside the species' historic range in response to population decline due to climate change. In the following discussion of some key literature we have used these terms interchangeably to maintain the original author's intentions.

McLachlan and colleagues (2007) highlighted the lack of scientifically based policy on assisted migration and suggested that the conservation community needed to consider it. They posed a framework that identified three positions: (1) aggressive assisted migration, (2) avoidance of assisted migration, and (3) constrained assisted migration. Aggressive assisted migration advocates believe that there is insufficient time for a species to adapt or move by itself, so there are no other options for the species' conservation. Followers of this position would also support using predictive modeling and perhaps broad-scale movement. At the opposite end of the spectrum are the critics of assisted migration, who think the practice should be avoided entirely. Those who focus solely on in situ conservation practices and have very little faith in predictive modeling hold this position. Constrained assisted migration balances the benefits and risks of the two previous positions. The risks of assisted migration are minimized by careful restrictions on planning, actions, monitoring, and adaptive management, while the potential benefit of preserving biodiversity is maintained. Additionally, McLachlan and colleagues (2007) suggested that the constrained assisted migration position be supported by proposals that would require rigorous planning and evidence of population decline attributable to climate change.

Hoegh-Guldberg and colleagues (2008) presented a decision assessment framework for evaluating the feasibility of assisted colonization. The framework was designed to assess the need to move species outside their historical range, but within the continent, as a mitigation measure to combat the loss of biodiversity from global climate change. This article advanced ideas about assisted migration from considering the potential options to considering the feasibility of its implementation.

Mueller and Hellman (2008) predicted that cases of detrimental invasion would be less likely for intracontinental movements of species than for intercontinental movements. They indicated that plants, as opposed to taxonomic groups such as fish, are at low risk as intracontinental invaders, and given the dispersal constraints of plants, assisted migration might be useful for this life form.

As with any controversial practice, the opponents of assisted migration have also emerged and argued their position. Ricciardi and Simberloff (2009a) made the case against assisted colonization as a viable conservation strategy. Their argument stems primarily from the invasive species literature, which includes many animal examples. The authors envision assisted migration being practiced as "large-scale transfers of species outside their natural ranges—in other words, planned invasions" (Ricciardi and Simberloff 2009a, p. 248). Schlaepfer and colleagues (2009), Sax and colleagues (2009), Schwartz and colleagues (2009), and Vitt and colleagues (2009) presented counterarguments. Among the counterarguments given are the need to weigh the risks of MR against the risk of extinction (Schlaepfer et al. 2009; Schwartz et al. 2009), the doubt that endangered species pose much risk of invasive behavior, the lack of evidence indicating that relocated species will cause extinctions (Sax et al. 2009), and the recognition that stakeholders may hold different values of outcomes (Schlaepfer et al. 2009). Vitt and colleagues (2009) implored practitioners to begin planning for the possibility of MR by making wise, genetically diverse ex situ collections now.

Seddon and colleagues (2009) warned against conservation practitioners prematurely embracing assisted colonization. As a proactive conservation measure, assisted colonization must be considered in light of the uncertainties of climate change predictions, consequent species responses, habitat needs, and the effects of translocations on ecosystem functions. Given these uncertainties, Seddon and colleagues emphasized a need to exercise caution against initiating a new era of ill-conceived species translocations.

Richardson and colleagues (2009) were the first to incorporate social aspects in their multidimensional decision-making framework for MR. Managed relocation should not be a last resort approach, they argued. Instead, MR should be an intervention strategy and part of a portfolio of options built around both ecological and social considerations (Richardson et al. 2009). Including social criteria, such as the cultural importance of a species and the financial costs of implementing an MR project in the decision-making process, is critically important to broadening the scope and examining the potential consequences of MR. Richardson and colleagues established another important point in the MR debate, which is that all the stakeholders in a potential MR project are unlikely to concur, making compromise a necessity.

Considering MR as a conservation tool is now expanding beyond whether or not it is a practical tool to whether or not it is an ethical tool. Minteer and Collins (2010) presented ethical and policy questions focused on candidate species, institutional context, authorization and oversight, motive, and environmental responsibility that are designed to make practitioners think about whether they should rather than could practice MR. Minteer and Collins (2010) then moved forward by suggesting that criteria for "good" and "bad" MR proposals be developed.

Like other conservation practices, MR occurs on a continuum. At the conservative end of this spectrum, MR projects would simply move rare species just beyond their historical range, where change in environmental variables is expected to be minimal. And at the radical or extreme end of the spectrum, MR projects would constitute constructing new ecological communities (Seddon 2010) using rare plant introductions as a starting point for ecosystem engineering (Jackson and Hobbs 2009). Seddon (2010, p. 800) proposed that practitioners "consider the possibility of adopting an ecological engineering perspective to use conservation translocations as a means to introduce species into suitable habitat outside their historical distribution range in order to contribute to the construction of new ecological communities." Furthermore, Seddon implored practitioners to consider both traditional and radical management options in times of pending ecological crisis. The suite of new concerns that would arise as a result of planning construction of a new community is beyond the scope of this chapter.

Unregulated, haphazard movement of plants in the commercial trade is also a rising concern (Shirey and Lamberti 2011). Currently some endangered species and hybrids intentionally cultivated from endangered species, are legally and readily available for sale over the global Internet. To prevent adverse effects of hybridization and genetic swamping of wild rare populations, Shirey and Lamberti called for a uniform coordinated rigorous enforceable policy, such as the Convention on International Trade in Endangered Species of Wild Fauna and Flora to control commerce of rare plants within the United States.

Managed Relocation: The Practitioner's View

The volume of work published recently on the topic of MR is testament to the deep feelings and emotions that this concept evokes (see table 13.1). Practitioners are torn between using a practice that goes beyond their formal training and comfort zone and the potential chance to save a species. The unknown costs of MR are a major debate.

In fall of 2009, the Center for Plant Conservation (CPC) held an international symposium titled "Evaluating Plant Reintroductions as a Plant Conservation Strategy: Two Decades of Evidence." In addition to presentations of the topics from this volume, attendees including horticulturists, conservation biologists, botanical garden scientists, land managers, regulating agency personnel, and academics participated in discussion related to MR. The following are their responses, which well represent the range of global opinions held about MR. In an effort to present a balanced perspective, we present the concerns, discussion, and related literature here. As is true with the published literature on this topic, the stakeholders at this symposium were not all in agreement, which made for lively debate. The value of such discussion is that when concerns are clarified and scrutinized, there is an opportunity to redress them.

What Are Your Feelings and Fears about the Practice of MR?

Fear 1: Important aspects of the target plant's ecology may be unknown and may be missing in the recipient environment. This fear stems from a general lack of knowledge about the basic ecology of many rare and threatened plant species. The solution, of course, is to conduct more research and to publish those results, be they statistically significant or not. The funding climate is changing almost as rapidly as the environmental climate, and in general it is not becoming more favorable. If traditional reintroduction projects from this point forward are conducted as experiments, then we can learn more about species' biology in a cost-effective manner. Some practitioners thought that before any MR, tests within range should be accomplished. This information can inform potential future MRs if necessary.

Fear 2: Traditional conservation practices (i.e., habitat conservation) will be abandoned or resources withdrawn in favor of MR. In addition to other conservation organizations, the CPC believes that MR should be used only as a last resort at this point and only in certain circumstances (see Maschinski et al. 2011 for examples). Furthermore, the practice of MR should follow appropriate planning and documentation (see appendix 1) in conjunction with ex situ propagule collection (Guerrant et al. 2004) and in situ habitat management. A multipronged management approach will improve the odds of conservation success.

Fear 3: Reintroduction success within the historical range is often low, so reintroduction into a novel site has an even lower chance at success. Perhaps one reason for low reintroduction success within historical range is that the climate has already changed and no longer supports optimal population growth of the taxon. Unfortunately, proving this impact on rare plant populations is most difficult. Using historical range as a guide to understanding plant distributions has other faults, in addition to changing climate. Seddon (2010) highlighted the fact that historical ranges are often based on arbitrary points, they are places where botanists prefer to visit and collect, and they are based on historical records that are often fraught with errors. In addition, other factors may change the suitability of any geographic space such that it is no longer suitable for maintaining a rare plant population with positive population growth rate (Knight, Maschinski et al. [chap. 7], this volume). Therefore, historical record data may not be the best indicator of where rare plant populations should occur now or in the changing future. Several studies have indicated that introductions to suitable habitats outside a known range can support persistent populations of rare species (Maschinski et al. [chap. 7], this volume). However, moving a species out of its current or recent range should be done with caution, long-term monitoring, and an exit strategy in case a problem occurs.

Fear 4: Climate modeling for future bioclimatic envelopes is not accurate enough to determine where MRs should occur. Climate models are useful tools for broadly categorizing fundamental niche space, and they are useful for planning experiments. For example, to determine whether *Lomatium* species were climate limited or dispersal limited, Marisco and Hellman (2009) used sites outside the current species ranges where climate models predicted sites with suitable future habitat. The researchers successfully used these modeled sites to determine that the *Lomatium* species were primarily dispersal limited, not climate limited

(Marisco and Hellman 2009). Like all disciplines, climate modeling is improving (Krause and Pennington, this volume), yet it still relies on accurate ecological data, which are often lacking for rare species.

Fear 5: MRs will lead to the spread of pests and disease. Davis and colleagues (2000, p. 531) argued, "The basic processes that admit exotic plant species are essentially the same as those that facilitate colonizations by native species or allow repeated regeneration at the same site." If Davis and colleagues are correct, then literature on invasive species can provide information on plant introductions and information to help minimize the risk of introduced plants becoming invasive. Indeed, this seems to be the case (Reichard et al., this volume). This fear further highlights the need for careful guidelines because introducing pests and disease can also occur with traditional reintroductions if precautions are not taken (appendix 1, #21). For example, every time soil is moved, there is a degree of risk in introducing harmful microbes to the recipient site. But there are also precautions that can be taken to reduce the potential for causing inadvertent introductions (Haskins and Pence, this volume).

Fear 6: The relocated species will become an invasive species, affecting all members of the recipient community. Rare species are rare for a reason. The traits that commonly characterize a rare species, such as poor competitive ability, low fecundity, strong site endemism, and weak dispersal mechanisms, are not the traits of an invasive species. In fact, Reichard and colleagues (this volume) and Gordon and Gantz (2008) found that intentionally introduced plants had a low probability of becoming invasive. Reichard and colleagues (this volume) suggest using weed risk assessment protocols to reduce fears related to invasive behavior of species proposed for MR.

Where Do We Go from Here?

The practice of moving plants has occurred for centuries (Sauer 1988), sometimes with highly detrimental results (e.g., kudzu, *Pueraria lobata*, and Brazilian pepper, *Schinus terebinthifolius*). The idea of moving plants for conservation purposes is more recent (see table 13.1), yet the associated risks with plant movement remain the same despite the noble intent. The light on the horizon is that progress is being made. Given the rigorous review of plant reintroduction practice provided in this volume, we are now faced with absorbing the latest knowledge and implementing the lessons we have learned to improve our future plant conservation efforts. The material, advice, and models suggested for traditional conservation practice (appendix 1) can and should be applied to justification, planning, implementing and monitoring MR populations.

What Would Make MR an Acceptable Practice?

By far, the most overwhelming response to this question was, "More research is needed to address the unknowns." Research in all areas, including the ecology of rare species and their mutualists, climate change models and predictions, historical occurrences, and political and ethical repercussions, is needed. A few studies have begun to address some of these issues. Using rare endemic plants and animals from the Florida Keys as case studies, Maschinski and colleagues (2011) proposed a systematic process for evaluating the effectiveness and risk of traditional conservation options in contrast to MR, with species characteristics, habitat-specific concerns, and legal issues in mind. Such individual assessments will be necessary for any MR proposal and are a wake-up call to the shortfalls of existing legislation to account for necessary changing roles of natural areas in our rapidly changing world. Additionally, Svenning and colleagues (2009) used regression modeling to determine whether potential introduction sites in northern Europe could withstand the addition of more species without suffering a loss in species richness. Their results suggested that many areas had not achieved maximum species richness and could potentially absorb additional species that were deemed suitable for MR (Svenning et al. 2009). Keel (2007) provided another example of the kind of proactive studies that are needed to address many questions that must be addressed before MR is attempted (see box 13.1).

In Which Cases Should MR Be Considered?

It bears repeating that well-educated practitioners are not promoting the use of MR for every rare species and never would advise that MR be used without careful study and planning of the introduction design. Hoegh-Guldberg and colleagues (2008) created a decision-making tool for determining whether a species is an appropriate candidate for MR, and we advise using this process as an initial step in MR planning. We are aware that extreme cases exist. Some species do not have other options available, such as coastal and island species threatened by sea-level rise (Maschinski et al. 2011) or species currently extant only in ex situ holdings (see Guerrant, this volume). Working with these types of species is a logical first step as their needs are most dire.

How Can One Build a Case for MR?

Rejecting MR as a viable strategy based on fears is akin to "putting one's head in the sand," according to Schwartz and colleagues (2009). Many public land

Contributed by: Brian G. Keel
Species Name: *Habenaria repens* Nuttall
Common Name(s): Water spider orchid
Family: Orchidaceae

Habenaria repens (Orchidaceae) is one of an estimated 25,000 orchid species worldwide (Dixon et al. 2003) and is found in open wet habitats in Mexico, the West Indies, Central America, and in the Gulf and Atlantic coastal plains in the United States (Brown 2002). In the wild, terrestrial and some epiphytic orchids need mycorrhizal fungi for seed germination and survival as adults (Dixon et al. 2003). Orchid seeds are dustlike and contain only a seed coat and an embryo. This minimal energy invested per seed increases plant fecundity and wind dispersal, but this tradeoff makes a fungal mycobiont necessary for survival. For *H. repens*, a North American terrestrial orchid, extended photoperiod and mycobiont distributions are two factors that will limit plant movement as the climate changes. These factors were examined to determine the managed relocation (MR) potential of this species.

To investigate mycobiont distributions, seed packets were constructed and used to sow seeds and capture fungi at three wild sites, with techniques adapted from Rasmussen and Whigham (1993). Fungi capable of germinating *H. repens* seed were captured at the site that historically contained *H. repens* in the 1960s and at the site presently containing a population of *H. repens*. However, sites that were beyond the present range of *H. repens* did not yield mycobiont fungi. These results do not preclude the possibility of mycobiont presence outside the range, but they do suggest the need to include mycorrhizal fungi along with the obligate plant in any MR. However, see Haskins and Pence (this volume) for cautions and advice on the movement of soil fungi.

Range expansions in a northerly direction would subject *H. repens* to shorter growing seasons but longer day lengths than they currently experience. Photoperiod is an important cue for breaking dormancy, resuming growth, initiating seed germination, and preparing for dormancy before the onset of climatically unfavorable conditions (Kimmins 1996). Keel subjected *H. repens* seedlings to three photoperiods representative of the seed source, a site 580 kilometers north of the seed source, and northern Quebec

---

BOX 13.1. CONTINUED

(table 13.2). Increased photoperiod was significantly and positively corre-lated with growth rate and plant size.

All orchids may not respond to photoperiod as *H. repens* did, and for some, an extended photoperiod may be cause to avoid northward MR. Other species may be capable of maintaining adequate growth and repro-duction in new locations. Despite this good news, the potential lack of fun-gal mycobionts is a critical missing piece for *H. repens* survival in this new environment. Thus, these findings emphasize the need to examine impor-tant ecological aspects of species targeted for MR.

TABLE 13.2

Habenaria repens *latitudes, corresponding photoperiods, and geographic locations.*

Latitude	Daylight at the Summer Solstice	Geographic Location
27°20′	14.75 hr	Avon Park, Florida, USA
39°44′	16.25 hr	Pennsylvania–Maryland border, USA
60°	20.00 hr	Northern Quebec, Canada

Avon Park was the project seed source. The Pennsylvania–Maryland border is 580 kilometers north of the northern range boundary of *H. repens.*

---

managing agencies are beginning to develop policies regarding MR (Camacho 2010), so it is important to move the debate to the next level. In response to the Minteer and Collins (2010) suggestion that we begin planning ways to critique MR proposals, we are providing suggestions for evaluating MR proposals for rare plants (table 13.3). We advise concerned conservationists first to document adher-ence to *CPC Best Reintroduction Practice Guidelines* (appendix 1). In addition, we recommend several steps: Document that climate is causing species decline, assess invasive risk, guarantee that adequate resources are available and have not been reallocated from protecting extant populations, find suitable recipient sites, provide experimental evidence that the recipient site has low probability of being harmed, ensure that the MR is legal and that recipient site managers approve, and justify all with a well-conceived, accurately documented plan.

TABLE 13.3

Criteria for ranking good and bad MR proposals.

Proposal Criteria	A Good Proposal Would Have . . .	A Bad Proposal Would Have . . .	Recommendations for Improved Practice
Justification	Reintroduction plan based on the CPC guidelines.	Reintroduction plan incomplete.	Follow the CPC guidelines and provide accurate documentation.
Climate change documentation	Species demographic data corroborated with climate data indicate decline.	No or inadequate species and climate data.	Install climate monitoring devices; monitor populations.
Negligible risk of invasiveness	Acquired passing score using WRA.	No support.	Conduct WRA.
Resource acquisition	Extramural funding and resources in hand.	Resources being pulled from other conservation projects.	Determine and seek sources for extramural funding, labor, and materials.
Legality ensured	Appropriate permits in hand; laws have been cross-checked; recipient site land managers approve.	Missing permits; unclear legal issues; no management support.	Research local, state, federal, and international law; negotiate with land managers of recipient sites; obtain permits.
Recipient site predicted and available	Used best available GCC models for site prediction and recipient site assessments.	Recent GCC models have not been used.	Work with modelers to improve predictions for your area; conduct common garden experiments to document feasibility of MR.

Developing a plan and providing documentation using the *CPC Best Reintroduction Practice Guidelines* (appendix 1) should always precede any MR proposal. MR = managed relocation; CPC – Center for Plant Conservation; WRA = weed risk assessment (Reichard et al., this volume); GCC = global climate change.

Summary

There remains little doubt that our world is changing, and the pace of that transformation is worrisome. Plants are not strangers to changing environments, but their ability to keep up with the speed of our changing climate is causing a conundrum about what to do for species at risk of extinction. Although most rare plant species cannot match the charisma of a panda bear, advocates for plant conservation are no less passionate about saving their species. Strong feelings about what to do for species in decline become highly apparent when we view the history of MR and how this conservation tool has evolved over the years. Two main camps exist. The first camp is resistant to moving plants based on a suite of fears. One of the biggest concerns associated with MR is the potential for introducing invasive species. Reichard and colleagues (this volume) address this major issue in the next chapter. The second camp sees the potential in moving plants to novel habitats to provide those species with a chance to live and persist in a natural habitat. The members of this camp are willing to explore the potential of MR in a controlled and scientific fashion. To this effect, we have proposed guidelines for preparing MR projects. We hope that discussion of these ideas will lead to the advancement of plant conservation options.

We thank attendees of the 2009 CPC Plant Reintroduction symposium for lively debate and exchange of ideas on the topic of MR. Additionally, K. Haskins would like to thank Matthew Albrecht for providing a constant flow of new and pertinent literature.

Chapter 14

Is Managed Relocation of Rare Plants Another Pathway for Biological Invasions?

SARAH REICHARD, HONG LIU, AND CHAD HUSBY

Why Use Managed Relocations?

Numerous studies have suggested that Earth's climate has changed and will continue to change in response to human activities. It is expected that average temperatures will increase, in some places as much as 4°C, between 1990 and 2040 (IPCC 2007). In addition, precipitation will probably change, becoming drier in some areas and wetter in others. The intensity of storms, such as hurricanes, may also increase in some areas (IPCC 2007). Biological organisms have physiological tolerances to climate extremes, which may decrease their ability to survive or be sufficiently competitive in the new climates, with different ranges. Although most animals and some plants may be able to migrate to more appropriate locations, many plants have limited seed dispersal capabilities and may be unable to reach suitable habitat without assistance.

Compounding the difficulty in dispersal is increased degradation and fragmentation of landscapes by humans. Natural barriers such as mountain ranges and large water bodies have always limited plant dispersal. These areas may represent large expanses of unsuitable habitat that may prevent population continuity, and most plants lack effective seed dispersal mechanisms to cross such barriers. Increasingly, human development of large land areas disrupts plant dispersal. According to the World Health Organization (Hagmann 2001), the human population is expected to grow to 9 billion by 2050, an increase of about 30% over the current numbers. Accommodating human population growth and economic expansion will lead to loss of habitat and a patchwork of intervening nonnatural environments that may prevent dispersal. Even lands that are preserved may be seriously degraded by agriculture, pollution, invasion by introduced species, pests, pathogens, and other factors that prevent successful colonization by native species. Moving threatened species beyond these influences may provide the only hope for a growing number of rare plants and animals.

Montane and island species are especially at risk because their habitats are isolated. Depending on the distance from mainland, islands may be isolated from potential habitat expansion. Mountains are islands of habitat surrounded by lower, warmer altitudes with numerous competitive species. Habitat at appropriate altitudes may be many kilometers away. Air and soil usually cool with increasing altitude, so some species may be able to relocate to higher zones, but species in mountainous areas may simply run out of higher locations for population persistence (Liu et al. 2010). Smaller populations, often the case in alpine areas, may be the most vulnerable.

Biologists are concerned that some species will become increasingly rare, or even extinct, because they cannot withstand local climate change, and they will be unable to disperse across landscape barriers to reach suitable habitat without human intervention (Vitt et al. 2010). Similar concerns apply to rare species displaced by human activities from the microclimates to which they are adapted, so these apprehensions are not limited to large-scale climate change (Wilby and Perry 2006). In general, rare species are known to have poor dispersal abilities (Gaston and Kunin 1997). Human-assisted movement of a rare species outside its historic range may allow its survival. The biological and philosophical debates around managed relocation (MR) are thus important and worthy of careful consideration (e.g., see Haskins and Keel, this volume).

Why Not Use MR?

It may seem initially an almost moral imperative that humans participate in MRs for species truly needing assistance. If climate change is mostly human induced, as most scientists think (IPCC 2007), should not humans also take responsibility for helping species that will be harmed by it? Many think so, and that has led some people to rush to move species northward or propagate and distribute them for garden use (Shirey and Lamberti 2010, 2011). Many species listed as endangered or threatened under the United States Endangered Species Act (ESA) of 1973 have a formal recovery plan that includes *reintroduction within the historic range only*. Of the forty-five states that have some form of state-level legislation to conserve endangered species, only twenty-nine include provisions for endangered plants, and most of these provisions are much weaker than the limited protection afforded by the ESA (George et al. 1998). However, Section 10(j) of the ESA allows experimental populations to be established, such as those proposed in MR plans, and specifies management requirements. Biologists assess many factors before a reintroduction, including the number of available seeds and plants, which seed sources are most appropriate for new locations, site suitability, and the stability of the management of the site. Most formal reintroductions are also done carefully to ensure that the material used is free from diseases and pests. Those less aware of these issues may do ad hoc reintroductions, and we do not support them.

Perhaps the greatest concern, however, is the potential that a species introduced outside its current or historical geographic distribution could become invasive, harming species native to the recipient site. Invasive species have a number of serious impacts. They may compete for resources with native plants (Brown et al. 2002), change soil chemistry (Dougherty and Reichard 2004) or hydrology (Gordon 1998), and alter food webs (Urgenson et al. 2009). After habitat destruction and degradation, invasions are the second leading cause of imperilment of species in the United States (Wilcove et al. 1998), although actual extinctions from invasions appear to be rare, especially for plants. In trying to establish a rare species to a new location for its recovery under climate change, there is concern for the recipient community (Ricciardi and Simberloff 2009a).

What Can We Learn from Invasion Biology?

Although Darwin (1859) mentioned the invasion of European plants in Argentina in his book *On the Origin of Species*, and agricultural scientists often noted weeds near crop plants, the field of invasion biology did not become a discipline of its own until the late 1900s (Reichard and White 2003). Furthermore, the complexity and diversity of natural communities may restrict patterns in invasion biology to local and contingent circumstances rather than broad, powerful generalizations (Hansson 2003; Simberloff 2009). However, concepts and theories in invasion biology are emerging that may be useful to examine when discussing managed relocation.

Biotic Resistance and Resilience

Biotic resistance is the ability of the native community to repel newly introduced organisms. At its core is the enemy release hypothesis (ERH), which has three essential points: (1) Natural enemies (or herbivores) regulate plant populations; (2) native enemies will affect native plant species more than introduced plant species, because the enemies have evolved with and may have chemical relationships with native plants; and (3) in the absence of enemies, introduced species are released from predation pressure, and their populations increase in comparison with those of native species (Keane and Crawley 2002). Any species in the community may be attacked by generalist herbivores. However, the likelihood that native species will have specialized pests present is greater than the likelihood that specialized pests from an introduced species' home range will also be introduced. The introduced species can reduce native species by replacing them in the community. The ERH has led to the development of some successful biological control programs. For instance, the release of three species of European weevils and beetles to control *Lythrum salicaria* (purple loosestrife) (Malecki et al. 1993) has led to a significant decrease of this invader in the United States.

Keane and Crawley (2002) argued that successful classic biological control is not proof that the release from natural enemies explains invasion, just as the absence of herbicide control in crops does not fully explain the infestations of weeds. They call for a better understanding of generalist enemies and their effects on native species to gain insight into the role of enemy release in plant invasions. Nevertheless, introduced plants are sometimes found to be released from specialist enemies (Liu and Stiling 2006; Liu et al. 2006, 2007), and such release allows some introduced plants to reinvest their energy in better defense against generalist herbivores and more vigorous growth (Joshi and Vrieling 2005). On the other hand, specialist herbivores can also catch up with their hosts, as has happened with *Eucalyptus* species in California, where at least twelve insect associates of eucalypts have established, some of which cause defoliation and outright mortality of trees (Hanks et al. 2000). Over time, native herbivores may also adapt to the invader (Siemann et al. 2006).

In contrast to the ERH, mutualisms may also influence invasive ability. Janzen (1985) identified four key types of mutualism: dispersal, pollination, nutrition, and protection. Lack of specialized mutualist pollinators and seed dispersers in an introduced range was generally considered an impediment to naturalization of introduced plants (Richardson et al. 2000). However, studies suggest that such impediments can be overcome with the gain of unexpected native partners (Valentine 1977; Gardner and Early 1996; Richardson and Higgins 1998) or the arrival of the plant's old partner (Nadel et al. 1992; Pemberton and Liu 2008a, 2008b). For example, *Ficus microcarpa* was long used in Florida horticulture but did not become invasive until its pollinator was introduced (Nadel et al. 1992).

Nutritional mutualisms may also influence invasibility. Many species have relationships with bacteria found in root nodules that change atmospheric nitrogen (N_2) into ammonia, the form of nitrogen that plants use (see Haskins and Pence, this volume, for more on microbial mutualists). More commonly, about 85% of all plant families have a relationship with fungi known as arbuscular mycorrhizas, whose fungal hyphae enter plant cell membranes. The arbuscules increase the surface area of the interface between the hyphae and cell cytoplasm to facilitate the transfer of nutrients. A less common form is found in about 10% of plant families, especially woody species. Ectomycorrhizas cover the root tips of plants and form an extensive network in the soil and leaf litter. Some evidence indicates that nutrients move between the plants connected in the network (Simard et al. 1997). Many of these nutritional mutualisms appear to have low specificity and often do not impede plant naturalization or invasion (Richardson et al. 2000). However, the influence of mycorrhizas on plant invasions may be underestimated and merits further study (Pringle et al. 2009).

Protection mutualisms are still understudied and may greatly expand our knowledge of invasions eventually. Endophytes, fungi that are found in the tissues of plants, may play important roles in increasing plant resistance to stress and herbivory. For instance, endophytes may increase tolerance to extreme temperatures (Redman et al. 2002). A grass that has an endophyte present may have up to 70% fewer arthropods and up to 20% less arthropod diversity (Rudgers and Clay 2008). Endophytes may be transmitted from the mother to the seed, so introduced plants may be carrying predation resistance into new locations.

Recent discoveries suggest that some plants may have mechanisms that allow them to be more competitive against novel species than those with which they occur in the native range. For instance, Russian knapweed (*Centaurea diffusa*) is uncommon in its native range in Russia but highly invasive in the American west. Callaway and Aschehoug (2000) grew the species in pots with both American and Eurasian grasses. American grasses produced 85% less root and leaf mass when planted with the knapweed, with no effect on the knapweed. When the knapweed was grown with Eurasian grasses, however, the grass biomass dropped by 50% and the knapweed also declined. The knapweed produced a chemical that reduced the American grasses' ability to take up phosphorus, an essential nutrient. Pots with the Eurasian grasses and knapweed produced 12% more biomass and took up 63% more phosphorus than those with the American grasses and knapweeds. Callaway and Aschehoug (2000) hypothesized that long-term associations of species may increase both productivity and resource use. There has been much discussion about these findings, and the argument for comparative studies of invasive plants in both invaded and native ranges is compelling (Hierro et al. 2005). There is also the suggestion that such chemical interactions may change over time, allowing native species to rebound (Lankau et al. 2009).

Community composition also can greatly affect whether a newly introduced species is likely to invade successfully. Unlike earlier hypotheses that species-rich communities tend to resist invasion (Elton 1958), most studies have found that the diversity of native species is not a good predictor of whether a community can be invaded (Tilman et al. 1996; Stohlgren et al. 1999; Stadler et al. 2000). Highly diverse communities tend to have more resources and therefore higher productivity

(Srivastava 1999; Stohlgren et al. 1999). The relationship between native diversity and habitat invasibility is scale dependent; that is, on the local scale, native diversity is often negatively correlated to invasibility. At the regional scale, however, the relationship is positive, with greater heterogeneity promoting both spread and coexistence of an introduced species that is not possible in homogeneous environments (Melbourne et al. 2007). After reviewing the literature on diversity and invasibility, Levine and D'Antonio (1999, p. 15) observed that "the consistent positive relationship between exotic species abundance and resident species diversity found in spatial pattern studies suggests that invaders and resident species are more similar than often believed."

Taxonomic relationships between native and introduced species may affect which species will invade and which will not. Darwin (1859) first argued that close relatives may overlap in resource use and be excluded. The ERH also suggests that natural enemies may shift more easily to relatives. Lockwood and colleagues (2001) and Ricciardi and Atkinson (2004) found that severe invasions were more likely when the invaders were genera not represented in the native community. A phylogenetic study of all grass species in California found that highly invasive grasses are less related to native grasses than are less invasive introduced grasses (Strauss et al. 2006). Other studies found that relatedness of introduced plants to natives played no role in naturalization and invasion (e.g., Daehler 2001; Duncan and Williams 2002; Dehnen-Schmutz et al. 2007b; Pemberton and Liu 2009), but these studies generally looked at all invasive species in a region, not those that have more serious impacts.

In addition to resisting invaders, communities may also respond to invasion with resilience. Resilient communities are able to recover from disturbances or changes such as the addition of new species without a change in function or structure. The concepts of ecological resilience emerged from studies of predator–prey interactions and the presence of multiple stable states for ecosystems (Holling 1973). Adaptive management strategies have been developed from these concepts, combining social and ecological theory to analyze and manage how systems function under uncertainty (Gunderson 2000). Each ecosystem will have different levels of resilience, with those that are more dynamic naturally responding to the introduction of new species without a major change to stability of structure or function. Of course, the characteristics of species will also affect the ability of the community to be resilient; very aggressive species or those that are functionally very different from natives may shift the community substantially. Rare species often lack aggressive traits (Gaston and Kunin 1997), and if they are introduced into a resilient community there may be little change to function or structure. The resilience of a community to invaders is often correlated with the levels of the other stressors present (Gunderson 2000).

Propagule Pressure, Residence Time, and Spatial Distribution

Few species become serious pests shortly after introduction. In general, a species may be present for many years before spread begins and impacts are noticed. This period between introduction and the onset of spread is usually called the lag phase. There may be many reasons for a lag phase, including a change in dispersal corridors, the introduction of a new pollen or seed disperser, or an episodic disturbance, such as a hurricane that temporarily decreases biotic resistance (Crooks 2005). In many cases, the lag phase may exist because there are insufficient numbers of people on the ground to observe and report the invasion, not because there was an actual lag. Because MRs are intentional, with subsequent monitoring and augmentation and with presumed adequate pollination, seed dispersal, and important mutualist provision, it is not expected that they will go through a lag phase associated with insufficient observation, and thus signs of invasiveness should be able to be detected quickly.

There appears to be a strong correlation between the number of introduced plants or seeds and the probability of invasion (Mulvaney 2001; Von Holle and Simberloff 2005; Pemberton and Liu 2009). This phenomenon has been called by various terms, including *propagule pressure*, *inoculation pressure*, or *infection pressure*. Having a larger founder population may increase the probability that seeds will land in spots appropriate for germination and ensure that despite stochastic processes that might lead to dips in the population size, a sufficient number of individuals will survive to reestablish the population. The number of separate introductions may also increase the probability that one or more events will encounter favorable environmental conditions and that a greater variety of genotypes will be introduced. Greater numbers of individuals planted during MR might increase the probability of the species behaving invasively or increase the probability of successful reintroductions. It is probably prudent to plant as many propagules as is feasible, with careful monitoring afterward.

Perhaps related to propagule pressure is the concept of residence time affecting invasion success. The longer a species is present in an area, the higher the probability it will move through a lag phase and express invasive tendencies. In Chile, invasive plants with a large geographic distribution had a longer residence time than those with a small distribution (Castro et al. 2005). A study evaluating the relative contributions of residence time, propagule pressure, and species traits found that a long residence time strongly contributed to escape from cultivation in the Czech Republic and to the probability of naturalization in Europe (Pyšek et al. 2009). Studies of the horticulture trade have also suggested that the longer a species has been available in the trade, the greater the likelihood of invasion (Dehnen-Schmutz et al. 2007a; Pemberton and Liu 2009). This suggests that very long-term monitoring of MRs may be necessary to determine establishment success and to ensure that the species does not become increasingly aggressive over time.

The spatial distribution of patches also influences the spread of species. Moody and Mack (1988) demonstrated that small patches of plants will lead to a larger population faster than a single large patch of the same total area. Spread from these multiple nascent foci, as they called the smaller patches, will increase population size rapidly even without geographic corridors to facilitate spread or barriers to prevent it. This approach may help an MR establish in a new area more quickly, but as with residence time it should be monitored to ensure that native plants are not overwhelmed.

These considerations highlight the importance of establishing criteria to distinguish between a successful MR and an invasion, because both phenomena will share certain characteristics such as successful reproduction and establishment. Thus, operational "damage criteria" (Warren 2007) that delineate a healthy community integration process from an invasion would be useful to determine before the MR is undertaken and these criteria used to focus the monitoring program.

Comparing Biological Traits of Invasive Plants and Rare Species

Most introduced plants, even though planted in large numbers over a long period of time, do not become invasive (Reichard 1997), and a number of studies have examined the biological attributes of invasive plants (Roy 1990; Rejmánek and Richardson 1996; Pyšek and Richardson 2007). In general, traits related to high reproduction, either sexual or vegetative, and high stress tolerance increase invasive ability. For instance, comparative studies between introduced invasive and noninvasive species have found that the invaders are significantly more likely to have consistent and high seed production, effective long-distance seed dispersal, and long flowering and fruiting periods and, if perennial, may also reproduce vegetatively through stolons, rhizomes, or other means (Reichard 1997; Sakai et al. 2001; Bass et al. 2006). Invaders may also be drought tolerant and retain their leaves for long periods of time, allowing them to photosynthesize and grow longer as well. In the western coastal United States a high level of semievergreen leaves or photosynthetic stems was found in the invaders, potentially allowing them to grow during the mild, wet winters of the region (Reichard 1997; Pyšek et al. 2009). Many invasive species in a given part of the world are also problems in other locations, and significantly fewer consistently noninvasive species become invaders anywhere, suggesting that biological traits contribute to postintroduction behavior (Reichard and Hamilton 1997). It is unclear whether rare species are less likely to be moved into new locations through horticulture or other efforts, so this predictor may be less useful in evaluating MR. Comparative studies on rare species show that they are quite different from invasive species (Gaston and Kunin 1997). Although there are perhaps fewer definitive traits shared by rare species, studies suggest that they have a lower reproductive investment and may, in fact, have minimal sexual reproduction or none at all. Rare species are often genetically impoverished by their small populations, which could lead to a reduction in genotypes capable of tolerating stress. One study also found that rare species tend to have bilaterally symmetric flowers, a trait related to specialized pollinators; thus, a lack of that mutualism could prevent seed production (Harper 1979). Some studies have also suggested that species with restricted ranges may have less efficient seed dispersal, but that, again, is not conclusive among the rare plants that have been studied (Gaston and Kunin 1997).

Weed Risk Assessment

Because invasive plants are environmentally and economically destructive (Mack et al. 2000; Pimentel et al. 2005) and because many are intentionally introduced (Reichard and White 2001), there have been several attempts to develop predictive methods to assess risk of invasion potential in introduced plants (e.g., Rejmánek and Richardson 1996; Reichard and Hamilton 1997; Pheloung et al. 1999). The Australian and New Zealand governments have adopted the weed risk assessment (WRA), developed in Australia (Pheloung et al. 1999), and a number of other countries have tested and modified it (Gordon et al. 2008a). It has also been used to evaluate the invasive potential of bioenergy feedstock crops (Davis et al. 2010). The WRA asks forty-nine questions about a species' distribution and suitable climates, domestication history, undesirable traits, reproduction, dispersal, and persistence attributes. Each answer is scored numerically between –3 and 5 points, although most are in the range of –1 to 1. The final score is the sum of all the questions. There are three outcomes: If the total score is less than 1, the species has a low probability of becoming invasive and is acceptable for importation; if it is greater than 6, it is rejected for being too high a risk; and a score between 1 and 6 is inconclusive and the species is recommended for additional evaluation.

To address the species needing further evaluation, a secondary screen was developed in Hawaii (Daehler et al. 2004). This is a short decision tree that emphasizes a few questions in the primary assessment about the biology of the species. The secondary screening gives more weight to certain traits in the assessment process. In a number of tests, the secondary screening has consistently reduced the number of species needing further evaluation (Gordon et al. 2008a).

Gordon and colleagues (2008b) evaluated the Australian WRA for use in Florida. They tested 158 annual and perennial plant species in six growth forms from fifty-two families in twenty-seven orders. Major invaders were paired with related species that had little or no invasion history in Florida. The WRA with the secondary screen correctly rejected 92% of the species that were known to be invasive in Florida and accepted 73% of noninvaders, with the remaining 19% falling into the evaluate further category. They concluded that the WRA is an effective screening mechanism for evaluating new introductions into Florida.

Using the WRA for MR

We compared the invasive risks of species introduced for ex situ conservation (hereafter called conservation species) with those of horticultural introductions (hereafter called horticultural species) at the Fairchild Tropical Botanic Garden (FTBG) and Montgomery Botanical Center in southern Florida. These species have been growing in these gardens for many years, and extensive data are available in the gardens' records to answer questions in the Australian WRA adapted for Florida (Gordon et al. 2008b).

We chose species from the current collection database of FTBG and from sale lists of horticultural distribution plants published in the *Fairchild Bulletin* from 1955 to 1979. This timeframe reflects a period when plants were distributed without regard to risk of invasiveness. We sampled twenty-two conservation species specifically denoted in the FTBG database as being introduced to the garden for conservation. They are narrow endemics in the Caribbean region or have International Union for the Conservation of Nature (IUCN) status of at least "Vulnerable." These species are categorized by IUCN as threatened, endangered, or critically endangered (table 14.1) and are part of an ex situ conservation collection at FTBG. For comparison, we selected a random sample of twenty-two species from Fairchild plant sale lists to match the distribution of plant habits (woody tree or shrub, palm, cycad, herbaceous plant) represented by the conservation species. Because two of these horticultural species had IUCN status of "Vulnerable," we moved them into the conservation category, yielding a final conservation species sample of twenty-four and a horticultural species sample of twenty.

We evaluated the invasive risk of each species using the Florida WRA (Gordon et al. 2008b), using data compiled from garden records at FTBG and Montgomery Botanical Center, web and library resources, and field observations. We followed available guidance on how to address the WRA questions (Gordon et al. 2010). Although three people were involved in the WRAs across species, final results were checked and corrected for consistency. If taxa fell into the "evaluate further" range (score of 1–6), we used a secondary screen proposed by Daehler and colleagues (2004), as has become common practice when using a WRA based on the Australian system (Gordon et al. 2008a).

TABLE 14.1

Conservation and horticulture species included in the weed risk assessment analysis and their conservation status.

Category	Species	Score	Outcome	2nd Screen	Introduction Year	IUCN Category	Years Since Introduction
Conservation	*Attalea crassispatha*	-6	Accept	Accept	1940	Critically Endangered	69
Conservation	*Brunfelsia densifolia*	-1	Accept	Accept	1971		38
Conservation	*Buxus vahlii*	1	Evaluate	Reject	1981	Critically Endangered	28
Conservation	*Calyptranthes thomasiana*	-4	Accept	Accept	1989	Endangered	20
Conservation	*Catesbaea melanocarpa*	-1	Accept	Accept	1994		15
Conservation	*Coccoloba pallida*	-2	Accept	Accept	1991		18
Conservation	*Copernicia ekmanii*	-3	Accept	Accept	1993	Endangered	16
Conservation	*Cordia rupicola*	-4	Accept	Accept	1991	Critically Endangered	18
Conservation	*Cornutia obovata*	-3	Accept	Accept	1950	Critically Endangered	59
Conservation	*Cresentia portoricensis*	-4	Accept	Accept	1989		20
Conservation	*Croton fishlockii*	-10	Accept	Accept	1995		14
Conservation	*Erythrina eggersii*	-5	Accept	Accept	1989	Endangered	20
Conservation	*Gaussia attenuata*	-5	Accept	Accept	1937	Vulnerable	72
Conservation	*Goetzea elegans*	-3	Accept	Accept	1981	Endangered	28
Conservation	*Maytenus elliptica*	-4	Accept	Accept	1991		18
Conservation	*Maytenus ponceana*	-5	Accept	Accept	1994	Vulnerable	15
Conservation	*Nashia inaguensis*	-3	Accept	Accept	1994		15
Conservation	*Roystonea borinquena*	-7	Accept	Accept	1947		62
Conservation	*Peperomia wheeleri*	-1	Accept	Accept	1991		18
Conservation	*Sida eggersii*	-3	Accept	Accept	1989		20
Conservation	*Stahlia monosperma*	-2	Accept	Accept	1989	Endangered	20
Conservation	*Zamia neurophyllidia*	-1	Accept	Accept	1978	Near Threatened	31
Conservation	*Delonix regia*	1	Evaluate	Accept	1953	Vulnerable	56
Conservation	*Palaquium philippense*	-3	Accept	Accept	1959	Vulnerable	50
Horticultural	*Ardisia guianensis*	0	Accept	Accept	1967		42

TABLE 14.1

Continued

Category	Species	Score	Outcome	2nd Screen	Introduction Year	IUCN Category	Years Since Introduction
Horticultural	*Cordia boissierii*	-3	Accept	Accept	1961		48
Horticultural	*Peltophorum adnatum*	-6	Accept	Accept	1954		55
Horticultural	*Clerodendrum minahassae*	6	Evaluate	Accept	1941		68
Horticultural	*Diospyros maritima*	4	Evaluate	Reject	1940		69
Horticultural	*Catalpa punctata*	1	Evaluate	Accept	1966		43
Horticultural	*Harpullia arborea*	2	Evaluate	Accept	1941		68
Horticultural	*Bulnesia arborea*	-4	Accept	Accept	1948		61
Horticultural	*Chorisia insignis*	-6	Accept	Accept	1967		42
Horticultural	*Sesbania punicea*	18	Reject	Reject	1954		55
Horticultural	*Lonchocarpus punctatus*	5	Evaluate	Accept	1960		49
Horticultural	*Plumbago indica var. coccinea*	4	Evaluate	Accept	1940		69
Horticultural	*Brosimum alicastrum*	1	Evaluate	Evaluate	1962		47
Horticultural	*Couroupita guianensis*	-2	Accept	Accept	1942	Lower Risk/Least Concern	67
Horticultural	*Saraca declinata*	-8	Accept	Accept	1942		67
Horticultural	*Butia capitata*	2	Evaluate	Evaluate	1935		74
Horticultural	*Encephalartos ferox*	-4	Accept	Accept	1957	Least Concern	52
Horticultural	*Carpentaria acuminata*	-3	Accept	Accept	1964		45
Horticultural	*Ptychosperma salomonense*	-4	Accept	Accept	1940		69
Horticultural	*Thrinax excelsa*	-5	Accept	Accept	1966		43

IUCN = International Union for the Conservation of Nature.

We performed statistical analyses using JMP 8.0 (SAS Institute, Inc.) and compared mean risk scores between the two plant categories using a Welch ANOVA for unequal variances due to significant differences in the variances across categories. We made pairwise comparisons between means using a Tukey HSD test. We compared the distribution of outcome categories associated with each score (accept, evaluate further, or reject) between samples using a chi-square test for homogeneity of a 2 × 3 contingency table (two plant categories and three outcome categories).

There was significant nonhomogeneity of variance between the outcome categories ($p = 0.001$, Levene's test), with scores of conservation species varying much less than those of horticultural species (0.594 vs. 32.35 in coefficient of variance). Using a Welch ANOVA, the average WRA score for conservation species was significantly lower (-3.25 ± 2.45 SD) than that for the horticultural species (-0.095 ± 5.74 SD) ($p = 0.0346$, fig. 14.1). A similar pattern was found for outcome (accept, evaluate, or reject) distributions (fig. 14.2). Specifically, 91.7% and 57.1% of the conservation and horticultural species were accepted, respectively. No conservation species were in the "reject" category, whereas one horticultural species was in this category. The outcome distribution of these plant groups was significantly nonhomogeneous ($p = 0.02$, for likelihood ratio test of homogeneity; fig. 14.2). After the secondary screen produced decisions for most of the "evaluate" categories, the outcome distribution no longer showed significant differences between the groups ($p = 0.13$, for likelihood ratio test of homogeneity). Scores aside, none of the conservation species have naturalized, despite having been in cultivation for

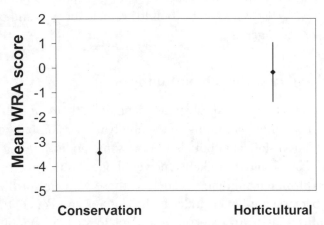

FIGURE 14.1. Mean weed risk assessment scores of species introduced to Fairchild Tropical Botanic Garden for ex situ conservation and horticultural introduction purposes. Vertical bars are ±1 standard error of the mean.

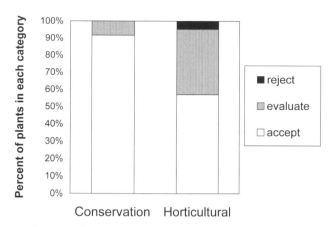

FIGURE 14.2. Distribution of weed risk assessment outcomes among species introduced for conservation and for horticultural introduction in Florida.

an average of 30.8 years. However, several of the horticultural species have naturalized, with one (*Sesbania punicea*) being among the worst invaders in Florida (FLEPPC 2009). *Sesbania punicea* also scored the highest among our forty-four samples. *Sesbania* was a popular landscape species for many years, so a combination of residence time and propagule pressure may have facilitated its current invasiveness. On average, the horticultural species have been in cultivation for 56.7 years, with the longest cultivation record being 74 years.

Despite their harm to the environment, the probability of an intentionally introduced plant becoming invasive is low (Gordon and Gantz 2008), and our findings are consistent, with most of the species being accepted as noninvasive. We recommend using a WRA as one of the tools to evaluate the pros and cons of managed relocation of endangered species.

Areas of Need and Research Opportunities

Although invasion biology is a rapidly progressing field, there are still more questions than answers, so there are limits on the ability of invasion biology to inform MR (table 14.2). For instance, although we have some understanding of biotic resistance, we still have limited predictive power about community resilience to a novel species. Nutritional mutualisms such as those with mycorrhizal fungi or nitrogen-fixing bacteria are better understood (Haskins and Pence, this volume), but how endophytes help invasive species adapt to stressful conditions is still largely unknown. In addition, protection mutualisms are increasingly appearing to be important (Callaway and Aschehoug 2000). How these interactions change

TABLE 14.2

Summary of hypotheses from invasion theory and how they might affect relocations.

Hypothesis	Summary of Evidence	Relevance to Traditional Restoration	Relevance to Managed Relocation
Enemy release	Successful biocontrol projects.	N/A.	Depends on how close to the home range.
Dispersal, pollination, nutrition, protection mutualisms	Need for specialized pollinators can stall invasions.	N/A.	Surveys should be done to ensure pollinator presence.
	Mycorrhizas may be needed for nutrition.	Mycorrhizal needs should be known.	Mycorrhizal needs should be known.
	Endophytes may provide stress resistance.	Endophytes are in leaf and seed tissue, so should be transferred.	Endophytes are in leaf and seed tissue, so should be transferred.
Community resilience	Studies show that some communities have multiple stable states.	Habitat heterogeneity increases spread and resilience of the recipient community.	Habitat heterogeneity enhances spread and resilience of the recipient community.
Propagule pressure	Probability of invasion increases with higher numbers of initial plantings.	Higher numbers of initial plants and repeated plantings will increase success.	Higher numbers of initial plants and repeated plantings will increase success and invasion potential.
Residence time	The longer a species is in an area, the greater the chance it will spread.	Older, established reintroductions will need less monitoring.	Older, established relocations should continue to be monitored to reduce possibility of invasion.
Spatial distribution	Grouping plantings in several small patches, instead of fewer larger, increases spread.	Plant small groupings throughout the site.	Plant small groupings throughout the site, but monitor to ensure that the species does not spread from the site.

over time and how both native and invasive species might evolve in response to them remain unclear (Lankau et al. 2009).

Although plant responses to increased temperature alone are straightforward to anticipate, response to other aspects of global climate change may be species-specific (Rawson 1992). For instance, increased CO_2 elevates levels of urushiol, the chemical in poison ivy (*Toxicodendron radicans*) that causes dermatitis (Mohan et al. 2006). Could climate change also alter the chemical ecologies of rare species, increasing allelopathy or other chemical interactions? Although research on the effects of increased CO_2 is ongoing, we are still far from anticipating how individual species will respond.

More comparative work between rare and common species, especially those that are related, will yield a greater understanding of how we might predict invasion risk. Although the WRA proved to be effective at distinguishing the conservation and horticulture species in our sample, it might be even better when adjusted for the purpose of rare species' MRs. For instance, unlike most invasions, where the geographic occurrence is unpredictable, the recipient community for an MR is known. Some questions might be added or changed to reflect that knowledge, including aspects of biotic resistance, propagule pressure, or spatial design of the outplanting.

Balancing Risks, Benefits, Uncertainties, and Complexities

Reintroductions should be done as closely as possible to the historic bioclimatic envelope of a rare species to reduce invasion probability and increase chances of successfully establishing the species in a new locality. Risks are inevitably present in any biological relocation and probably increase with distance from the historic range of the species. The risk that the MR species may become invasive must be balanced against the risk of extinction if nothing is done. On one hand, competition from invading species appears unlikely to lead to extinction of long-term resident species, especially among plants (Davis 2003; Gurevitch and Padilla 2004; Schlaepfer et al. 2009). This is probably even more true when an MR species is being moved within a continent. On the other hand, many species are in dire threat of extinction or genetic erosion. If moving these species can significantly reduce their chances of extinction or severe genetic impoverishment, the benefit of this action seems likely to outweigh the risks of MR, especially if some precautions, such as WRA, are undertaken and some long-term monitoring is conducted.

Furthermore, emphasis on communities as essentially static entities can obscure the knowledge that communities are able to accommodate a certain degree of compositional change without lasting biodiversity loss (Rosenzweig 2001; Thompson et al. 2007; Lugo 2009). The factors surrounding invasiveness and nativity are far from absolute (Warren 2007), suggesting that it is important to focus on case-by-case assessment of the benefits, risks, and complexities of individual MR cases rather than placing excessive emphasis on generalizations (Slobodkin 2001).

Cautions for Appropriate Use of Managed Relocations

Sites for MR should be carefully selected, with biotic resistance and resilience of the recipient community considered. Locating the rare species in an area with no close relatives would be prudent to reduce risk of hybridization. Although an analysis of chemical ecology is probably not possible in most cases, if a rare species is believed to be allelopathic to other species, the MR should proceed with great caution. In addition, sites as close to the historic range as possible, while still meeting expected climate change scenarios, will potentially decrease the risk of invasion (Mueller and Hellmann 2008).

Because the idea of managed relocations is so new, it is still unclear how sites will be selected for introductions (but see Krause and Pennington, Maschinski et al. [chap. 7], appendix 1, #22, this volume). Presumably, most will be undertaken near native range, but at higher altitudes or more northerly latitudes. If the introduction site is within the same continental region as the native sites, it appears less likely that invasive behavior should be expected. Mueller and Hellmann (2008) studied a database of United States invasive species and their geographic origins to determine whether intracontinental movement of species carried less risk of invasion than intercontinental movement. They thought that given Darwin's hypothesis mentioned previously, species of intracontinental origin would be less likely to become problematic because they would be more likely to have closer relatives in the greater native flora than intercontinental species. Overall, they found that intracontinental species were far less likely to be invasive than species from other continents but that species that became invasive were equally severe regardless of origin. However, the results were somewhat skewed by the particularly high representation of fish and crustaceans in the data. In fact, only 7.5% of the invading plants were intracontinental invaders. This suggests that if relocations are done as closely as possible within the historic bioclimatic envelope, invasion probability may be reduced. This likelihood, along with the findings of Lockwood and colleagues (2001), Ricciardi and Atkinson (2004), and Strauss and colleagues (2006), suggest that more work examining the interactions of invasives with sympatric versus allopatric species (e.g., Callaway and Aschehoug 2000) should be done to gain greater understanding of why these patterns exist.

Planting in a nascent foci pattern is likely to increase the probability of establishment (appendix 1, #29) but also potentially will increase the risk that it might spread too aggressively. Similarly, planting as many individuals as possible increases the probability that more will survive stresses to reproduce. In both cases, because planting large numbers in scattered foci is known to promote invasions, regular monitoring is especially important. This will be especially true if the species has high reproductive rates and stress-tolerating abilities, although it appears this is seldom the case for rare species (Gaston and Kunin 1997). Because a lengthy "residence time" is correlated with invasive ability (Pyšek et al. 2009), the monitoring should be planned for several decades. However, given the large amount of resources needed for long-term monitoring, the benefits must be weighed against the use of these resources to help save other rare species from extinction. This is especially the case because risk is essentially unavoidable, with human-caused impacts imperiling many rare species even if the human activities do not include MR. If most resources are focused on risk mitigation, even in the cases where available information suggests very little risk, then few resources will be available to attempt and actually rescue the growing numbers of rare species in need of assistance.

Not all species are suitable candidates for MR. Hoegh-Guldberg and colleagues (2008) provided a framework to determine whether a species should be considered. The decision tree presents choices to be made, such as the level of extinction risks and technical difficulties, which then feed into options. Six options are possible, with one of them being MR. We recommend that the WRA should also be used before any plant undergoes MR. Because it appears to be robust in places as diverse as Hawaii and the Czech Republic (Gordon et al. 2008a), it should be useful in any location with proper modifications relevant to the geographic and climatic features of the recipient region. The secondary screen may also be used to reduce the number of species that fall into the "evaluate further" category (Gordon et al. 2008a). If the species is in the "evaluate further" category after the secondary screening, it should probably be kept in a controlled ex situ collection and the MR done only under the direst scenarios of imminent extinction. Under those circumstances, it is extremely important that the site conditions are thoroughly considered and an extended monitoring plan is in place. If the species has a WRA score of 6 or higher, then MR should not be carried out under any circumstance.

Climate envelope modeling may also assist in determining potential dispersal range of a species that might be considered a risk (Krause and Pennington, this volume). Risky species would be those that needed "further evaluation" even after the secondary screening and those that have higher reproductive rates. This will provide some guidelines for future introduction locations.

The MR species may become invasive, but the outplanting may also introduce other invasive organisms. As with all reintroductions, every plant or seed used must be carefully scrutinized for insects and pathogens to avoid spreading pests. All infected plants should be treated or destroyed. If the MR plants are rooted in soil, that soil should also be treated to reduce the probability of unintentionally moving species, or the plants should be moved bare-root and reestablished in a nursery at the recipient site. Vehicles used to transport the plants and all equipment should be similarly scrutinized and cleaned before visiting the recipient site.

Finally, these recommendations are for reintroduction of plants only. Although some principles of invasion theory may also inform reintroductions of imperiled mammals, fish, or insects, we have considered only plants in this evaluation.

Summary

The first preference of any conservation practitioner will always be to protect and preserve the ecosystems where rare species naturally occur. However, it is an unavoidable reality that some imperiled species are likely to suffer interactions of climate change and habitat loss or fragmentation that will lead to extinctions and range contractions (Parmesan 1996; McLaughlin et al. 2002) and often genetic erosion. These species may be candidates for MR after careful consideration of many factors (Hoegh-Goldberg et al. 2008). Managed relocation will always carry risks. Those risks must be considered against the risk of species extinction. No species should be moved outside its historic range without a careful analysis of its invasive potential and its potential to carry pests, using the recommendations in this chapter and the *Center for Plant Conservation Best Reintroduction Practice Guidelines* (appendix 1). Even with analysis of risk, invasion theory suggests that the introduction sites should be as close as possible to the historic range.

Section 10(j) of the ESA allows listed species planted outside their range to be designated as "non-essential experimental populations." Although weaker under the law, this designation is particularly important because it allows the population to be removed if it becomes ecologically harmful (Shirey and Lamberti 2010), and should be used for all managed relocations, regardless of the outcome of the risk analysis. Species not listed under the ESA should also be planted with the idea that they can be eliminated if harmful. Because there is always the potential that MR species might need to be removed or that MR populations will not successfully establish, it is especially critical that reserve propagules of viable seed or other irreplaceable rare species germplasm be collected and stored in a secure location (Guerrant et al. 2004a; appendix 1, #20).

We thank Jason Downing for his help with the initial phase of the weed risk assessment. We also thank Kelly Foundation's Montgomery Botanical Research Fellows program for funding support for the initial phase of the project. We gratefully acknowledge the key assistance of Mary Collins and Marilyn Griffiths of Fairchild Tropical Botanic Garden and Arantza Strader of Montgomery Botanical Center in retrieving plant database records. Doria Gordon and an anonymous reviewer provided thoughtful comments on an earlier draft. Finally, we appreciate the invitation of Joyce Maschinski and Kristin Haskins to consider this subject and the participants attending the Center for Plant Conservation symposium for their thoughtful comments.

PART IV

Synthesis and Appendices

"Like music and art, love of nature is a common language that can transcend political or social boundaries."

—*Jimmy Carter*

The growing information and experience in the field of plant conservation science provided an opportunity to pause and take stock of what we have done, why we have done it, and how we can do it better. This final part of the book summarizes the earlier sections and highlights key points for best practice (chap. 15). As with all young science, it is important to reflect on past paradigms and incorporate current findings to chart a new course. Kennedy and colleagues (this volume) do this as they clarify the risks and potential for managed relocation and make specific recommendations for collaborative, sound reintroduction science in the future. Examples of reintroductions that have been well designed and executed are provided as models for future studies.

We encourage our readers to provide feedback about the contents of this volume at the national office of the Center for Plant Conservation (http://www .centerforplantconservation.org/ or cpc@mobot.org). Furthermore, we implore practitioners to use the *CPC Best Reintroduction Practice Guidelines* (appendix 1) to plan any new reintroduction. These guidelines draw from past protocols, the chapters in this volume, comments from conference attendees, and experience of practitioners to refine reintroduction planning. As more reintroductions are implemented, particularly in light of the research agenda advised in this volume, more rigorous meta-analyses may be possible, the practice of reintroduction can be further refined, and recovery of endangered plant species can be accomplished.

Chapter 15

Synthesis and Future Directions

Kathryn Kennedy, Matthew A. Albrecht,
Edward O. Guerrant Jr., Sarah E. Dalrymple,
Joyce Maschinski, and Kristin E. Haskins

Reintroduction work must continue and should be expanded because it is an important tool to stabilize and restore vulnerable declining species. Reintroduction can play a vital role in keeping species present in our landscape through climate change, but this will be possible only through careful planning, research, modeling, and priority setting. In this chapter we review the insights emerging from the sections of this volume: the meta-analyses of plant reintroductions, the science and practice of reintroduction, and managed relocation (MR). We provide examples of well-conceived reintroduction projects to serve as models for planning future reintroductions. Furthermore, we make suggestions for improving plant reintroduction science and practice, preparing for climate change, and moving forward to best conserve biodiversity.

Insights from the Meta-Analyses

Reintroduction science is a young and rapidly growing discipline, and increased sophistication in its practice is emerging. Our meta-analyses revealed broad-scale patterns in plant reintroductions and showed that there is an increasing trend toward reintroduction projects designed to test explicit hypotheses (Albrecht and Maschinski, Dalrymple et al., Guerrant, this volume). In the 1990s plant reintroductions began to emerge as a serious conservation tool; early guidelines emphasized the use of replicated experiments designed to evaluate specific factors that limit or facilitate population establishment and persistence. Many promising studies have been conducted (CPCIRR 2009; appendix 2).

Our reviews of rare plant reintroduction projects revealed trends in successful and failed plant reintroductions. Of the forty-nine projects reported in the *CPC*

International Reintroduction Registry initiated in 1985–2008 for which the fates in 2009 were known, 92% were still extant; 76% had attained reproductive adulthood, 33% produced next-generation offspring, and 16% of the second generation had achieved reproductive maturity. Similarly, Dalrymple and colleagues (this volume), analyzing a different data set, found that the majority of young projects were extant. However, see the caveat later in this chapter about interpreting young projects.

Two fundamental constraints on the available data limited our ability to detect causes for reintroduction project success or failure: (1) The time elapsed since inception for most reintroduction projects was too brief to determine population trajectory with great certainty, and (2) many projects initiated more than 5 years ago lacked long-term monitoring and subsequent documentation of population status. Indeed, the status was unknown in the majority of reviewed projects that had been initiated more than 5 years before the review (Dalrymple et al., this volume), creating challenges in determining the efficacy of reintroduction as a conservation tool. Similarly, drawing inferences from more recent projects can be misleading because the short-term demographic dynamics of reintroduced populations have sometimes been poorly correlated with their long-term persistence (e.g., Maschinski 2006; Albrecht and McCue 2010; Monks et al., this volume). The timeframe for many projects was simply not long enough to determine whether reintroductions were self-sustaining. This was especially true for long-lived taxa that may take years or decades before they reach demographic benchmarks (i.e., sexual reproduction and recruitment of next generation; Monks et al., this volume). Although young projects have not yet accumulated enough data to inform our work fully, they provide a good baseline for future evaluations. More important, these reintroduced populations increase the total number of individuals surviving on the landscape and the spatial occupancy of the species, thereby reducing extinction risk. A critical focus of future work should include monitoring and analyzing reintroduced population dynamics, particularly individual transplant survival, recruitment of the next generation, and the spatial spread within the recipient site or dispersal to unoccupied sites (Pavlik 1996). Emphasizing research and analysis on long-term monitoring efforts will help refine strategies that can improve reintroduction as a conservation tool.

In many instances small sample sizes limited our ability to detect significant differences between groups we wanted to test. For example, Dalrymple and colleagues (this volume) found that wild-sourced propagules achieved higher recruitment levels than ex situ propagules, but the small sample sizes restricted robust statistical support. Similarly, comparisons of the success of projects where the cause of decline was removed or not and of those conducted inside and outside of range suffered from small sample sizes, and the findings should be considered with this in mind. Propagule type and plant life history influenced the success of the reintroduction, but uneven representation of groups influenced the statistical power. Seeds were used as the source material more often for annuals than for perennials, which confounded propagule success with life history (Dalrymple et al., Albrecht and Maschinski, this volume). The majority of reintroduction projects involved perennial species. Among all projects analyzed, whole plant propagules had the greatest overall survival (Dalrymple et al., Albrecht and Maschinski, this volume).

We discovered gaps in our knowledge of reintroduction practice for diverse plant taxa, life forms, and life histories (Guerrant, this volume). Consequently, generalizations drawn from tested groups may not apply across all plant taxa. There are certainly rare and endangered taxa within these life history and life form categories, but they were not reported in the CPCIRR (2009). Additional exploration with untested groups is needed to inform us about the reliability of generalizations within and between plant groups.

Careful and well-designed long-term monitoring efforts would allow us to determine how much management and intervention are necessary before reintroduced populations can be considered self-sustaining. Without long-term monitoring we will not be able to know whether long-lived plant species are able to reach demographic benchmarks and form self-sustaining populations in the wild. Vital rates of reintroduced populations should be compared with those of natural or reference populations to evaluate reintroduction success (Bell et al. 2003; Menges 2008), and of course, to be considered sustainable, reintroduced populations must have positive population growth (Knight, this volume).

Few failures were reported in the CPCIRR (Guerrant, this volume) or in the published literature, a phenomenon also observed by Godefroid et al. (2011), who conducted an independent review of plant reintroductions. This bias influenced our ability to detect reasons for reintroduction success and failure because failed projects were underrepresented. We encourage practitioners to share both unsuccessful and successful projects. Publication of all reintroduction projects would improve our overall knowledge base. Recently journals requiring peer review, such as *Restoration Ecology*, have encouraged publishing setbacks and surprises, realizing the value of sharing these experiences. Even if trials result in nonsignificant data or fail to establish plants in the long term, this information will contribute to reintroduction practice, especially if conditions related to the reintroduction are reported. Furthermore, we encourage project entry into databases such as the *CPC International Reintroduction Registry*. Monitoring periods longer than 5 years will yield valuable information for land managing agencies and practitioners. Although traditional demographic monitoring yields important information, alternative measures also can help evaluate the health of reintroduced populations (Monks et al., this volume).

Our reviews revealed another area where improved practice could help inform the science of reintroduction. Nonexperimental reintroductions, projects done without monitoring, or projects lacking published results have limited value in documenting success or evaluating causes of failures and are not recommended as best practice. Such studies restrict the inferences that can be drawn about why reintroduced populations failed to establish (or persist) and provide little insight into the basic ecology of the focal species, which could further inform future conservation efforts. Furthermore, they are subject to misinterpretation, and they represent lost opportunities to learn and refine small population management.

Consequently, we have carefully selected several well-designed and executed reintroduction projects that can serve as useful models for planning and implementing future reintroduction experiments (table 15.1). We purposely selected studies across a broad range of life histories, propagule types, and experimental designs. Furthermore, many of the selected studies, although by no means exhaustive, exemplify how to select genetically appropriate source material, which can influence the long-term persistence of reintroduced plant populations (see Neale, this volume). Although some of the selected studies can be viewed as unsuccessful in reintroducing viable plant populations, their robust experimental designs im-

TABLE 15.1

Plant reintroduction studies that exemplify well-replicated experimental designs and can serve as models for planning and implementing future reintroduction projects.

Study	Life History	Founder Stage	Experimental Factors
Holl and Hayes (2006)	Annual forb	Seeds and seedlings	Competition (varying frequencies of vegetation removal), litter and soil disturbance, and grazing
Pavlik et al. (1993)	Annual forb	Seeds	Competitor exclusion (herbicide vs. clipping) and fire
Maschinski et al. (2004a)	Perennial shrub	Seeds	Topographic moisture gradient and nurse plants
Sinclair and Catling (2004)	Perennial herb	Seeds and seedlings	Soil disturbance and fertilization
Smith et al. (2009)	Perennial herb	Adult transplants	Presence of mycorrhizal fungi, soil aeration, and seasonal timing

part valuable information on the species' ecology, help advance reintroduction theory, and can be easily replicated in other systems.

We recognize that not all rare plant reintroduction projects can undertake well-replicated experiments because of the extreme rarity of some species or for logistical reasons. In these instances, initiating intervention rather than accepting inevitable loss is understandable. However, even in these cases close quantitative monitoring will help support adaptive management action to increase the probability of population persistence (see Kawelo et al., this volume). Registering and reporting the fate of these projects is also valuable for practitioners facing similar challenges.

Science and Practice

Evidence is mounting that we are making progress to understand both important broad principles governing plant reintroduction success and underlying species characteristics that can facilitate success (see appendix 1). The reviews of reintroduction practice in this section quantified public impact on rare plant recovery, described new horticultural techniques that emphasize an improved understanding of community elements, and extended applications of ecological theory to advance reintroduction practice.

Careful attention to choosing appropriate source material to incorporate adequate representative genetic diversity in the reintroduced population is advised (Falk and Holsinger 1991; Falk et al. 1996; Guerrant et al. 2004a; Neale, this volume). Although genetic analyses are not always possible or necessary before conservation intervention begins (depending on population characteristics), comprehensive genetic evaluation of species and populations can provide valuable information revealing evidence of genetic divergence or convergence between geographically and ecologically separate populations. Additionally, genetic evaluation can guide ex situ collection strategies for maintaining genetically representative collections and can help identify limiting conditions such as lack of compatible alleles needed for reproductive success (Weekley et al. 2002). Although "normal" levels of population diversity vary greatly for different taxa, understanding the molecular and adaptive quantitative genetic diversity of a species, the profile of declining populations, and the diversity of potential source materials for reintroduction can help plan new populations with suitable levels of diversity and reduced risk of deleterious genetic processes (Neale, this volume). Kawelo and colleagues (this volume) demonstrate how these techniques are implemented. Genetic analyses play an important role in evaluating success of reintroductions. Differences in levels of genetic diversity and changes in allele frequencies between reintroduced and wild populations can be assessed over time as another measure of success (Monks et al., Neale, this volume).

New horticultural techniques may improve propagule health and reintroduction success. Advances for in vitro acclimatization methods, such as using liquid culture techniques (see box 6.1) combined with customizing soil microbial inoculum development (Haskins and Pence, this volume) are new contributions to improve rare plant propagation and potential outplanting success. Examining the ecology of rare plants, including the belowground soil mutualists and how these important soil microbes promise to improve acclimatization processes in the laboratory, greenhouse, and field, will be essential preparation for plant reintroduction in a changing climate.

In comparison with *Restoring Diversity* (Falk et al. 1996), the previous reintroduction review by the CPC, this volume has a greater emphasis on population-level dynamics and models. Models can help in planning the number and type of propagules needed to establish a sustainable population (Albrecht and Maschinski, Knight, this volume) and are an essential tool for documenting the extinction risk of reintroduced populations (Monks et al., this volume). Several suggestions are given to improve selection of recipient sites for reintroductions (Maschinski et al. [chap. 7], this volume), as site conditions play a critical role in population growth rates (Knight, this volume) and reintroduction success (Dalrymple et al., this volume). Experimental reintroductions have helped elucidate the intricate link between specific sites and population growth rates. Testing conditions in the habitat such as biotic interactions (herbivory; Knight, this volume) or abiotic factors (light; Possley et al. 2009) help elucidate the degree to which such factors affect reintroduction success. Dalrymple and colleagues (this volume) remind us that critical factors are not always apparent; they stress that current climatic information should be incorporated into reintroduction planning. Intentional use of an explicitly experimental approach provides practitioners with the greatest chance to discern the important factors for their species.

Extensive and intensive measures in the context of integrated conservation strategies, including ex situ, intersitu, and reintroduction, may be needed to save our rarest species (Kawelo et al., this volume). Unfortunately, models examining future conditions project dim trajectories for many endemic species (Krause and Pennington, this volume). This should be a call to action to ameliorate the impacts of climate change now, while there is still time. The fate of the rarest of the rare species remains unclear, but thoughtful planning and proactive measures from collecting ex situ source material and ecological work to assess species niches to modeling projected future conditions in planning sites for reintroduction work can improve population restoration success and reduce extinction risk in the near and long term.

The law will always influence reintroduction options. Kawelo and colleagues (this volume) provide an excellent example of how government regulation of an important pesticide is hampering reintroduction success. In preparation for oncoming changes in climate, it will be necessary not only to understand and respond to existing laws and policies but to become engaged in supplying the rigorous science that will support feasible, appropriate policies and regulations for the future (Maschinski et al. 2011).

Collaborations, public outreach, and information sharing must be cultivated to advance reintroduction science (Maschinski et al. [chap. 4], this volume). We anticipate that future collaborations between modelers, statisticians, ecologists, policymakers, lawyers, and economists will be needed to advance plant reintroduction science. As plant reintroduction science grows, so must the efforts to implement these projects on the ground. Enlisting the public as a resource for education, communication, planting, and monitoring will continue to be essential (see box 4.2). The guidelines we offer practitioners (appendix 1) are based largely on our experiences and are intended to help improve reintroduction practice and science. We encourage practitioners to follow these or other published guidelines (e.g., Vallee et al. 2004) and to provide feedback to the authors and the Center for Plant Conservation at cpc@mobot.org. Through professional exchange with agencies such as the International Union for Conservation of Nature (IUCN) we are exploring ways to expand the availability of these guidelines to the world community. This will require overcoming language barriers and publicizing the *CPC International Reintroduction Registry*.

Facilitating the accumulation and synthesis of reintroduction information will allow even more targeted and powerful metadata analyses to clarify significant principles and best practices. Building comparative data sets over time and distance will help identify conditions directly and indirectly associated with climate change and further our understanding of the requirements for persistence in the face of those challenges. In addition, practitioners can benefit from reviewing reintroductions that have been conducted by others. As the conservation community develops and supports a culture of information sharing (Dalrymple et al., Guerrant, this volume), collaborative syntheses can aid research on reintroduction science and improve our understanding of the impacts of climate change on rare plant populations.

Managed Relocation

Experience over the last 20 years of conservation reintroductions has demonstrated the difficulty of reintroduction even in the best circumstances. Now, the conservation community has to seek solutions for rare species reintroductions at an ever-growing pace to keep up with the effects of climate change. A few of these solutions, such as MR, are controversial (Haskins and Keel, this volume), yet managed relocation is an approach that may assist in maintaining the rarest species on the landscape. There are obvious parallels between traditional within-range reintroductions, introduction outside a taxon's historic range, and MR, as all involve the establishment of new populations. Unfortunately, insufficient rigorous field studies have been conducted comparing reintroductions within range of current habitat with efforts outside the known habitat to evaluate either the risk of failure (and unintended damage) from MR or its potential for success (Dalrymple et al., Guerrant, this volume, but see Marsico and Hellmann 2009). The path forward is confused by extreme views that have been presented by some authors. For example, one scenario is to assist populations through incremental introductions into suitable habitats near existing populations (see box 7.1), and more radical scenarios envision relocation to much more extreme ecological or geographic distances into completely new habitats and communities (Barlow and Martin 2005, but see Schwartz 2005) or moving entire communities (Seddon 2010).

Our experience, techniques, and approaches with traditional reintroductions can help inform future work to advance the science of reintroduction, evaluate and test the potential of MR, and develop responsible guides and protocols that may help achieve the ultimate goal of sustainable reintroduced populations. In contrast, the scenario of relocation over great ecological and geographic distances presents challenges on many levels and is not supported by CPC. The conservation community is defining areas of concern and inquiry regarding MR and is examining its potential risks and benefits. Long-distance relocations into dissimilar habitat present high risks and may potentially harm recipient communities (Davidson and Simkanin 2008; Ricciardi and Simberloff 2009a, 2009b; Seddon et al. 2009). The challenges inherent in MR require that we define and address broad research needs and establish interdisciplinary partnerships to focus work on the highest priorities in science and practice. A good consensus that has examined the potential and risks comprehensively must be a result of broad interdisciplinary work. While we strive to keep species from disappearing from the landscape, we also cannot proceed recklessly without a reasoned, stepwise approach that will be informative and conscientious.

The approaches and cautions developed in conservation science and practice in reintroduction work to date remain important, especially as the focus of debate in the conservation literature centers on MR. Previous recommendations by CPC and other conservation organizations (Falk et al. 1996; IUCN 1998a; Society for Ecological Restoration Science & Policy Working Group 2002; Vallee et al. 2004) stressed that preparation of explicit reintroduction plans, hypothesis-driven experimental approaches, quantitative data collection, and good long-term monitoring are essential to eventual success. We emphasize following these guidelines, along with our updated *CPC Best Reintroduction Practice Guidelines* (appendix 1), to problem-solve for specific cases and inform plans to test the promise and risks of MR and other conservation options.

Preparation for an unknown future climate requires an extension of traditional conservation practice. As has been demonstrated (Knight, Krause and Pennington, Maschinski et al. [chap. 7], Monks et al., this volume), more rigorous ecological and demographic data are needed to understand rare plant population growth, to build accurate models, and to validate them. These same needs apply to efforts to predict and evaluate climate change and to inform research and support decisions for MR. Future MR work will require a detailed understanding of species biology, population dynamics, and ecological relationships. While the species and systems are still functional and available to us, ecological characterization is urgent work that must be done now.

To date our research on reintroduction site selection has focused largely on the rare introduced plant population and not the recipient community. There is a great need to understand the effects of reintroduced populations on recipient communities. Possible impacts include modified environmental gradients (e.g., nutrient and moisture availability) or modified biotic interactions (e.g., attracting predators, pollinators, and dispersers). It is critical to increase our knowledge of these relationships because this information contributes to successful evaluations of recipient sites and any plans for creating new populations. Methods suggested for selecting traditional reintroduction sites still apply to selecting MR sites (Maschinski et al. [chap. 7], this volume). Reintroduction biology must be integrated with ecosystem restoration, which will require more collaboration between reintroduction biologists and restoration ecologists. In addition, we emphasize consideration of the human community. Community involvement for activities affecting both donor and recipient sites will be important where projects are initiated (Parker 2008).

To provide adequate, diverse source material for reintroductions and MRs, there is a need to be proactive and increase ex situ genetic resources (Vitt et al. 2010). Research and development are needed on storage protocols for more diverse kinds of material and for seed-increasing techniques that maximize wild-adapted traits. Similarly, developing propagation and production techniques to provide sufficient material for establishing large populations quickly will promote success of populations reintroduced to changing landscapes. There is a clear need to increase ex situ work by securing wild plant genomes with broad genetic diversity, and it may be necessary to target extreme traits. Populations at the edge of range may have traits that would allow a species to persist and adapt in a changing climate.

Carefully documented reintroductions can provide significant genetic and ecological information for modeling, monitoring, and guidance for developing new management approaches such as MR. Long-term monitoring on a landscape scale is needed to detect climate-driven changes in populations. Increasing the numbers of responsible, well-designed reintroduction research projects in current habitats can help clarify species' strengths or vulnerabilities and support adaptive management to maximize resistance and resilience of rare species to stochastic events while supporting natural adaptation and migration processes.

The CPC's collective experience, techniques, and approaches can inform future experimental designs to advance the science of reintroduction, to evaluate the potential of MR, and to develop responsible guides and protocols that may improve success. This process will require establishing interdisciplinary partnerships to focus on the highest priorities in science and practice. Current funding reality requires that we carefully examine and allocate resources across priorities for personnel, research, and active conservation management as our native habitats experience climate change–related stress.

The challenges in defining and evaluating the potential of MR are not just biological. We will also need to develop public policy and professional guidance that will provide a framework to implement active ethical management of suites of species. Engagement of the scientific community, agencies, and public outreach specialists is needed. For a good consensus policy to be formulated, information gaps and rigorous science guiding recommended practice must be clear. Benefits and risks must be comprehensively and broadly reviewed. Practical protocols must be developed, and professional ethics must be elaborated. Agency and public policies need to be written, and international cooperation must be gained. Our work ahead is demanding and abundant but highly necessary to preserve biodiversity. The fate of many rare species rests in the hands of policymakers and land managers.

As is true for traditional reintroduction science, developing the theory and practice for MR will be best served by taking a similar path of carefully planned hypothesis-driven experiments. Strategic collaborative work is vital to provide informative multidisciplinary and complementary studies that can be compared within and between systems. Test cases should be selected in both simple and complex systems. We suggest that there is a high priority for integrated planning to identify species that present valuable research models and determine how and when to initiate work along a gradient of expected change. Identifying the most urgent cases of decline or projected imminent habitat loss, such as the Key tree cactus, *Pilocereus robinii* in Florida (Maschinski and Goodman 2008), alpine species (Grace et al. 2002; Raven, this volume), and *Banksia montana* in South West Australia (Cochrane and Barrett 2009), and initiating multidisciplinary work with them is important. Pilot studies need not focus solely on population establishment issues. They can also examine related issues such as producing and validating population and climate models, assessing genetic change, and conducting risk and cost analyses. Assessing risks and monitoring for the appearance of adverse impacts is critical to building confidence that we can minimize negative impacts on recipient communities (Reichard et al., this volume). New studies should provide context by comparing performance and population dynamics in current and new localities while investigating multivariate and interactive effects. A conceptual effort to guide the evaluation of proposed MR studies is offered by Haskins and Keel (this volume).

We need to proceed as a scientific community by design rather than by chance. Ensuring that we conduct well-designed pilot studies of sufficient depth and breadth in a comprehensive and rigorous scientific manner is not a simple matter. Future studies must be supported by planning and policy and backed by significant funding. Academic peer review and agency review in initial study proposals is critical to steady, productive progress.

Summary

Accepting the premise that maintaining the current level of plant diversity is in the best interest of a sustainable planet and the quality of human life, our charge remains to prevent the extinction of species and minimize the loss of diversity. Reintroduction research and clarification of the practical potential for proposed MR techniques are critical to natural resource management through climate change. Defining and implementing the agenda for research and practice will require coordinated regional, national, and even international efforts across agencies and conservation scientists in theoretical and applied sciences. The CPC will continue to work to support the partnerships and strategic initiatives that will be needed to do so. The work ahead is challenging and exciting. Our aim for plant conservation is to move forward carefully, using good science, strategic thinking, and purposefully collaborative work to identify and resolve the issues at hand.

Center for Plant Conservation Best Reintroduction Practice Guidelines

Joyce Maschinski, Matthew A. Albrecht, Leonie Monks, and Kristin E. Haskins

The ultimate goal of rare plant conservation is to ensure that unique taxa experience continued evolution in a natural context. Over the past 20 years conservation officers working with the Center for Plant Conservation (CPC) have conducted plant reintroductions of many species in many habitats. In this appendix we provide our *CPC Best Reintroduction Practice Guidelines*, which refine reintroduction planning based on a review of past protocols, the experience of CPC practitioners, findings presented in this volume, and comments from conference attendees. The science and practice of rare plant reintroduction are expanding, and these guidelines represent the state of the art.

Our goal is to provide a quick reference for practitioners to use when planning and executing rare plant reintroductions (fig. A1.1). The term *reintroduction* in this appendix implies any attempt to introduce propagules to an unoccupied patch, including augmentations, introductions, and translocations. Managed relocations would require following these same guidelines in addition to the points presented by Haskins and Keel (this volume) and the modeling, interdisciplinary, multiagency, and potentially international collaborations cautioned by Kennedy and colleagues (this volume). The sections are intended to help practitioners do the following: justify the decision to conduct a reintroduction; prepare the reintroduction design with legal, funding, species biology, horticulture, and recipient site considerations in mind; implement the reintroduction; conduct project aftercare; and design monitoring to document long-term establishment of the rare population. All phases of the reintroduction process should include opportunities for public involvement. In addition, we suggest a template to use for documenting all aspects of the reintroduction that can be found on the North Carolina Botanical Garden website (North Carolina Plant Conservation Program Scientific Committee 2005).

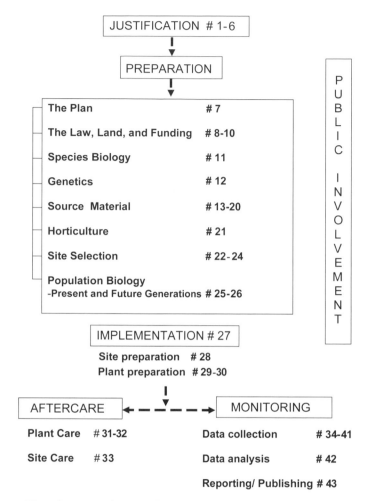

FIGURE A1.1. Flow diagram of reintroduction justification, preparation, implementation, aftercare, and monitoring.

In comparison to previous guidelines, these offer suggestions based on the meta-analyses described in this volume. Thorough examinations of existing populations are recommended to help determine the trajectory of population growth and guide selection of recipient sites. When possible we advise linking the ecology to the demography of the species. In addition, we provide suggestions for improving the possibility of creating a sustainable population in a changing climate.

To support our recommended best practices, we reference appropriate sections of the text of this volume. Additional literature can be found in the chapters. For more in-depth details about some of the sections, we refer readers to previous publications with reintroduction guidelines: *Restoring Diversity* (Falk et al. 1996), *IUCN Guidelines for Reintroductions* (IUCN 1998a), *The SER Primer on Eco-*

logical Restoration (Society for Ecological Restoration Science & Policy Working Group 2002), and *Guidelines for the Translocation of Threatened Plants in Australia* (Vallee et al. 2004).

It is our hope that these guidelines will improve recovery of endangered species and will leave a lasting impression on all those who are concerned with saving biodiversity. We welcome feedback on the guidelines and encourage practitioners to report any reintroductions to the *CPC International Reintroduction Registry*. CPC practitioners throughout the United States can be contacted through the CPC national office (http://www.centerforplantconservation.org/ or cpc@mobot.org).

I. Justifying and Deciding Whether to Conduct a Reintroduction

We do not support or promote reintroduction as an alternative to in situ ecosystem protection. All those working in plant conservation firmly agree that the priority is to conserve species in situ and to preserve their wild populations in natural habitats in as many locations as possible. Reintroduction is never the first action to take for a critically endangered species, even when crisis is imminent. First steps for species in dire straits must be ex situ collection, threat control, and habitat management (Guerrant et al. 2004a; Bruegmann et al. 2008).

Before any reintroduction is conducted, thorough status surveys and careful review of rarity status and threats should be undertaken. Reintroduction should be considered only if habitat protection is not possible or if the taxon is critically imperiled and appropriate sites and propagule source materials are available. We recognize that in the very near future introductions may need to be used as a tool to mitigate the impacts of climate change, because some in situ rare plant populations will be unsustainable in their current historical ranges.

To determine whether a species should be considered for reintroduction, it should meet the criteria described in the checklist box (box A1.1). If the species does not meet these criteria, a reintroduction should not be attempted at this time. If conditions change in the future, a second evaluation could be done. For some taxa, it may *never* be appropriate to conduct reintroductions. For others, changed conditions and improved horticultural, genetic, and ecological knowledge may make it feasible to conduct a reintroduction in the future.

1. DOCUMENT THE SPECIES' STATUS AND DISTRIBUTION.

- Conduct surveys, create maps, and obtain population distribution information.
- Assess habitat-specific population information (Knight, this volume). In each population, count the number or estimate the percentage of

---

Box A1.1. Justification for Reintroduction

A reintroduction may be justified if:

_____ Species is extinct in the wild OR

_____ The distribution of the species is known and there are few, small, and declining populations; AND

_____ Alternative management options have been considered and conducted yet have been judged to be inadequate for long-term conservation of the species; AND

_____ Threats have been identified; AND

_____ Threats from habitat destruction, invasive species, land conversion and/or climate change are imminent and are uncontrollable. Species has high risk of extinction if only managed in situ.

If the species meets any one of the following criteria, then do *not* proceed with reintroduction. Consider ex situ conservation practices (Guerrant et al. 2004b). If the unmet criterion is resolved in future, then reevaluate.

_____ Reintroduction will undermine the imperative to protect existing sites.

_____ Previous tests indicate that it has not been possible to propagate plants or germinate seeds.

_____ High-quality, diverse source material is not available.

_____ Existing threats have not been minimized or managed.

_____ The reintroduced species may potentially negatively affect species in the recipient site via competition, hybridization, or contamination.

_____ There is evidence that the reintroduced taxon would harm other threatened and endangered species or conflict with their management.

_____ The reintroduction is not supported legally, administratively, or socially.

_____ Suitable habitat is not available or not understood.

---

reproductive, juvenile, and seedling stages and, if possible, measure growth and reproduction.

• Note abiotic and biotic conditions in occupied patches. Whenever possible, quantify these factors (e.g., near adults and seedlings, record the canopy cover, associated species, plant density, soil moisture, light, and other factors).

2. Ascertain threats and, when possible, take action to remove, control, or manage them.

- Note specific abiotic and biotic factors that may be causing the population decline. Realize that threats may be direct or indirect and will be best observed over time (Dalrymple et al., this volume).
- If stochastic processes (e.g., wildfires, storms, or random events) have occurred and have decreased the number of individuals in the population, we advise augmenting the population.

3. Engage land managers in discussion about options for the species conservation.

- Attempt or consider all feasible alternative management options before considering reintroduction.
- Ensure that the population will have long-term protection and management (e.g., invasive species removal, controlled burns).

4. If you cannot justify a reintroduction, do not proceed. Use other conservation options.

5. Consider whether your reintroduction will do any harm to the recipient community or to existing wild populations. If so, consider alternative conservation strategies.

- Determine whether the potential collateral impacts of the species in the recipient site are negligible. Is there a threat of hybridization, invasion, or contamination?
- The reintroduction should not undermine the imperative to protect existing populations and their habitats.

6. Determine that the reintroduction is feasible legally, logistically, and socially.

- Laws governing rare species protection vary by location and jurisdiction; therefore, it is essential to discuss any plans for a reintroduction with authorities.
- Determine whether the species has a legal document such as a recovery plan or a conservation action plan, wherein reintroduction has been identified as an important step for preserving the species.

- Hold public meetings to review reasons for the reintroduction and solicit support or involvement.
- Document that the recipient site landowner (public or private) is committed to protecting the reintroduced population.

II. Preparing the Reintroduction

Although it is impossible to say definitively, we believe that many failed reintroductions could have succeeded if appropriate preparation had been undertaken. Being prepared for a reintroduction requires a good plan coupled with large investments of time and resources. This demands commodities that are often in short supply in our rapidly changing world: patience and persistence. It may not be possible to know all of the factors we describe here, but the more that is known, the higher the likelihood of success, and practitioners should at least be aware of the gaps in their knowledge.

Reviewing your reintroduction plan by addressing the following questions will allow you to assess your degree of preparedness (box A1.2). Once knowledge gaps are identified, there is an opportunity to weigh whether there is adequate information to proceed. The risk of proceeding without the knowledge can be assessed along with the risk of taking no action and losing the species. We strongly recommend that reintroductions be conducted as experiments precisely designed to address these knowledge gaps. In this way, each reintroduction can not only help future reintroductions of the practitioner's target species but also help others doing plant reintroductions around the world.

Previous CPC publications have addressed detailed preparations for reintroductions with regard to demography, genetics, and horticultural practice (Falk and Holsinger 1991; Falk et al. 1996; Guerrant 1996a). Specific guidance for ex situ collection and management is essential preparation for reintroductions (see Guerrant et al. 2004a). Our aim here is to provide guidance for establishing sustainable populations in the wild where they may have opportunities for adaptation, evolution, and interactions within a natural ecosystem. Although it is necessary to describe the steps of the plan sequentially, often several steps are conducted simultaneously.

The Plan

7. DEVELOP A REINTRODUCTION PLAN. WHENEVER POSSIBLE, DESIGN THE REINTRODUCTION AS AN EXPERIMENT AND SEEK PEER REVIEW.

- Identify the project leader and key collaborators, who will be responsible for planning, supporting, implementing, site management, monitoring, and reporting findings.

Box A1.2. Questions to Consider When Planning a Reintroduction
(Falk et al. 1996; Vallee et al. 2004)

_____ Is this an augmentation (reinforcement), reintroduction, or introduction (see Glossary)?

_____ Have you considered legal issues, logistics, and land management
(McDonald 1996)?

_____ Are the biology and ecology of the species understood (Menges
2008; Maschinski et al. [chap. 7], this volume)?

_____ Are genetic studies needed (Neale, this volume)?

_____ Have germination protocol and propagation methods been determined (Guerrant 1996a; Guerrant et al. 2004a; Haskins and Pence,
this volume)?

_____ Has a suitable recipient site been identified, and are land managers supportive (Fiedler and Laven 1996; Maschinski et al. [chap. 7],
this volume)?

_____ Are pollinators known and present?

_____ Are plants susceptible to herbivory? Will they be protected?

_____ Have threats been reduced or eliminated?

_____ How many plants or seeds are available, and how many are needed
(Guerrant 1996a; Albrecht and Maschinski, Knight, this volume)?

_____ What question is being addressed, and does your experiment answer the question?

_____ How will success be measured (Pavlik 1996; Monks et al., this
volume)?

_____ What kind of aftercare for plant and site management will be
needed and how frequently?

_____ What is the involvement of the land manager or owner?

_____ What are the monitoring design and plan for reporting results?

_____ In what ways will you involve the public in your project (Maschinski et al. [chap. 4], this volume)?

- Identify areas of expertise needed to execute the reintroduction. If they are
not represented in the collaborative group, then seek outside experts to join
the team. For example, enlist the help of a scientist with experience in experimental design and statistical analysis to ensure that you have adequate
replication to answer your research question. Consider addressing theoretical questions (box A1.3).

- Plan the reintroduction based on the best scientific information available.
Rely on peers to review your reintroduction plan and provide feedback and

Box A1.3. Questions to Consider When Designing Reintroduction Experiments

_____ What additional knowledge is needed about the species' biology or other factors?

_____ What is the question being asked? Does your experimental design answer the question?

_____ How much replication is needed for adequate statistical power? How will the study be analyzed?

_____ Who will conduct the data analyses?

_____ Have you considered testing aspects of ecological theory, such as founder events, small population dynamics, establishment phase competition, dispersal and disturbance ecology, succession, metapopulation dynamics, patch dynamics on population persistence, resilience, and stability over time?

_____ Using the reintroduced population as a cohort, will you examine natural variation in survival, mortality, and recruitment and tie these to environmental factors?

_____ Will the reintroduction test key habitat gradients of light, moisture, elevation, or temperature?

_____ Will the underlying environmental drivers of λ be measured (Knight, this volume)?

_____ Will genetic factors be part of the experimental design? If so, how will they be analyzed?

_____ Will the reintroduction further our knowledge of key principles related to rare species' ability to cope with climate change?

_____ Are you testing factors within a single site or across multiple sites?

_____ Has a monitoring plan been developed? How long will monitoring be conducted? Have you considered an adaptive monitoring plan? What will the duration of the experiment be?

_____ Have you developed a clear unambiguous datasheet to track reintroduced plant growth, reproduction, and survival? If the monitoring persists for decades, will your successors be able to interpret the data you have collected?

_____ Will the data be housed within your institution or elsewhere so that your successors will able to use it?

_____ How will the plants be mapped and marked or numbered?

_____ If plants are susceptible to herbivory, will their response be included in the design, or should the plants be protected?

Sources: Falk et al. (1996); Vallee et al. (2004).

> ### Box A1.4. Potential Reviewers for Reintroduction Plans
>
> In some regions, there are panels of plant conservation experts who review reintroduction plans as a part of an ongoing legislative process. For example, the scientific advisory committee of the North Carolina Plant Conservation Program requests and evaluates reintroduction plans as part of the process of granting legal permission to proceed with a plant reintroduction in the state of North Carolina (North Carolina Department of Agriculture and Consumer Services 2010).
>
> Experts operating in different areas of the world are also available. *The CPC International Reintroduction Registry* (CPC 2009) provides a resource to learn about reintroductions that have been done and is a source for potential peer reviewers.
>
> The IUCN Reintroduction Specialist Group (IUCN 1998b) has a *Reintroduction Practitioner's Directory 1998* intended to facilitate communication between individuals and institutions undertaking animal and plant reintroductions.
>
> The Global Restoration Network (Society for Ecological Restoration 2009) provides a web-based information hub linking research, projects, and practitioners.

alternative points of view. Rely on the global conservation community to assist you (see suggested reviewers in box A1.4).

- Train and carefully manage all personnel and volunteers who are involved.
- Define goals of reintroduction related to the recovery of the species. Set objectives.
- Develop methods, select which plant and population attributes will be measured, and determine monitoring protocol, frequency, and duration.

The Law, the Land, and Funding

8. Obtain legal permission to conduct the reintroduction.

- In some locations you may be required to obtain one or many permits before conducting a reintroduction (e.g., from the land owner or manager and local, regional, and national authorities). A reintroduction plan is often required for permit acquisition.
- Note the expiration date of all permits involved. Also note the requirements of permits, such as periodic reports or updates to the permitting agency.

- If the reintroduction is done as mitigation, it is critical that all preliminary planning steps be taken within legal parameters. (See Falk et al. 1996 for extensive discussion of mitigation.)

9. ENSURE THAT LANDOWNERS AND MANAGERS ARE SUPPORTIVE OF THE PROJECT AND CAN ACCOUNT FOR POSSIBLE CHANGES IN THE FUTURE.

- Discuss the long-term support and management of the recipient habitat with land managers.
- Develop a written agreement outlining who will be responsible for what action and any special protocols that need to be followed by parties working on the site.
- Set a schedule to meet periodically with the recovery team to assess the species' condition and the status of the reintroduction.
- If future changes warrant intervention, determine a process for evaluating impacts on the reintroduced species. For some agencies, it may be necessary to develop a protocol or decision tree to trigger management action.
- Develop a mechanism for handling any conflicts that may arise (e.g., management for one species is detrimental to another species).

10. SECURE ADEQUATE FUNDING TO SUPPORT THE PROJECT.

- Ideally, funding should be garnered for implementation and management for several years, if not decades, after the installation. At the very least, parties proposing to reintroduce a species should be committed to seek long-term funding support for the project. Committed partners who are willing to provide in-kind services or volunteer citizens who are willing to monitor the reintroduction will help make this step feasible.
- Determining the outcome of a reintroduction takes much more time than we thought. Expect to devote more than 10 years to monitoring to determine whether a population is sustainable (Monks et al., this volume).

Species Biology

The design of your reintroduction will benefit from knowing the biology and ecology of your taxon. We advise gathering information from the literature on your target taxon and closely related congeners. If there are gaps in your knowledge, use the reintroduction as an opportunity to learn more about the species and its ecology. See documentation section (p. 306).

11. KNOW THE SPECIES' BIOLOGY AND ECOLOGY.

- Knowing the mating system will determine whether source material should come from a single population or from mixed populations. For example, because remnant populations lacked compatible alleles for successful reproduction, reintroductions done with Florida ziziphus required carefully selecting compatible individuals from more than one location to achieve reproductive success (Weekley et al. 1999, 2002). In contrast, the facultatively autogamous *Schiedea obovata* requires keeping all outplantings separate (Kawelo et al., this volume).
- Because some taxa need symbionts to germinate or grow (Ogura-Tsujita and Yukawa 2008; Janes 2009; Haskins and Pence, this volume), knowing whether there are obligate mutualists will influence reintroduction success. Attempts to germinate or grow such species without their obligate mutualists will fail.
- If a species is dioecious, the spatial design of plantings should place male and female plants in close proximity (e.g., *Zanthoxylum coriaceum* in Maschinski et al. 2010).
- Species or conditions that may require special techniques for growing and implementing a reintroduction include edaphic endemics, species with specialist pollinators, and species that need symbionts for germination and growth.

Genetics

Ideally, the genetic composition of the reintroduced material is a balance between representing the local gene pool and creating a new, genetically diverse population. Reviewing your current knowledge of wild population genetics will facilitate decisions about appropriate locations for collecting source material, confirming whether hybridization may be a potential problem, or confirming the species taxonomy (Falk and Holsinger 1991; Falk et al. 1996; Neale, this volume; see boxes A1.3 and A1.5). For example, you may want to pursue genetic studies before your reintroduction if you suspect there are hybridization problems, if the morphology of the species looks different in different locations, if one or more populations of the species has distinct ecology from the majority of populations, or if it is difficult to distinguish this species from a congener. Using genetically heterogeneous founders will improve the ability of propagules to cope with varying environmental conditions (Falk et al. 1996; Guerrant et al. 2004a; Neale, this volume). Theoretically, high levels of genetic diversity will equip the new population with the adaptive potential needed to withstand stochastic and deterministic

---

Box A1.5. Questions Related to Wild Populations (McKay et al. 2005; Neale, this volume)

_____ What is the genetic structure of the wild populations?

_____ What is the dispersal capability of the species?

_____ If hybridization is a concern, what are the ploidy levels of the wild populations (McKay et al. 2005)?

_____ Does the species show symptoms of inbreeding depression?

_____ Is there evidence of outbreeding depression?

_____ Based on special ecology, unique morphology, or spatial disconnection from other populations, do you suspect that a population has local adaptation?

_____ Based on the presence of a congener in the wild population or variable morphology, do you suspect that the species is hybridizing with a congener?

---

events, including climate change, and can defend against potential genetic pitfalls of small populations such as founder effects and inbreeding depression.

Working with local geneticists at universities or government facilities to do the genetic studies may be necessary. Adequate funding must be garnered for proper genetic work. But also be aware that there are alternatives to genetic studies. These include hand pollination studies, common garden experiments, and reciprocal transplant studies. Each has advantages and disadvantages.

12. Ascertain whether genetic studies are needed before conducting the reintroduction and, if possible, conduct studies to measure genetic structure of the focal species (Neale, this volume).

- A genetic assessment of wild populations is advised before a reintroduction if the species meets any of the following criteria (S. Wagenius, personal communication).

  The population has fewer than fifty individuals flowering and setting fruit.

  The species has highly fragmented and isolated populations.

  No pollinators are present.

No viable seed is being set.

There are high levels of herbivory, especially on flowers, seeds, and fruits.

The morphology of the species looks different in different locations.

One or more populations of the species have distinct ecology from the majority of populations.

It is difficult to distinguish this species from a congener.

There is recent disagreement about the taxonomy, and a reintroduction may create the undesired opportunity for hybridization.

- In the absence of genetic data, it is valuable to use information on species life history traits, such as habit and breeding system, to inform reintroduction decisions (Neale, this volume).

Source Material and Horticulture

The source material used for any reintroduction may determine its fate. To give the new population a chance at success and a buffer against future stochastic or catastrophic events, it is important to use plants that are genetically diverse and vigorous.

13. SELECT APPROPRIATE SOURCE MATERIAL.

- Collect source material from a location that has similar climatic and environmental conditions to the restoration site(s).
- Minimize artificial selection during seed increases or augmentation of natural populations by resisting the temptation to use abundantly available, vigorously growing maternal lines that may skew the diversity of the population, but rather attempt to maintain even family line representation for a reintroduction (Guerrant et al. 2004a; McKay et al. 2005).
- Traditionally it is recommended to use a single source unless adequate information is available about mating system, dispersal, and genetic structure to justify mixing source material. Justifications for mixing source material include a lack of concern about disrupting local adaptation and evidence of inbreeding depression (Dalrymple et al., Neale, this volume).
- Consider the genetics of the reintroduced population in the context of the wild populations (box A1.6). For example, if the species is an obligate outcrosser and is locally adapted to a site, then breeding with natural populations may lead to outbreeding depression (Neale, this volume).

> ### Box A1.6. Questions to Consider about the Genetics of Source Material
>
> _____ From which wild population(s) should the material be collected for use in the reintroduction?
>
> _____ What is the basis for collecting source material from a particular location?
>
> _____ How will the source material be sampled?
>
> _____ What is the genetic composition of the reintroduced material?
>
> _____ Should material come from an ex situ source, only one wild source population, or mixed population sources?

14. USE EX SITU SOURCE MATERIAL BEFORE COLLECTING NEW MATERIAL FROM THE WILD UNLESS THE EX SITU PROPAGULES YOU HAVE AVAILABLE ARE NOT GENETICALLY DIVERSE OR THERE IS A MORE APPROPRIATE WILD SOURCE POPULATION THAT CAN WITHSTAND COLLECTION (GUERRANT ET AL. 2004A).

- Ex situ samples are not immortal, and they degrade over time. Consider using ex situ material first, and then replenish ex situ stock.
- As a precaution favoring wild population integrity, we recommend using ex situ propagules despite some evidence that wild-sourced propagules tended to achieve higher levels of recruitment than ex situ propagules (Dalrymple et al., this volume). The comparative advantage of wild-collected over ex situ propagules may be related to greater plant age or size of wild-collected propagules. For example, an introduction of wild source and ex situ propagules of _Amorpha herbacea_ var. _crenulata_ showed that the largest plants had greatest survival (Wendelberger et al. 2008). The propagule origin was a less critical factor influencing transplant survival than was plant size.
- Bulking up ex situ collections through vegetative reproduction is recommended if feasible.
- If ex situ material is not available, collect no more than 10% of seed produced in any year from wild populations to avoid harm to the wild populations with more than fifty plants. Collect from all individuals within the population if there are fifty or fewer plants. Capturing broad genetic diversity may require collecting in different years and across the range of the fruiting season. See Guerrant et al. (2004a) for specific guidance on ex situ collection and management.

15. FOR LONG-LIVED SPECIES, REINTRODUCE PLANTS OF VARYING SIZES AND LIFE STAGES TO ACCOUNT FOR VARIABLE SUCCESS OF STAGES IN DIFFERENT MICROSITES (ALBRECHT AND MASCHINSKI, THIS VOLUME).

- The key is to provide heterogeneity. For example, use juveniles and reproductive plants in your reintroduction. Sometimes the two will have different microsite needs. Using different-stage plants will result in a more diverse population structure in the present and future and will increase your probability of finding the optimal conditions for the whole population to grow.

16. USE LARGE, MATURE FOUNDERS TO INCREASE THE LIKELIHOOD OF ESTABLISHING A PERSISTENT POPULATION (GUERRANT ET AL. 2004A; ALBRECHT AND MASCHINSKI, THIS VOLUME); USE WHOLE PLANTS RATHER THAN SEEDS UNLESS THERE ARE COMPELLING CIRCUMSTANCES (E.G., ROCK OUTCROP HABITATS) WHERE SEEDS ARE NECESSARY.

- Grow plants as large as is feasible to manage for transport to the reintroduction site and planting.
- Develop a demographic model for the species to determine the optimum founder plant and population size (Knight, this volume).
- To maximize the number of plants that will be available for the reintroduction, particularly when few seeds are available, we recommend germinating seeds under controlled nursery conditions and transplanting whole plants to the reintroduction site (Albrecht and Maschinski, this volume). A sample of 100 seeds may yield 95 plants if germinated in a greenhouse, whereas only a single seedling may emerge in the field.
- When seeds are the only option (e.g., annuals) we recommend using an experimental protocol that involves irrigation in the field until seeds germinate and become established, a practice often used with long-lived perennials. Also consider protecting seeds from herbivory or providing conditions that will decrease the probability of desiccation (e.g., Bainbridge 2007).

17. CONFIRM THAT THE SPECIES CAN BE SUCCESSFULLY PROPAGATED AND THAT AN ADEQUATE AMOUNT OF HIGH-QUALITY, HEALTHY, GENETICALLY DIVERSE SOURCE MATERIAL IS AVAILABLE.

- A critical step to accomplish before reintroduction is mastering the art of propagating large numbers of the species, acclimatizing them, and growing them ex situ. A declining species that cannot be propagated ex situ is simply not a good candidate for reintroduction. Acknowledge that you are not ready to proceed if you have not mastered this step.

18. ALLOW ENOUGH TIME TO GENERATE AN ADEQUATE AMOUNT OF SOURCE MATERIAL BEFORE INITIATING THE REINTRODUCTION, KNOWING THIS COULD TAKE MONTHS OR YEARS.

19. KEEP DETAILED DOCUMENTATION ON ALL SOURCE MATERIAL USED TO RESTORE POPULATIONS. THIS DOCUMENTATION SHOULD BE LINKED TO PERMANENT PLANT LABELS OR ID TAGS ATTACHED TO THE REINTRODUCED PLANTS. STORE THESE DATA IN MULTIPLE LOCATIONS.

20. DO NOT USE ALL YOUR SOURCE MATERIAL FOR THE REINTRODUCTION.

- Genetically diverse source material should be safely backed up in an ex situ location so that regardless of whether the reintroduction succeeds or fails, there is still germplasm conserved.

21. USE GOOD HORTICULTURAL PRACTICE.

- Acclimate plants to novel conditions (Haskins and Pence, this volume). Transitions from culture medium to soil and from greenhouse to outdoors will require a period of adjustment to reduce the chance of shock. If using propagules that were derived from tissue culture, we recommend gradually decreasing humidity while subjecting cultures to ventilation or air exchanges before transfer to soil. Alternatively, methods could include increasing ambient CO_2, decreasing sugar levels in the cultures, or treating with growth regulators to increase stress tolerance.
- Take phytosanitary precautions to ensure that diseases will not be transmitted.
- Using native soils from the recipient site is advised for nursery production to provide necessary microbial mutualists. Native soils may need augmentation with sterile perlite or vermiculite to achieve consistency necessary for container growth. The possibility of transferring pathogens with native soil should be considered, and good nursery hygiene practices must be followed. If the use of native soil is impractical, then microbial inoculum can be purchased or self-cultured (Brundrett et al. 1996; Dumroese et al. 2009). Note that microbial additions involve translocating multiple species, and therefore all the considerations discussed in these guidelines must be considered for the microbes as well.
- Remove weeds from pots containing reintroduction propagules.

Site Selection

A recipient site should be chosen with great care and intention. Several conditions influence a species' ability to colonize a new site, including functional ecosystem

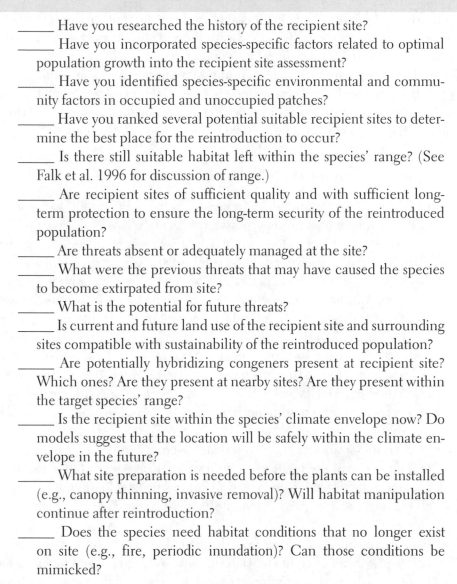

Box A1.7. Questions to Ask about Recipient Site or Reintroduction Location

_____ Have you researched the history of the recipient site?

_____ Have you incorporated species-specific factors related to optimal population growth into the recipient site assessment?

_____ Have you identified species-specific environmental and community factors in occupied and unoccupied patches?

_____ Have you ranked several potential suitable recipient sites to determine the best place for the reintroduction to occur?

_____ Is there still suitable habitat left within the species' range? (See Falk et al. 1996 for discussion of range.)

_____ Are recipient sites of sufficient quality and with sufficient long-term protection to ensure the long-term security of the reintroduced population?

_____ Are threats absent or adequately managed at the site?

_____ What were the previous threats that may have caused the species to become extirpated from site?

_____ What is the potential for future threats?

_____ Is current and future land use of the recipient site and surrounding sites compatible with sustainability of the reintroduced population?

_____ Are potentially hybridizing congeners present at recipient site? Which ones? Are they present at nearby sites? Are they present within the target species' range?

_____ Is the recipient site within the species' climate envelope now? Do models suggest that the location will be safely within the climate envelope in the future?

_____ What site preparation is needed before the plants can be installed (e.g., canopy thinning, invasive removal)? Will habitat manipulation continue after reintroduction?

_____ Does the species need habitat conditions that no longer exist on site (e.g., fire, periodic inundation)? Can those conditions be mimicked?

processes, appropriate associated species, and ongoing management to remove threats and maintain ecosystem health. Review what is known about a proposed recipient site (box A1.7). Seek a recipient site with great similarity to the place where the rare species is thriving. Understanding the site history may help explain existing conditions. Although it is impossible to predict with certainty what a site will become in the future, as much as possible practitioners should try to imagine the future conditions the reintroduced population will face. Ongoing management and threat abatement are essential for maintaining conditions conducive to population sustainability.

In addition, it is important to think about any recipient site in the context of the species' whole distribution. Because corridors may facilitate migration and dispersal between patches, especially with the onset of climate change (Noss 2001), a reintroduced population can serve an important function of connecting existing populations by forming a stepping stone between patches or expanding the size of existing patches. Connecting fifteen or more patches will improve chances for the entire metapopulation capacity (see Hanski and Ovaskainen 2000).

22. CHOOSE A SUITABLE RECIPIENT SITE.

- Evaluate potential reintroduction sites using the recipient site assessment or other quantitative assessment (Maschinski et al. [chap. 7], this volume). Base your evaluation on the natural habitat where a population has positive (or at least stable) growth rate (Dalrymple et al., Knight, this volume).
- To choose between several potential sites, rank reintroduction sites incorporating logistics or ease of implementation, quality of habitat, and management influencing the species' ability to persist at a site (table 7.1; Maschinski et al. [chap. 7], this volume).
- Consider landscape-level phenomena. Evaluating the landscape from the perspectives of topography, ecosystem dynamics, and patterns of possible restoration trajectories will help determine the locations with greatest likelihood of sustaining a reintroduced population (Maschinski et al. [chap. 7], this volume).
- To account for uncertainty, incorporate heterogeneity into the reintroduction plan. Use multiple sites and multiple microsites (even outside your expectations) to test heterogeneity of conditions needed for optimal growth for all life stages of a species (Dalrymple et al., Maschinski et al. [chap. 7], this volume).
- Because the fine-scale needs for individual plant growth and optimal population growth are often unknown, using microsite as an experimental factor is good practice. Measure abiotic conditions (e.g., soil, precipitation, temperature) and biotic conditions (e.g., predators, mutualists, invasive species) at the reintroduction site that are associated with plant performance and population growth (Knight, Maschinski et al. [chap. 7], this volume). Ensure that there are adequate areas for population expansion (e.g., microsites are available within the recipient site and adjacent suitable habitat is available outside of the recipient site).

- Realize that if environments conducive to positive population growth are rare or nonexistent, additional activities, beyond simply reintroducing propagules, will be necessary (Knight, Maschinski et al. [chap. 7], this volume).

23. NOTE THAT USING EXTANT POPULATIONS AND THEIR HABITAT CONDITIONS AS REFERENCE POINTS FOR REINTRODUCTIONS WILL NOT ALWAYS BE APPROPRIATE IF THE SPECIES DOES NOT HAVE POSITIVE GROWTH RATE AT THESE LOCATIONS (POSSLEY ET AL. 2009; DALRYMPLE ET AL., KNIGHT, MASCHINSKI ET AL. [CHAP. 7], THIS VOLUME).

- An experimental context is essential to determine factors necessary for positive population growth.
- Reference points may not be available within core habitat under climate change conditions (Dalrymple et al., this volume). Similarly, geographic distribution may not be a good reference for fundamental niche space. For this reason, known historic range may not necessarily be the only guide to assess optimal habitats for successful reintroduction (Maschinski et al. [chap. 7], this volume).

24. INCREASE THE PROBABILITY OF CREATING A SUSTAINABLE POPULATION BY CHOOSING RECIPIENT SITES THAT HAVE CONNECTIVITY AND INCREASE THE PROBABILITY OF DISPERSAL TO ADDITIONAL LOCATIONS (MASCHINSKI ET AL. [CHAP. 7], THIS VOLUME; BOX A1.8).

- Recipient sites in close proximity to wild or reintroduced populations may have a higher probability of gene exchange.
- Recipient sites with adequate suitable habitat have a higher probability of providing space for population expansion.

Population Biology: Present and Future Generations

Guerrant (1996a, p. 194) suggested that the "founding population should be as large as possible, with the ceiling set primarily by practical and other strategic considerations." With this in mind, it is important to introduce enough individuals (seeds or juveniles) to break through demographic and environmental stochasticity of low populations to achieve a viable population (Knight, this volume). A good reintroduction plan will address population biology questions (box A1.9).

BOX A1.8. QUESTIONS RELATED TO HABITAT OR LANDSCAPE-LEVEL CONSIDERATIONS

_____ Does the recipient site contribute to natural patterns of heterogeneity in the species' distribution?

_____ Have you considered habitat connectivity? Is healthy suitable habitat nearby that will allow the reintroduced population to expand in area and number of individuals? Is adjacent property suitable habitat? Is adjacent property protected?

_____ Are there metapopulation possibilities? Have you accounted for between-site factors as well as within-site factors? Is the site located close to extant populations or other reintroduced populations?

_____ What are the distances between the proposed reintroduction and nearby wild populations? What advantages or disadvantages do the nearby sites give the reintroduced population?

BOX A1.9. QUESTIONS RELATED TO POPULATION BIOLOGY CONSIDERATIONS

_____ What founder population size will be used? (Albrecht and Maschinski, Knight, this volume)

_____ What size and stage structure of plants will be used?

_____ How will the founding population be spatially configured to favor demographic persistence and stability?

_____ What is known about population growth, recruitment, and survivorship in wild habitats, and what environmental or community factors are correlated with population growth rates?

_____ How will population growth, recruitment, and survivorship be monitored in the reintroduced population? And by whom?

---

25. USE AT LEAST FIFTY PLANTS FOR A REINTRODUCTION (ALBRECHT AND MASCHINSKI, THIS VOLUME).

- When working with perennial herbs and sites in highly competitive environments such as grasslands, founder population sizes will need to be larger than fifty.
- We recommend developing a demographic model for the species to determine the optimum founder size (see Knight, this volume).

26. Seek or develop growing conditions with the intention of improving germination, establishment, and survival of next-generation seedlings (Albrecht and Maschinski, this volume).

- Implementing techniques or manipulating site conditions, such as using nurse plants, drip irrigation, or sculpting microcatchments (Bainbridge 2007) to improve success of field germination and seedling establishment, is a critical part of creating a sustainable population. More attention should be paid to this step in the reintroduction process.

III. Implementing the Reintroduction

To use our limited conservation resources to the fullest extent, all reintroductions should be viewed as opportunities to learn about the species, either through experimentation or through documented observation. Even when there is reasonably good information about the environmental attributes associated with the species and its occurrence, test plantings can show which microhabitat conditions are optimal for growth, survival, and long-term population growth (Maschinski et al. [chap. 7], this volume). Effective implementation entails considering logistics and design (box A1.10).

27. Determine the time, materials, personnel, and logistics needed to implement the reintroduction.

- Ensure that you have enough help to treat the site and install plants.
- This is a wonderful opportunity for student and citizen volunteers of all ages. Ensure that they are provided with adequate training, supervision, water, and snacks.

28. If necessary, remove invasive species or thin canopy to improve site conditions for the reintroduced species.

- Site preparation will take time before and after the reintroduction.
- Multiple treatments (e.g., irrigation, soil amendment) may be needed to ensure ideal conditions for reintroduced plants.

29. Place plants in a spatial pattern that will promote effective pollination, seed production, and recruitment.

- Plant density strongly influences variation in outcrossing (or selfing) among plants, so plant in a spatial pattern that will encourage appropriate breeding for your species (Monks et al., this volume).

**BOX A1.10. QUESTIONS ABOUT IMPLEMENTATION LOGISTICS
(VALLEE ET AL. 2004)**

_____ What is the best season to transplant or sow seeds? Keep in mind that best season for rainfall may also be that hottest time of the year, and plants may need more attention.

_____ Have you invited participation from enough staff, volunteers, community members, agency members, and landowners or land managers to execute the reintroduction?

_____ Are permits acquired and up to date?

_____ How will you ensure that plants will be able to be tracked for many years in the future? Are plants tagged and coordinates recorded?

_____ How will you transport plants to the recipient site? Do you have necessary off-road equipment for transport away from roadways?

_____ What is the planting layout design?

_____ Have you notified the press or arranged for photos to be taken of the event? (Note that there may be circumstances in which the exact location of the reintroduction must not be publicized to prevent unauthorized collection of the taxon; however, good conservation news with general descriptions of the reintroduction can be used to engender public enthusiasm for plant conservation. If you are uncertain, talk to your regulatory agency before notifying the press.)

- Planting individuals in small clusters throughout the recipient area, instead of a few large clusters, may increase spread of the population (Reichard et al., this volume).
- Understanding a target species' tolerance for competition and disturbance, as well as habitat composition and structure, can help inform spatial and temporal placement of any reintroduction (Maschinski et al. [chap. 7], this volume). For example, if the target species is not a good competitor, planting into open spaces with few other species present is advised.

30. USE A SYSTEM SUCH AS COLOR CODING TO DISTINGUISH PLANTS IN DIFFERENT EXPERIMENTAL TREATMENTS EASILY. SELECT DURABLE, LONG-LASTING TAGS FOR LABELING PLANTS AND PLOTS.

- If you have a large number of plants and a large number of people helping with the installation of the reintroduction, it is important to be able to distinguish plants from different treatments. For example, if you are testing plants that received mycorrhizal fungal inoculum and those that did not, clearly mark plants before getting to the field and clearly mark the location at the site where plants of each group should be planted.

IV. Conduct Aftercare of the Reintroduction

After the reintroduction is installed, it will need additional care. Success cannot be assumed just because plants or seeds are in the ground. The first few weeks are often most crucial in ensuring that the species survives in its new home. Practitioners should take care to consider these activities in time and cost estimates (box A1.11).

31. WATER PLANTS AND SEEDS UNTIL ESTABLISHED.

32. PERIODICALLY REMOVE WEEDS NEARBY UNTIL PLANTS ARE WELL ESTABLISHED.

33. ONGOING SITE MANAGEMENT IS IMPORTANT. COLLABORATORS SHOULD REVIEW THE STATUS OF THE SITE PERIODICALLY TO ASCERTAIN WHETHER FURTHER MANAGEMENT IS NEEDED.

- Control overabundant herbivores. Cage plants, if necessary.
- Restore historical disturbance regimes such as fire.
- It may be necessary to control competing native and exotic vegetation over the long term, especially if fire cannot be restored to the recipient site.
- Periodically survey the site to detect unforeseen problems (e.g., trampling, theft, herbivory, pest insects, vandalism, maintenance personnel abuse of plants). Take appropriate action to protect the reintroduced population.

BOX A1.11. POST-PLANTING QUESTIONS TO CONSIDER (VALLEE ET AL. 2004)

_____ What aftercare will be needed, and how frequently will plants need attention?

_____ What habitat management and threat abatement are needed? How frequently?

_____ Has a monitoring plan been prepared and reviewed?

_____ Are sufficient funds available for aftercare?

_____ Do permits cover aftercare activities?

V. Design Appropriate Monitoring Plans

A well-designed monitoring plan is an essential component of any reintroduction program. To ensure the long-term persistence of a species in the face of environmental change, a long-term monitoring plan is needed to evaluate how reintroduced populations respond to infrequent events (e.g., drought) and to detect changes in the population that might take years to express (e.g., inbreeding depression in long-lived perennials, replenishing of the soil seedbank). Our goal in this section is not to provide an exhaustive review of how to monitor plant populations but rather to provide standards for the minimum amount of information needed to evaluate the long-term fate of reintroduced populations. Although all monitoring plans must be tailored to individual projects in order to obtain relevant data, all reintroduction monitoring plans include basic components needed to provide information relevant to species' biology and techniques for managing rare plant populations (table A1.1). A long-term monitoring strategy will depend on a number of factors, including the trajectory of population growth, the life history of the focal species, monitoring resources, and the goals of the experimental components of the project. See Elzinga and colleagues (1998) for more details.

34. DEVELOP A MONITORING PLAN.

- A well-designed monitoring plan with clear objectives provides information on the species' biology and techniques for managing rare plant populations. It should be easily understood by your successors; record details as if you are writing for institutional memory.
- If any changes are made to the monitoring plan, then document changes in detail.

35. GATHER DEMOGRAPHIC DATA ON THE REINTRODUCED POPULATION UNLESS IT IS NOT APPROPRIATE FOR THE LIFE HISTORY OF THE TARGET SPECIES (MORRIS AND DOAK 2002; SEE #37).

- Determine the stages of your population and count them. Most commonly, this will be seedlings, juveniles, nonreproductive adults, and reproductive adults.
- We recommend measuring survival, growth, and reproduction on each plant, preferably over multiple generations (Monks et al., this volume).
- If you plan to develop and compare population viability analysis models for the reintroduced population and natural populations, then the frequency of monitoring will need to be at a rate that accurately charts movement of an individual from one life stage to another (table A1.1).

List of actions essential to monitoring plans for reintroduced plant populations.

Action	Description
1. Develop clear monitoring objectives.	Take into account the life history of the focal species, propagule stages planted, and biological and project goals (Pavlik 1996).
2. Define sample units.	Use individuals or transplants for demographic monitoring or plot- or transect-based methods for monitoring demographic structure. All transplants and plots must be permanently marked and mapped, preferably with GPS.
3. Determine appropriate monitoring frequency.	Monitoring period should match key phenological phases (e.g., peak fruiting and flowering) and life history of the focal species.
4. Monitor vital rates.	Follow the fates (survival, growth, fecundity, and recruitment) of transplanted individuals and their progeny or quantitatively track abundance of stage classes (seedling, juvenile, nonreproductive adult, reproductive adult).
5. Evaluate fecundity.	Measure seed production by counting the number of fruits per plant and estimate the number of seeds per fruit through subsampling. Compare results to reference or natural populations.
6. Survey new habitat patches for dispersal and spread.	Search for seedlings at each census, both near and far from sample units. Add new recruits to demographic studies; subsample if recruitment densities are large. Conduct searches for the focal species in suitable habitat patches within and beyond the initial planting site. Establish new sample units to monitor the growth and development of new patches or populations.
7. Monitor wild reference populations.	Simultaneously monitor reintroduced and natural populations to gain insight into key factors that underlie restoration success. Natural populations should be monitored across several sites and during the same years to capture variation in vital rates for comparison to reintroduced populations.
8. Monitor threats.	Document evidence of changes in exotic species distribution and abundance, successional patterns, hydrology, disturbance regimes, land management, herbivores, predators, and disease.
9. Prepare backup plan to relocate lost sample units.	Document all sites and plots with GPS and supplement with precise directions that include compass directions and measured distance from permanent visible landmarks (Elzinga et al. 1998). Produce geographic information system layers and maps if possible.
10. Archive monitoring data and provide metadata.	Enter, store, and back up all monitoring data in digital files. A minimum of two copies of raw datasheets should be kept on paper file, preferably in separate locations. One copy should be accessible to take into the field during subsequent monitoring events. Metadata should be assembled during the project and continually updated.

GPS = Global Positioning System.

• Define the boundaries of your search area to determine dispersal of new recruits and survey these as needed. Realize that these boundaries may need to be expanded or changed over time.

36. WHEN POSSIBLE, MONITOR MULTIPLE WILD REFERENCE POPULATIONS TO COMPARE TO THE REINTRODUCED POPULATION (BELL ET AL. 2003; COLAS ET AL. 2008; MENGES 2008).

• Reference populations will give context for spatial and temporal variation in the reintroduced population's vital rates (table A1.1) and aid in identifying the vital rates that are driving population trends (Morris and Doak 2002).
• In augmentations, the fate of augmented individuals and naturally occurring ones should be distinguished in demographic or quantitative censuses whenever possible to determine whether transplants are performing differently from naturally occurring individuals in the population.

37. ADOPT A MONITORING STRATEGY THAT IS APPROPRIATE FOR THE LIFE HISTORY OF YOUR TARGET SPECIES AND THE FOUNDING PROPAGULE USED.

a. For long-lived perennial plants, monitoring plans will need to accommodate changes in population structure over time.

• Note when transplants transition into larger size classes and sexually reproduce.
• Tag new seedlings as they recruit into the population.
• Most perennial plants will need to be monitored each year to obtain annual vital rates, but some long-lived species (e.g., trees) with slow growth and low reproduction may need less frequent monitoring.
• Time monitoring visits with peak seasonal activity of fecundity and seed germination.
• Searches beyond the transplant plots or transects will need to be conducted to document dispersal, seedling recruitment, and metapopulation dynamics adequately.

b. For short-lived plants, such as annuals, whose populations are often spatially and temporally variable, seed will most often be used to found reintroduced populations (Albrecht and Maschinski, Dalrymple et al., this volume). We recommend sowing seed into permanently marked and mapped plots or transects.

• In annual species, dormancy and germination are often driven by climatic cues that vary from year to year, resulting in wide annual fluctuations in distribution and abundance. As subsequent generations disperse seed, restricting the census to the original sown plots would fail to capture local dispersal. It will be important to note which microsites are suitable for germination and survival.

- Regular counts of individuals within grids or belt transects that cover broad areas within the habitat may be needed to capture changes in the complete spatial distribution and abundance over the long term and to assess population trends effectively (Young et al. 2008).

c. The method used to monitor seeds will depend on the sample unit.

- When sample sizes are small, seeds can be tracked individually. In most cases, however, seeds are sown directly into plots and cohorts are followed.

d. If demographic monitoring of individuals is not possible, monitor stages or size classes that are most important in maintaining population growth.

- If the importance of the vital rates is known for your species, you can concentrate on the most important vital rate and note changes across years to understand population trends.
- If populations begin to decline, then monitoring individuals in all stage classes may be needed to understand mechanisms that are driving the decline and to determine what management actions are needed to reverse the decline.

e. When the target species has characteristics or traits such as clonal reproduction, seed or plant dormancy, or cryptic life history stages (e.g., orchid germinants), all of which make demographic monitoring of marked individuals difficult or impractical, we recommend doing census counts of all or key life history stages to detect population trends (Menges and Gordon 1996).

38. MONITOR FOR AT LEAST 3 YEARS AND IF POSSIBLE FOR 10 YEARS OR MORE (FALK ET AL. 1996; DALRYMPLE ET AL., THIS VOLUME).

- Long-term monitoring provides information necessary to evaluate how reintroduced populations respond to events (e.g., drought) that were infrequent or nonexistent during the early phase of population establishment. It can reveal genetic problems that might play out only after multiple generations (e.g., inbreeding). The importance of these data cannot be overemphasized.
- To develop population viability models and predict population trajectories, a minimum of 3 years of monitoring data is needed. To predict long-term trends (10–100 years) and determine whether a reintroduced population is

potentially self-sustaining under current environmental conditions, extended periods of monitoring are necessary.

- Enlist the help of volunteers to accomplish long-term monitoring (Maschinski et al. [chap. 4], this volume). When possible, include land managers in the monitoring process to foster a close connection between project members and the reintroduced population.

39. IT IS SAFE TO ASSUME THAT SOME OF THE SAMPLE UNITS WILL BE LOST OVER TIME. USE MULTIPLE PERMANENT MARKERS AND MAP PLANTS AND PLOTS WITH A GLOBAL POSITIONING SYSTEM DEVICE TO HELP PREVENT THE LOSS OF VALUABLE DATA.

- Realize that over time, natural or anthropogenic disturbances can impede access to sites or complicate relocating sample units. Plots and transect boundaries or demographic markers may be lost due to fire, flood, downfalls, burial, vandalism, animal impacts, and so on.
- Losses can be mitigated with a good insurance plan, which can be used to reestablish or relocate the boundaries of sample units or tagged individuals when necessary. Whether through plot-based methods or monitoring of individuals, there are several ways to ensure the accurate relocation of lost plot markers, transects, and tagged individuals. See pages 190–191 in Elzinga and colleagues (1998) for more details.

40. DETERMINE HOW SUCCESS WILL BE MEASURED AND HAVE REALISTIC GOALS.

- Identify and define short-, mid-, and long-term goals and determine how you will assess whether those goals have been met.
- Consider project success and biological success (Pavlik 1996).
- Consider population, genetic, and reproductive attributes as indicators of success (Monks et al., this volume).

41. AS SHORT-TERM GOALS ARE ACHIEVED IN A REINTRODUCTION PROGRAM, MONITORING INTENSITY MAY CHANGE FROM EXPERIMENTAL TO OBSERVATIONAL.

- For example, when reintroducing the perennial forb *Helenium virginicum* to sinkhole ponds in the Ozarks, Rimer and McCue (2005) initially set out to determine how planting position and maternal lines affected establishment rates of transplants over a 2-year period. Individuals of the species were grown ex situ and transplanted in a replicated experimental design, and then the fates of transplants were followed demographically. After meeting the initial goals of the reintroduction, the populations grew rapidly due to vegetative reproduction and successful seedling recruitment, making it impractical to differentiate demographically between transplants and new recruits in subsequent censuses. Because the short-term goals of the experimental design were accomplished, the populations grew rapidly, and the species was capable of completing its life cycle in this location, the monitoring protocol switched to count estimates and surveys for new threats rather than full-scale demographic monitoring of individuals. Likewise, transitioning to observational monitoring may lead to less frequent data collection (e.g., annual rather than quarterly) than was needed during the more intense experimental stage.

42. ANALYZE DATA IN A TIMELY FASHION. DISCUSS YOUR ANALYSES WITH PEERS AND STATISTICIANS.

43. REPORT RESULTS BY PUBLISHING OR PUBLICIZING VIA THE POPULAR MEDIA, NEWSLETTERS, AND WEBSITES. ENTER DATA INTO RELEVANT DATABASES FOR GLOBAL ACCESS.

Documentation

Because documentation is an essential component of reintroduction (box A1.12), we encourage careful documentation so that the reintroduction project is justified, good decisions are made about preparedness before the reintroduction event, appropriate monitoring is implemented, the data are analyzed, and the project is published and made available to others in one form or another. These steps are important to represent the reintroduction accurately from a legal and scientific perspective (see Dalrymple et al., this volume). A documentation form is available on the North Carolina Botanical Garden website (North Carolina Plant Conservation Program Scientific Committee 2005).

Box A1.12. Documentation Needed to Justify and Decide Whether to Conduct a Reintroduction

_____ Survey and status updates are complete. Status includes degree of protection, threats, and management options for the extant populations.

_____ Specific information on the number of populations has been collated within the last 18 months.

_____ Counts or estimates of the number of individuals in each population have been done.

_____ The age structure of the populations is known.

_____ The relationship of populations in a metapopulation context is compiled.

_____ Surveys identifying suitable habitat are complete.

_____ Suitable recipient sites have been assessed and ranked.

_____ Long-term protection and management plans are documented for suitable recipient sites.

_____ Sufficient money is secured to conduct the reintroduction.

Studies Used for Meta-Analyses

EDWARD O. GUERRANT JR., MATTHEW A. ALBRECHT,
AND SARAH E. DALRYMPLE

The reintroductions that were used in the meta-analyses reported by Albrecht and Maschinski, Dalrymple and colleagues, and Guerrant (this volume) are listed here. Included are the species epithet, family, country where the work was conducted, life history or life form, reintroduction type, year of first attempt, the authors who used this species in their meta-analysis, and the references. In some cases the reintroductions have been published in peer-reviewed literature, but many are reported either in gray literature or in the *CPC International Reintroduction Registry* (2009).

Species Name	Family	Country	Life History or Form	Reintroduction Type	First Year	Dataset Identifier	References
Abronia umbellata ssp. *breviflora*	Nyctaginaceae	USA	Ann, SLMP	Int WHR	1995	A, D, G	(CPCIRR: Thorpe et al. 2008a; McGlaughlin et al. 2002)
Acacia aprica	Mimosaceae	Australia	WP, LLPP	Int OHR, Int WHR	1998	A, D, G	(CPCIRR: L. Monks; Monks 2002)
Acacia cochlocarpa ssp. *cochlocarpa*	Mimosaceae	Australia	WP	Int WHR	1999	A, G	(CPCIRR: L. Monks)
Acacia cretacea	Mimosaceae	Australia	WP	Aug	1992	A, D, G	(CPCIRR: Jusaitis 1997, 2005; Jusaitis and Val 1997)
Acacia whibleyana	Mimosaceae	Australia	WP	Int WHR, Aug	1996	A, D, G	(CPCIRR: Jusaitis 2005; Jusaitis and Sorensen 2007; Jusaitis and Polomka 2008)
Acanthomintha duttonii	Lamiaceae	USA	Ann.	Int WHR	1991	D	Pavlik and Espeland 1998
Acianthera saundersiana	Orchidaceae	Brazil	HP	Trans, Int WHR	2001	D	Jasper et al. 2005
Acianthera sonderana	Orchidaceae	Brazil	HP	Trans, Int WHR	2001	D	Jasper et al. 2005
Acomastylis rossii	Rosaceae	USA	HP, ULPP	Trans	1973	G	(CPCIRR: May et al. 1982)
Aconitum noveboracense	Ranunculaceae	USA	HP, ULPP	Aug	2009	G	(CPCIRR: V. Pence)
Aechmea calyculata	Bromeliaceae	Brazil	HP	Trans, Int WHR	2001	D	Jasper et al. 2005
Aechmea recurvata	Bromeliaceae	Brazil	HP	Trans, Int WHR	2001	D	Jasper et al. 2005
Agave arizonica	Agavaceae	USA	LLMP	Aug	1989	G	(CPCIRR: K. Rice; Ecker 1990; Gentry 1982)
Agrimonia incisa	Rosaceae	USA	HP	Trans, Int WHR	1998	D	Glitzenstein et al. 2001
Aldrovanda vesiculosa	Droseraceae	Czech Republic	HP	Int WHR	1995	A, D, G	(CPCIRR: Adamec and Lev 1999); Adamec 2005
Amorpha herbacea var. *crenulata*	Fabaceae	USA	WP	Reint, Int WHR, Int OHR	1995	A, D, G	(CPCIRR: Maschinski et al. 2006, 2007; Wendelberger et al. 2008)
Amsinckia grandiflora	Boraginaceae	USA	Ann.	Int WHR	1989	A, D, G	Pavlik 1991; (CPCIRR: Pavlik et al. 1993; Pavlik 1996)
Antennaria flagellaris	Asteraceae	USA	HP	Trans	1983	D	Fiedler and Laven 1996
Apium repens	Apiaceae	UK	HP	Int WHR	1996	D	McDonald and Lambrick 2006
Aquilegia canadensis	Ranunculaceae	USA	HP	Int WHR	1994	D	Drayton and Primack 2000

Species	Family	Country	Life form	Type	Year	Outcome	Reference
Arabis koehleri var. koehleri	Brassicaceae	USA	ULPP	Aug	2001	A, G	(CPCIRR: Guerrant 1996a; Yandell 1997; Guerrant and Kaye 2007)
Aralia racemosa	Araliaceae	USA	P	Int WHR	1994	D	Drayton and Primack 2000
Arenaria cumberlandensis	Caryophyllaceae	USA	HP	Int WHR	2005	G	(CPCIRR: V. Pence)
Arenaria grandiflora	Caryophyllaceae	France	HP	Reint	1999	A, G	Adamec and Lev 1999; (CPCIRR: Bottin et al. 2007)
Argusia argentea	Boraginaceae	Australia	WP	Reint	1999	D	McDonald 2005
Aristida beyrichiana	Poaceae	USA	HP	Int WHR	1993	D	Glitzenstein et al. 2001
Asclepias meadii	Asclepiadaceae	USA	HP, LLPP	Int WHR	1994	A, G	(CPCIRR: Tecic et al. 1998; Bowles et al. 1998, 2001; Hayworth et al. 2001; Bell et al. 2003)
Aster linosyris	Asteraceae	UK	HP	NS	NS	D	BSBI Introductions Database
Astragalus bibullatus	Fabaceae	USA	HP	Int WHR	2001	A, G	(CPCIRR: Albrecht and McCue 2010)
Astragalus cremnophylax var. cremnophylax	Fabaceae	USA	HP	Aug	1990	A, G	(CPCIRR: Maschinski and Rutman 1993)
Banksia anatona	Proteaceae	Australia	WP	Int OHR	2007	A, G	CPCIRR: L. Monks
Banksia brownii	Proteaceae	Australia	WP	Int OHR	2007	A, G	CPCIRR: L. Monks
Barbosella cogniauxiana	Orchidaceae	Brazil	HP	Trans, Int WHR	2001	D	Jasper et al. 2005
Billbergia nutans	Bromeliaceae	Brazil	HP	Trans, Int WHR	2001	D	Jasper et al. 2005
Bletia urbana	Orchidaceae	Mexico	HP	Int WHR	1986	A, D	(CPCIRR: Rubluo et al. 1989)
Brachycome muelleri	Asteraceae	Australia	Ann.	Aug, Int OHR	1996	A, D, G	(CPCIRR: Jusaitis et al. 2004)
Caladenia arenicola	Orchidaceae	Australia	HP	Int OHR	1996	D	Batty et al. 2006a
Caltha palustris	Ranunculaceae	USA	HP	Int WHR, Trans	1994	D	Drayton and Primack 2000
Campylocentrum burchellii	Orchidaceae	Brazil	HP	Trans, Int WHR	2001	D	Jasper et al. 2005
Capanemia micromera	Orchidaceae	Brazil	HP	Trans, Int WHR	2001	D	Jasper et al. 2005
Capanemia superflua	Orchidaceae	Brazil	HP	Trans, Int WHR	2001	D	Jasper et al. 2005
Carex pyrenaica	Cyperaceae	USA	HP, ULPP	Trans	1973	G	(CPCIRR: May et al. 1982)
Carex rupestris	Cyperaceae	USA	HP, ULPP	Trans	1973	G	(CPCIRR: May et al. 1982)
Carex vulpina	Cyperaceae	UK	HP	Aug	NS	D	Porley 2005

Species Name	Family	Country	Life History or Form	Reintroduction Type	First Year	Dataset Identifier	References
Carpinus caroliniana	Betulaceae	Mexico	WP	Reint	2000	G	(CPCIRR: Alvarez-Aquino et al. 2004)
Centaurea corymbosa	Asteraceae	France	HP	Int	1994	A	Kirchner et al. 2006
Cerastium nigrescens	Caryophyllaceae	UK	HP	Reint	1995	D	BSBI Introductions Database
Chamaesyce skottsbergii var. *skottsbergii*	Euphorbiaceae	USA	WP	Int WHR	1979	D	Mehrhoff 1996
Chrysopsis floridana	Asteraceae	USA	SLPP	Int WHR	2008	G	(CPCIRR: C.L. Peterson)
Cimicifuga elata	Ranunculaceae	USA	HP	Aug	2001	G	(CPCIRR: Kaye 2001)
Cirsium pitcheri	Asteraceae	USA	HP	Reint, Int WHR	1991	A, D, G	(CPCIRR: Bowles et al. 1993; McEachern et al. 1994; Bowles and McBride 1996; Bell et al. 2003)
Cirsium tuberosum	Asteraceae	UK	HP	Reint	1989	A, D	(CPCIRR: Pigott 1988)
Cochlearia polonica	Brassicaceae	Poland	HP	Trans	1970	A	Cieslak et al. 2007
Conradina glabra	Lamiaceae	USA	WP	Int	1991	A, D, G	(CPCIRR: Gordon 1996a, 1996b)
Consolea corallicola	Cactaceae	USA	LLPP	Aug, Int WHR	1996	A, D, G	(CPCIRR: Maschinski et al. 2004b; Stiling et al. 2000)
Cordylanthus maritimus ssp. *maritimus*	Scrophulariaceae	USA	Ann.	Int WHR	1995	A, G	(CPCIRR: Parsons and Zedler 1997)
Ctenium aromaticum	Poaceae	USA	HP	Int WHR	1993	D	Glitzenstein et al. 2001
Cycnea superba ssp. *superba*	Campanulaceae	USA	LLPP	Reint	1997	G	(CPCIRR: M. Kier; USFWS 2005)
Damasonium alisma	Alismataceae	UK	Ann.	Int WHR	NS	D	Plantlife data, Wheeler 2001
Daviesia bursarioides	Fabaceae	Australia	WP	Aug	1997	D	Cochrane et al. 2000
Decalepis arayalpathra	Periplocaceae	India	HP	Reint	1998	D	Gangaprasad et al. 2005
Delissea rhytidosperma	Campanulaceae	USA	ULPP	Reint	NS	G	(CPCIRR: D. Bender)
Deschampsia caepitosa	Poaceae	USA	HP, ULPP	Trans	1973	G	(CPCIRR: May et al. 1982)
Diuris fragrantissima	Orchidaceae	Australia	HP	Int	2004	A	Smith et al. 2009
Diuris magnifica	Orchidaceae	Australia	HP	Int WHR	NS	D	Batty et al. 2006a
Diuris micrantha	Orchidaceae	Australia	HP	Int WHR	NS	D	Batty et al. 2006a

Species	Family	Country	LLPP	Trans	Year	Code	Reference
Dodonaea subglandulifera	Sapindaceae	Australia	LLPP	Trans	1991	G	(CPCIRR: Jusaitis 1997; Jusaitis and Val 1997); Moritz and Bickerton 2010
Echinacea laevigata	Asteraceae	USA	HP, LLPP	Int WHR	2000	A, D, G	Alley and Affolter 2004; (CPCIRR: Alley et al. 2008)
Echinacea tennesseensis	Asteraceae	USA	SLPP	Reint	NS	G	(CPCIRR: A. Bishop)
Erigeron parishii	Asteraceae	USA	HP	Int WHR	1991	A, D, G	(CPCIRR: Mistretta and White 2001)
Erigonum ovalifolium var. vineum	Polygonaceae	USA	HP	Int WHR	1991	A, D, G	(CPCIRR: Mistretta and White 2001)
Erodium macrophyllum	Geraniaceae	USA	Ann.	Int	2001	A, G	(CPCIRR: Gillespie and Allen 2008)
Erysinium menziesii	Brassicaceae	USA	HP	Int	1985	A	Ferreira and Smith 1987
Eurystyles cotyledon	Orchidaceae	Brazil	HP	Trans, Int WHR	2001	D	Jasper et al. 2005
Fagus grandifolia var. mexicana	Fagaceae	Mexico	WP, LLPP	Reint	2000	G	(CPCIRR: Alvarez-Aquino et al. 2004)
Filago gallica	Asteraceae	UK	Ann.	Reint/Int WHR	1994	A, D, G	(CPCIRR: Rich et al. 1999)
Gladiolus imbricatus	Iridaceae	Estonia	HP	Int	2003	A, G	(CPCIRR: Jogar and Moora 2008)
Gomesa crispa	Orchidaceae	Brazil	HP	Trans, Int WHR	2001	D	Jasper et al. 2005
Grevillea calliantha	Proteaceae	Australia	WP	Int WHR	1997	D	Cochrane et al. 2000
Grevillea humifusa	Proteaceae	Australia	WP	Int OHR	2003	A, G	(CPCIRR: L. Monks)
Grevillea scapigera	Protoaceae	Australia	WP	Int OHR	1996	D, G	Dixon and Krauss 2001; Dixon 2004; (CPCIRR: Dixon and Krauss 2008)
Haloragis eyreana	Haloragaceae	Australia	HP	Trans	NS	G	(CPCIRR: M. Jusaitis)
Hedyotis caerulea	Rubiaceae	USA	HP	Int WHR, Trans	1994	D	Drayton and Primack 2000
Helenium virginicum	Asteraceae	USA	HP, ULPP	Int WHR	2003	A, D, G	(CPCIRR: Rimer and McCue 2005)
Helianthemum apenninum	Cistaceae	UK	P	NS	1955	D	BSBI Introductions Database
Hibiscus waimeae ssp. hannerae	Malvaceae	USA	WP, LLPP	Int	1997	A, G	(CPCIRR: D. Bender)
Hieracium attenuatifolium	Asteraceae	UK	P	Aug	1999	D	BSBI Introductions Database
Holocarpha macradenia	Asteraceae	USA	Ann.	Int	1999	A	Holl and Hayes 2006

Species Name	Family	Country	Life History or Form	Reintroduction Type	First Year	Dataset Identifier	References
Hurmboldtia smithiana	Orchidaceae	Brazil	HP	Trans, Int WHR	2001	D	Jasper et al. 2005
Hutera rupestris	Brassicaceae	Spain	SLMP	NS	1976	D	Sainz-Ollero and Hernandez-Bermejo 1979
Ipomopsis sancti-spiritus	Polemoniaceae	USA	SLMP	Int OHR	1998	A, G	(CPCIRR: Sivinski and Tonne 2008)
Ipsea malabarica	Orchidaceae	India	HP	Int WHR, Reint	1995	A, D, G	CPCIRR: Gangaprasad et al. 1999; Martin 2003
Iris lacustris	Iridaceae	USA	HP	Int	1989	A	Simonich and Morgan 1990
Isoetes louisianensis	Isoetaceae	USA	HP	Aug, Int WHR	2008	G	(CPCIRR: A. Tiller, P. Faulkner)
Isotria medeoloides	Orchidaceae	USA	HP	Trans	1986	A, G	(CPCIRR: Brumback and Fyler 1996)
Ixiolaena specis nova	Asteraceae	Australia	HP	Int	1993	A	Morgan 1999
Jacquemontia reclinata	Convolvulaceae	USA	HP	Int WHR, Reint	2001	A, G, D	(CPCIRR: Maschinski et al. 2003; Thornton and Wright 2003; Wright 2003a, 2003b; Wright and Fidelibus 2004; Maschinski et al. 2004b; Maschinski and Wright 2006; Maschinski et al. 2006, 2007)
Kobresia myosuroides	Cyperaceae	USA	HP, ULPP	Trans	1973	G	(CPCIRR: May et al. 1982)
Lambertia echinata ssp. *echinata*	Proteaceae	Australia	WP	Aug	1997	D	Cochrane et al. 2000
Lambertia fairallii	Proteaceae	Australia	WP	Int OHR	2007	A, G	(CPCIRR: L. Monks)
Lambertia orbifolia ssp. *orbifolia*	Proteaceae	Australia	WP	Int WHR	1997	A, D, G	(CPCIRR: L. Monks); Cochrane et al. 2000
Lantana canescens	Verbenaceae	USA	WP	Int WHR, Reint	2005	A, G	(CPCIRR: Possley et al. 2009)
Lasthenia conjugens	Asteraceae	USA	Ann.	Reint	1999	A	Ramp et al. 2006
Leionema equestre	Rutaceae	Australia	LLPP	Trans	1992	G	(CPCIRR: Jusaitis 1991, 1996, 1997)
Lepanthes eltoroensis	Orchidaceae	USA	HP	Trans	1999	A	Tremblay 2008
Lepidium hyssopifolium	Brassicaceae	Australia	HP	Int	1993	A	Morgan 1999

Species	Family	Country	Life form	Method	Year	Code	Reference
Lepismium cruciforme	Cactaceae	Brazil	P	Trans, Int WHR	2001	D	Jasper et al. 2005
Lepismium houlletianum	Cactaceae	Brazil	P	Trans, Int WHR	2001	D	Jasper et al. 2005
Lepismium lumbricoides	Cactaceae	Brazil	P	Trans, Int WHR	2001	D	Jasper et al. 2005
Lepismium warmingianum	Cactaceae	Brazil	P	Trans, Int WHR	2001	D	Jasper et al. 2005
Lilaeopsis schaffneriana ssp. recurva	Apiaceae (alt. Umbelliferae)	USA	HP	Reint	2003	A, G	(CPCIRR: Titus and Titus 2008)
Lilium occidentale	Liliaceae	USA	HP, LLPP	Int WHR	1996	A, G	CPCIRR: Guerrant 1996c, 2001; Guerrant and Fiedler 2004; Yandell 1997
Lindera melissifolia	Lauraceae	USA	P	Int WHR	1990	D	Smith 2003
Linnaea borealis	Caprifoliaceae	UK	HP	Int WHR	1999	A, D	Kohn and Lusby 2004
Linum carteri var. carteri	Linaceae	USA	HP	Aug	2006	A, G	(CPCIRR: Maschinski et al. 2007)
Liparis loeselii	Orchidaceae	UK	HP	Int WHR, Trans	2005	D	Land, pers. comm.
Lobelia cardinalis	Campanulaceae	USA	HP	Int WHR	1994	D	Drayton and Primack 2000
Lobelia urens	Campanulaceae	UK	HP	NS	1968	D	BSBI Introductions Database
Lupinus guadalupensis	Fabaceae	USA	Ann.	Int OHR	1995	A, C	(CPCIRR: Helenurm 1998)
Lupinus sulphureus ssp. kincaidii	Fabaceae	USA	HP	Int	1997	A, G	(CPCIRR: Thorpe et al. 2008b)
Luronium natans	Alismataceae	UK	P	NS	1983	D	BSBI Introductions Database
Lysimachia asperulifolia	Primulaceae	USA	HP	Trans	2004	A, C	(CPCIRR: M. Kunz)
Maxillaria ferdinandiana	Orchidaceae	Brazil	HP	Trans, Int WHR	2001	D	Jasper et al. 2005
Maxillaria juergensii	Orchidaceae	Brazil	HP	Trans, Int WHR	2001	D	Jasper et al. 2005
Maxillaria picta	Orchidaceae	Brazil	HP	Trans, Int WHR	2001	D	Jasper et al. 2005
Melampyrum pratense	Scrophulariaceae	UK	Ann.	Trans, Int WHR	1985	D	Walter 2005
Melampyrum sylvaticum	Scrophulariaceae	UK	Ann.	Int WHR	2005	D, G	(CPCIRR: Dalrymple et al. 2008)
Myricaria laxifolia	Myricaceae	China	WP	Trans, Int WHR	2002	D	Chen et al. 2005
Nepeta rtanjensis	Lamiaceae	Serbia	HP	Int WHR	2004	A, D	Misic et al. 2005
Okenia hypogaea	Nyctaginaceae	USA	Ann.	Aug	2003	A, G	(CPCIRR: Maschinski et al. 2004b; Pipoly et al. 2006)
Oncidium flexuosum	Orchidaceae	Brazil	HP	Trans, Int WHR	2001	D	Jasper et al. 2005
Oncidium macronix	Orchidaceae	Brazil	HP	Trans, Int WHR	2001	D	Jasper et al. 2005
Oncidium riograndense	Orchidaceae	Brazil	HP	Trans, Int WHR	2001	D	Jasper et al. 2005

Species Name	Family	Country	Life History or Form	Reintroduction Type	First Year	Dataset Identifier	References
Oryza rufipogon	Poaceae	China	HP	Reint	1993	D	Liu et al. 2004
Osmorhiza claytonii	Apiaceae	USA	HP	Int WHR, Trans	1994	D	Drayton and Primack 2000
Parnassia caroliniana	Saxifragaceae	USA	HP	Int OHR	1995	D	Glitzenstein et al. 2001
Passiflora sexflora	Passifloraceae	USA	HP	Reint	2006	A, G	(CPCIRR: Possley et al. 2007; Possley and Maschinski 2009)
Pediocactus knowltonii	Cactaceae	USA	WP	Reint	1991	A, G	(CPCIRR: Cully 1996; Sivinski 2008)
Phebalium glandulosum ssp. glandulosum	Rutaceae	Australia	LLPP	Trans	NS	G	(CPCIRR: Jusaitis 1996, 1997)
Phcrodendron rubrum	Viscaceae	USA	WP	Aug	2001	A, G	(CPCIRR: J. Duquesnel)
Phymatidium delicatulum	Orchidaceae	Brazil	HP	Trans, Int WHR	2001	D	Jasper et al. 2005
Pinus torreyana	Pinaceae	USA	WP	Reint	1994	D	Ledig 1996
Plantago sparsiflora	Plantaginaceae	USA	HP	Int WHR	1998	D	Glitzenstein et al. 2001
Platanthera leucophaea	Orchidaceae	USA	HP	Int WHR	1981	A, G	(CPCIRR: T. Bittner); Packard 1991
Pleurothallis aveniformis	Orchidaceae	Brazil	HP	Trans, Int WHR	2001	D	Jasper et al. 2005
Polystachya estrellensis	Orchidaceae	Brazil	HP	Trans, Int WHR	2001	D	Jasper et al. 2005
Posidonia australis	Posidonaceae	Australia	HP	Aug	1999	D	Meehan and West 2002
Potentilla robbinsiana	Rosaceae	USA	HP	Reint	1986	A, G	(CPCIRR: Brumback et al. 2004)
Primula vulgaris	Primulaceae	Poland	HP	Reint	1993	D	Kucharczyk and Teske 1996
Prostanthera eurybioides	Lamiaceae	Australia	WP	Int WHR	1996	A, D	Jusaitis 2005
Prostanthera eurybioides	Labiatae	Australia	LLPP	Aug	1996	G	(CPCIRR: Jusaitis 2005)
Pseudophoenix sargentii	Arecaceae	USA	WP	Aug, Int WHR	1991	A, D, G	(CPCIRR: Maschinski and Duquesnel 2007)
Psoralea tenax	Fabaceae	Australia	HP	Intro	1993	A	Morgan 1999
Pterostylis arenicola	Orchidaceae	Australia	HP	Trans	1995	G	(CPCIRR: M. Jusaitis)
Ptilimnium nodosum	Apiaceae	USA	HP	Int WHR	2006	A, G	(CPCIRR: Box 2.1, this volume)
Ptilotus erubescens	Amaranthaceae	Australia	HP	Int	1993	A	Morgan 1999
Pultenaea trichophylla	Leguminosae	Australia	LLPP	Aug	1991	G	(CPCIRR: Jusaitis 1997; Jusaitis and Val 1997)

Species	Family	Country			Year		Reference
Purshia subintegra	Rosaceae	USA	WP	Aug, Int WHR	1997	A, G	(CPCIRR: Baggs and Maschinski 2000; Maschinski et al. 2004a)
Quercus acutifolia	Fagaceae	Mexico	WP, LLPP	Reint	2000	G	(CPCIRR: Alvarez-Aquino et al. 2004)
Ranunculus aestivalis	Ranunculaceae	USA	HP	Aug, Int WHR	2007	G	(CPCIRR: Pence, Murray et al. 2008)
Remirea maritima	Cyperaceae	USA	HP	Aug	2003	A, G	(CPCIRR: Pipoly et al. 2005)
Rhipsalis cereuscula	Cactaceae	Brazil	P	Trans, Int WHR	2001	D	Jasper et al. 2005
Rhipsalis floccosa	Cactaceae	Brazil	P	Trans, Int WHR	2001	D	Jasper et al. 2005
Rhipsalis teres	Cactaceae	Brazil	P	Trans, Int WHR	2001	D	Jasper et al. 2005
Rhus michauxii	Anacardiaceae	USA	WP	Reint, Aug	1998	G	(CPCIRR: Braham et al. 2006)
Rutidosis leptorrhynchoides	Asteraceae	Australia	HP	Aug, Int WHR	1987	A, G	Morgan 1999; (CPCIRR: Morgan 2000)
Salix arizonica	Salicaceae	USA	WP	Aug	1995	A, G	(CPCIRR: Maschinski 2001)
Salix lapponum	Salicaceae	UK	WP	Aug	1991	D	Mardon 2003
Salix myrsinifolia	Salicaceae	UK	WP	Aug	1991	D	Mardon 2003
Salvia pratensis	Lamiaceae	UK	HP	Aug	1999	D	BSBI Introductions Database
Sanguinaria canadensis	Papaveraceae	USA	HP	Int WHR, Trans	1994	D	Drayton and Primack 2000
Sarracenia flava	Sarraceniaceae	USA	HP	Int WHR	1998	D	Sheridan and Penick 2000
Saxifraga virginiensis	Saxifragaceae	USA	HP	Int WHR, Trans	1994	D	Drayton and Primack 2000
Schiedea spergulina var. leiopoda	Caryophyllaceae	USA	HP	Aug	2007	A, G	(CPCIRR: D. Bender)
Schwalbea americana	Scrophulariaceae	USA	HP	Aug	1995	A	Obee and Cartica 1997
Scleranthus perennis ssp. prostratus	Caryophyllaceae	UK	HP	Reint	1995	D	Leonard 2006a, 2006b
Senecio paludosus	Asteraceae	UK	HP		1996	D	BSBI Introductions Database
Senecio macrocarpus	Asteraceae	Australia	HP	Int	1993	A	Morgan 1999
Sibbaldia procumbens	Rosaceae	USA	HP, ULPP	Trans	1973	G	(CPCIRR: May et al. 1982)
Silene douglasii var. oraria	Caryophyllaceae	USA	HP, SLPP	Aug, Int WHR	1998	A, D, G	(CPCIRR: Kephart 2004; Lofflin and Kephart 2005)
Silene regia	Caryophyllaceae	USA	SLMP	NS	2004	G	(CPCIRR: H. Alley)
Sorghastrum nutans	Poaceae	USA	HP	Int WHR	1997	D	Glitzenstein et al. 2001
Specklinia malmeana	Orchidaceae	Brazil	HP	Trans, Int WHR	2001	D	Jasper et al. 2005

Species Name	Family	Country	Life History or Form	Reintroduction Type	First Year	Dataset Identifier	References
Specklinia pabstii	Orchidaceae	Brazil	HP	Trans, Int WHR	2001	D	Jasper et al. 2005
Stephanomeria malheurensis	Asteraceae	USA	Ann.	Reint, Int OHR	1987	A, G	(CPCIRR: Brauner 1988; Parenti and Guerrant 1990; Guerrant 1996b; Guerrant and Pavlik 1998; Currin et al. 2007; Currin and Meinke 2008)
Symonanthus bancroftii	Solanaceae	Australia	HP	Reint	2002	G	(CPCIRR: Bunn and Dixon 2008)
Symplocos coccinea	Symplocaceae	Mexico	LLPP	Reint	2000	G	(CPCIRR: Alvarez-Aquino et al. 2004)
Syzygium travancoricum	Myrtaceae	India	WP	Aug	1999	D	Anand et al. 2004
Taraxacum palustre	Asteraceae	UK	HP	Reint	1998	D	BSBI Introductions Database
Tephrosia angustissima var. *corallicola*	Fabaceae	USA	HP	Int WHR	2003	A, G	(CPCIRR: Wendelberger and Maschinski 2006)
Tetraneuris herbacea	Asteraceae	USA	HP	Int	1988	A	McClain and Ebinger 2008
Teucrium scordium	Lamiaceae	UK	HP	Int	1998	A	Beecroft et al. 2007
Thlaspi perfoliatum	Brassicaceae	UK	Ann.	NS	NS	D	Rich et al. 1998, Plantlife data
Tillandsia geminiflora	Bromeliaceae	Brazil	HP	Trans, Int WHR	2001	D	Jasper et al. 2005
Tillandsia stricta	Bromeliaceae	Brazil	HP	Trans, Int WHR	2001	D	Jasper et al. 2005
Tillandsia tenuifolia	Bromeliaceae	Brazil	HP	Trans, Int WHR	2001	D	Jasper et al. 2005
Tillandsia usneoides	Bromeliaceae	Brazil	HP	Trans, Int WHR	2001	D	Jasper et al. 2005
Trichocentrum pumilum	Orchidaceae	Brazil	HP	Trans, Int WHR	2001	D	Jasper et al. 2005
Trifolium stoloniferum	Fabaceae	USA	HP	Trans, Int	1990	A, G	Smith 1998; (CPCIRR: D. White, D. Osbourne)
Trinia glauca	Apiaceae	UK	HP	NS	1955	D	BSBI Introductions Database
Veronica spicata	Plantaginaceae	UK	HP	NS	1955	D	BSBI Introductions Database
Verticordia albida	Myrtaceae	Australia	WP	Int WHR	2004	A, G	(CPCIRR: L. Monks)
Vitis vinifera	Vitaceae	France	WP	NS	1992	D	Arnold et al. 2005
Vriesea friburgensis	Bromeliaceae	Brazil	HP	Trans, Int WHR	2001	D	Jasper et al. 2005
Vriesea platynema	Bromeliaceae	Brazil	HP	Trans, Int WHR	2001	D	Jasper et al. 2005

Widdringtonia cederbergensis	Cupressaceae	South Africa	WP	D	1987	Reint	Mustart et al. 1995
Woodsia ilvensis	Dryopteridaceae	Estonia, UK	HP	D	1996	Reint	Aguraiuja, pers. comm., McHaffie 2005, 2006; Lusby et al. 2002
Zanthoxylum coriaceum	Rutaceae	USA	WP	A, G	1998	Aug, Int WHR	(CPCIRR: Maschinski et al. 2005, 2006, 2007)
Zanthoxylum flavum	Rutaceae	USA	WP	A, G	1999	Aug	(CPCIRR: J. Duquesnel)
Zizania texana	Poaceae	USA	HP	G	1992	Reint	(CPCIRR: Power 1995)
Ziziphus celata	Rhamnaceae	USA	WP	A, G	2002	Int WHR	(CPCIRR: Weekley and Race 1999, 2001; Weekley et al. 2002; Weekley 2004; Weekley and Menges 2008)
Zygostates alleniana	Orchidaceae	Brazil	HP	D	2001	Trans, Int WHR	Jasper et al. 2005

Life history or form abbreviations: Ann. = annual, HP = herbaceous perennial, LLMP = long lived (more than 5 years) monocarpic perennial, LLPP = long lived (more than 10 years) polycarpic perennial, P = perennial, SLMP = short-lived monocarpic perennial, SLPP = short-lived (less than 10 years) polycarpic perennial, U = unknown life form or history, ULPP = unknown longevity polycarpic perennial, WP = woody perennial.

Reintroduction type abbreviations: Aug = augmentation, Int OHR = introduction outside historic range, Int WHR = introduction within historic range, Reint = reintroduction, Trans = translocation, NS = not specified.

First year is the year of outplanting or the earliest year of an outplanting if multiple years were reported; NS = not specified.

Dataset identifier indicates the authors who used this study in their analysis: A = Albrecht and Maschinski, D = Dalrymple and colleagues, G = Guerrant.

References that used entries from the *CPC International Reintroduction Registry* (CPCIRR) are shown in parentheses with the principal investigator's name and publication. Studies used from peer-reviewed publications that were not reported in the CPCIRR are noted outside parentheses.

Acclimatization: The habituation of a plant's physiological response to environmental conditions (Begon et al. 1990).

Adaptation: Changes in the morphology or physiology of a plant via natural selection.

Adaptive management: A systematic process of continually improving management policies and practices by learning from the outcomes of existing programs (IUCN 1998a).

Assisted colonization: See *Managed relocation*.

Assisted migration: See *Managed relocation*.

Augmentation: The addition of individuals to an existing population, with the aim of increasing population size or diversity and thereby improving viability. Also called enhancement, reinforcement, or restocking (Falk et al. 1996).

Best practice: A superior or innovative method that contributes to the improved performance of an organization and is usually recognized as best by peer organizations. It implies accumulating and applying knowledge about what works and what does not work in different situations and contexts, including learning from experience, in a continuing process of learning, feedback, reflection, and analysis on what works, how, and why (IUCN 1998a).

Bioclimatic envelope: Typically derived by examining statistical correlations between existing species distributions and environmental variables to define a species' tolerance. Envelopes of tolerance are then drawn around existing ranges. With temperature, rainfall, and salinity forecasts, new range boundaries can be predicted.

CPC: Center for Plant Conservation, an organization dedicated to the conservation and restoration of imperiled native plants of the United States.

Dioecious: Having male and female reproductive organs on separate plants.

Endemic: A species native to and restricted to a particular geographic region. Highly endemic species are especially vulnerable to extinction if their natural habitat is eliminated or significantly disturbed (IUCN 1998a).

Enhancement: See *Augmentation*.

Evolution: Changes in the frequency of genes in a population over time; descent with modification.

Ex situ: The conservation of components of biological diversity outside their natural habitats (IUCN 1998a).

Fitness: The relative contribution an individual makes to the gene pool of the next generation (Begon et al. 1990).

Fundamental niche: The potential range of all biotic and abiotic conditions under which an organism can have a positive population growth rate. The Hutchinsonian fundamental niche can be conceptualized as the *n*-dimensional hypervolume.

Geitonogamous: Reproducing through self-pollination; one flower is pollinated by pollen from another flower on the same plant.

Gene flow: The spread of genes across and between populations as a result of cross fertilization or seed introductions (Begon et al. 1990).

Genetic drift: Random changes in gene frequency within a population resulting from sampling effects rather than natural selection (Begon et al. 1990).

Hermaphrodite: A plant that has perfect flowers and can self-pollinate.

Historic range: The geographic area where a species was known or believed to occur within historic time (USFWS 1999).

Inbreeding depression: A loss of vigor among offspring occurring when closely related individuals are crossed, resulting from the expression of deleterious genes in the homozygous state and from a low level of heterozygosity (Begon et al. 1990).

Introduction: The intentional or accidental dispersal by human agency of a living organism outside its historically known native range (IUCN 1998a).

Invasive species: Introduced species that increases in abundance at the expense of native species (Primack 2006).

Iteroparous: Capable of reproducing more than once (Silvertown 1982).

IUCN: International Union for Conservation of Nature, the world's oldest and largest global environmental network, focused on sustainable development and the environment.

Lambda (λ): Annual population growth rate or $\lambda_t = N_{t+1}/N_t$

Managed relocation: The deliberate introduction of organisms outside their native ranges to counteract the negative effects of climate change. Goals of managed relocation include reducing extinction risk, increasing evolutionary potential, and enhancing ecosystem services (Hellmann et al. 2008; Managed Relocation Working Group 2008).

Metapopulation: A system of connected, spatially distinct subpopulations (IUCN 1998a).

Mitigation: An action that is intended to offset environmental damage (SER 2002).

Monoecious: Having female and male reproductive parts on the same plant.

n-dimensional hypervolume: All aspects of the environment, physical and biological, are included in the niche (e.g., temperature tolerance, water requirements, competition, predation). Hutchinson (1957) mathematically described the hypervolume in n-dimensional space along n axes corresponding to environmental variables that permit a species' population growth rate to be positive indefinitely.

Native plant: A species that occurs naturally in an area.

Natural range: The geographic area within which a species can be found. Sometimes a distinction is made between a species' natural range and the places to which it has been introduced by human agency (deliberately or accidentally), as well as where it has been reintroduced after extirpation.

Niche: See *Fundamental niche* and *Realized niche*.

Outbreeding depression: A reduction in vigor or fertility (fitness) resulting from hybridization between genetically distinct individuals or populations of the same species. The loss in vigor is thought to be caused by breaking up co-adapted gene complexes.

Outplanting: Movement of plants from an ex situ location to an in situ location (Falk et al. 1996).

Phytosanitary: Any measure applied (a) to protect human, animal, or plant life or health (within a Member's Territory) from the entry establishment or spread of pests, diseases, or disease-carrying organisms; or (b) to prevent or limit other damage (within the Member's Territory) from the entry, establishment, or spread of pests (IUCN 1998a).

Population: A group of individuals of the same species that have the ability to genetically interact and inhabit a defined geographic area.

Population growth rate: Change in population size from one time to another. A positive population growth rate indicates an increasing population, whereas a negative population growth rate indicates a declining population. See *Lambda*.

Practitioner: A person involved with all aspects of plant reintroduction, including planning stages and actual placement of plants in the ground.

Raunkiaer plant life forms: A system for categorizing plants using life form categories, particularly related to locations of perennating buds, devised by Christen C. Raunkiaer (1934).

Realized niche: The subset of a fundamental niche remaining after competitive exclusion (Hutchinson 1957). The niche is separate from but can be mapped onto the physical space where an organism lives.

Rehabilitation: Reestablishment of part of the productivity, structure, function, and processes of the original ecosystem (IUCN 1998a).

Reinforcement: See *Augmentation.*

Reintroduction: The release of individuals into a formerly occupied area after the native population has been lost or become extinct. Also known as reestablishment (IUCN 1998a).

Relative risk ratio: In statistics and mathematical epidemiology, relative risk (RR) is the risk of an event (or of developing a disease) relative to exposure. Relative risk is a ratio of the probability of the event occurring in the exposed group versus a group that was not exposed.

Resilience: The ability of an ecosystem to regain structural and functional attributes that have suffered harm from stress or disturbance (SER 2002).

Resistance: An ecosystem's ability to maintain its structural and functional attributes in the face of stress and disturbances (SER 2002).

Restocking: See *Augmentation.*

Restoration: The process of assisting the recovery of an ecosystem that has been degraded, damaged, or destroyed (SER 2002).

SER: Society for Ecological Restoration, an organization providing a source for expertise on restoration science, practice, and policy.

Stability: The ability of an ecosystem to maintain its given trajectory despite stress; it denotes dynamic equilibrium rather than stasis. Stability is achieved in part on the basis of an ecosystem's capacity for resistance and resilience (SER 2002).

Translocation: The deliberate and mediated movement of wild individuals or populations from one part of their range to another (IUCN 1998a).

Transplanting: See *Outplanting.*

USFWS: The US Fish and Wildlife Service, a federal agency charged with working with others to conserve, protect, and improve fish, wildlife, and plants and their habitats for the continuing benefit of the American people.

Vital rate: The rate of change in factors such as fecundity, growth, and survivorship in a population. Even when population numbers are stable, there may be changes in the vital rates.

Abbott, I., and N. Burrows, eds. 2003. *Fire in Ecosystems of South-West Western Australia: Impacts and Management*. Leiden, The Netherlands: Backhuys.

Ackerly, D. D. 2003. Community assembly, niche conservatism, and adaptive evolution in changing environments. *International Journal of Plant Science* 164:S163–84.

Adamec, L. 2005. Ten years after the introduction of *Aldrovanda vesiculosa* to the Czech Republic. *Acta Botanica Gallica* 152:239–45.

Adamec, L., and J. Lev. 1999. The introduction of the aquatic carnivorous plant *Aldrovanda vesiculosa* to new potential sites in the Czech Republic: A five-year investigation. *Folia Geobotanica* 34:299–305.

Adesemoye, A. O., H. A. Torbert, and J. W. Kloepper. 2008. Enhanced plant nutrient use efficiency with PGPR and AMF in an integrated nutrient management system. *Canadian Journal of Microbiology* 54:876–86.

Akasaka, M., and S. Tsuyuzaki. 2005. Tree seedling performance in microhabitats along an elevational gradient on Mount Koma, Japan. *Journal of Vegetation Science* 16:647–54.

Akeroyd, J., and P. Wyse Jackson. 1995. *A Handbook for Botanic Gardens on the Reintroduction of Plants to the Wild*. Richmond, Surrey, UK: Botanic Gardens Conservation International.

Albrecht, M. A., and K. A. McCue. 2010. Changes in demographic processes over long time scales reveal the challenge of restoring an endangered plant. *Restoration Ecology* 18:235–43.

Allen, E. B., and M. F. Allen. 1986. Water relations of xeric grasses in the field: Interactions of mycorrhizae and competition. *New Phytologist* 104:559–71.

Allen, M. F. 1991. *The Ecology of Mycorrhizae*. Cambridge, England: Cambridge University Press.

Alley, H., and J. M. Affolter. 2004. Experimental comparison of reintroduction methods for the endangered *Echinacea laevigata* (Boynton and Beadle) Blake. *Natural Areas Journal* 24:345–50.

Alley, H., J. M. Affolter, and J. F. Ceska. 2008. Recovery of smooth cornflower in the Chattahoochee National Forest, Georgia, USA. In *Global Re-introduction Perspectives: Reintroduction Case-Studies from around the World*, edited by P. S. Soorae, 244–48. Abu Dhabi, UAE: IUCN/SSC Re-introduction Specialist Group.

Alvarez-Aquino, C., G. Williams-Linera, and A. C. Newton. 2004. Experimental native tree seedling establishment for the restoration of a Mexican cloud forest. *Restoration Ecology* 12:412–18.

Alward, R. D., J. K. Detling, and D. G. Milchunas. 1999. Grassland vegetation changes and nocturnal global warming. *Science* 283:229–31.

Anand, A., C. S. Rao, P. Eganathan, N. A. Kumar, and M. S. Swaminathan. 2004. Saving an endemic and endangered taxon: *Syzygium travancoricum* gamble (Myrtacae)—A case study focussing on its genetic diversity, and reintroduction. *Physiology and Molecular Biology of Plants* 10:233–42.

Anonymous. 1931. Establishing pines: Preliminary observations on the effect of soil inoculation. *Rhodesian Agricultural Journal* 28:185–87.

Antonelli, A., J. A. A. Nylander, C. Persson, and I. Sanmartin. 2009. Tracing the impact of the Andean uplift on Neotropical plant evolution. *Proceedings of the National Academy of Sciences of the United States of America* 106:9749–54.

Apostolo, N. M., C. B. Brutti, and B. E. Llorente. 2005. Leaf anatomy of *Cynara scolymus* L. in successive micropropagation stages. *In Vitro Cellular and Developmental Biology: Plant* 41:307–13.

Aracama, C. V., M. E. Kane, S. B. Wilson, and N. L. Philman. 2008. Comparative growth, morphology, and anatomy of easy- and difficult-to-acclimatize sea oats (*Uniola paniculata*) genotypes during in vitro culture and ex vitro acclimatization. *Journal of the American Society for Horticultural Science* 133:830–43.

Araujo, M., and A. Guisan. 2006. Five (or so) challenges for species distribution modeling. *Journal of Biogeography* 33:1677–88.

Arbor Day Foundation. *Arborday.org Hardiness Zones*. Accessed November 1, 2010, http://www.arborday.org/media/map_change.cfm.

Armstrong, D. P., and P. J. Seddon. 2007. Directions in reintroduction biology. *Trends in Ecology & Evolution* 23:20–25.

Arnold, C., A. Schnitzler, A. Douard, R. Peter, and F. Gillet. 2005. Is there a future for wild grapevine (*Vitis vinifera* subsp. *silvestris*) in the Rhine Valley? *Biodiversity and Conservation* 14:1507–23.

Asai, T. 1944. Über die Mykorrhizenbildung der leguminosen Pflanzen. *Japanese Journal of Botany* 13:463–85.

Athens, J. S. 2009. *Rattus exulans* and the catastrophic disappearance of Hawai'i's native lowland forest. *Biological Invasions* 11:1489–1501.

Athens, J. S., H. D. Tuggle, J. V. Ward, and D. J. Welch. 2002. Avifaunal extinctions, vegetation change, and Polynesian impacts in prehistoric Hawai'i. *Archaeology in Oceania* 37:57–78.

Atkin, O. K., and M. G. Tjoelker. 2003. Thermal acclimation and the dynamic response of plant respiration to temperature. *Trends in Plant Science* 8(7):343–51.

Augé, R. 2001. Water relations, drought and VA mycorrhizal symbiosis. *Mycorrhiza* 11:3–42.

Augustine, D. J., and L. E. Frelich. 1998. Effects of white-tailed deer on populations of an understory forb in fragmented deciduous forests. *Conservation Biology* 12:995–1004.

Australian Network for Plant Conservation Translocation Working Group. 1997. *Guidelines for the Translocation of Threatened Plants in Australia*. Canberra: Australian Network for Plant Conservation.

Baggs, J. E., and J. Maschinski. 2000. From the greenhouse to the field: Cultivation requirements of Arizona cliffrose. In *Southwestern Rare and Endangered Plants*, edited by J. Maschinski and L. Holter, 176–85. Flagstaff, AZ: US Department of Agriculture, Forest Service, Rocky Mountain Research Station.

Bainbridge, D. A. 2007. *A Guide for Desert and Dryland Restoration: New Hope for Arid Lands*. Washington, DC: Island Press.

Baraloto, C., and D. E. Goldberg. 2004. Microhabitat associations and seedling bank dynamics in a neotropical forest. *Oecologia* 141:701–12.

Barlow, C. 2000. *The Ghosts of Evolution*. New York: Basic Books.

Barlow, C., and P. S. Martin. 2005. Bring *Torreya taxifolia* north now. *Wild Earth* Fall/Winter:52–56.

Barrett, S. C. H., and J. R. Kohn. 1991. Genetic and evolutionary consequences of small population size in plants: Implications for conservation. In *Genetics and Conservation of Rare Plants*, edited by D. A. Falk and K. E. Holsinger, 3–30. New York: Oxford University Press.

Barroetavena, C., S. D. Gisler, D. L. Luoma, and R. J. Meinke. 1998. Mycorrhizal status of the endangered species *Astragalus applegatei* Peck as determined from a soil bioassay. *Mycorrhiza* 8:117–19.

Bartel, R. A., and J. O. Sexton. 2009. Monitoring habitat dynamics for rare and endangered species using satellite images and niche-based models. *Ecography* 32:888–96.

Bartlein, P. J., C. Whitlock, and S. L. Shafer. 1997. Future climate in the Yellowstone National Park region and its potential impact on vegetation. *Conservation Biology* 11:782–92.

Baskin, C. C., and J. M. Baskin. 1998. *Seeds: Ecology, Biogeography, and Evolution of Dormancy and Germination*. San Diego, CA: Academic Press.

Bass, D. A., N. D. Crossman, S. L. Lawrie, and M. R. Lethbridge. 2006. The importance of population growth, seed dispersal, and habitat suitability in determining plant invasiveness. *Euphytica* 148:97–109.

Batagin, K. D., C. V. de Almeida, F. A. Ossamu Tanaka, and M. de Almeida. 2009. Morphological alterations in leaves of micropropagated pineapple plants cv IAC "Gomo-de-mel" acclimatized in different conditions of luminosity. *Acta Botanica Brasilica* 23:85–92.

Batllori, E., J. J. Camarero, J. M. Ninot, and E. Gutiérrez. 2009. Seedling recruitment, survival and facilitation in alpine *Pinus uncinata* tree line ecotones: Implications and potential responses to climate warming. *Global Ecology and Biogeography* 18:460–72.

Batty, A. L., M. C. Brundrett, K. W. Dixon, and K. Sivasithamparam. 2006a. In situ symbiotic seed germination and propagation of terrestrial orchid seedlings for establishment at field sites. *Australian Journal of Botany* 54:375–81.

Batty, A. L., M. C. Brundrett, K. W. Dixon, and K. Sivasithamparam. 2006b. New methods to improve symbiotic propagation of temperate terrestrial orchid seedlings from axenic culture to soil. *Australian Journal of Botany* 54:367–74.

BCCR. 2005. IPCC DDC AR4 BCCR-BCM2.0 SRESB1 run1. World Data Center for Climate. CERA-DB "BCCR_BCM2.0_SRESB1_1" http://cera-www.dkrz.de/WDCC /ui/Compact.jsp?acronym=BCCR_BCM2.0_SRESB1_1.

BCCR. 2006a. IPCC DDC AR4 BCCR_BCM2.0 1PCTTO2X run1. World Data Center for Climate. CERA-DB "BCCR_BCM2.0_1PCTTO2X_1" http://cera-www.dkrz.de /WDCC/ui/Compact.jsp?acronym=BCCR_BCM2.0_1PCTTO2X_1.

BCCR. 2006b. IPCC DDC AR4 BCCR_BCM2.0 SRESA1B run1. World Data Center for Climate. CERA-DB "BCCR_BCM2.0_SRESA1B_1" http://cera-www.dkrz.de /WDCC/ui/Compact.jsp?acronym=BCCR_BCM2.0_SRESA1B_1.

Beardmore, J., and G. F. Pegg. 1981. A technique for the establishment of mycorrhizal infection in orchid tissue grown in aseptic culture. *New Phytologist* 87:527–35.

Beckage, B., W. J. Patt, M. G. Slocum, and B. Panko. 2003. Influence of the El Niño southern oscillation on fire regimes in the Florida Everglades. *Ecology* 84:3124–30.

Becker, U., G. Colling, P. Dostal, A. Jakobsson, and D. Matthies. 2006. Local adaptation in the monocarpic perennial *Carlina vulgaris* at different spatial scales across Europe. *Oecologia* 150:506–18.

Beecroft, R. C., C. J. Cadbury, and J. O. Mountford. 2007. Water germander *Teucrium scordium* L. in Cambridgeshire: Back from the brink of extinction. *Watsonia* 26:303–16.

Begon, M., J. L. Harper, and C. R. Townsend. 1990. *Ecology: Individuals, Populations and Communities*. 2nd ed. Boston: Blackwell Scientific Publications.

Bell, S., M. Marzano, J. Cent, H. Koberiska, D. Podjed, D. Vandzinskaite, H. Reinert, A. Armaitiene, M. Grodzinska-Jurczak, and R. Mursic. 2008. What counts? Volunteers and their organizations in the recording and monitoring of biodiversity. *Biodiversity Conservation* 17:3443–54.

Bell, T. J., M. L. Bowles, and A. K. McEachern. 2003. Projecting the success of plant population restoration with viability analysis. In *Population Viability in Plants: Conservation, Management, and Modeling of Rare Plants*, edited by C. A. Brigham and M. W. Schwartz, 313–48. Berlin: Springer Verlag.

Bever, J. D. 1994. Feedback between plants and their soil communities in an old-field community. *Ecology* 75(7):1965–77.

Beyra Matos, A. 1998. Las leguminosas (Fabaceae) de Cuba, II. Tribus Crotalarieae, Aeschynoimeneae, Millettieae y Robinieae. Institute Botanica. *Collectanea Botanica* 24:263.

Bond, W. J., and J. E. Keeley. 2005. Fire as a global "herbivore": The ecology and evolution of flammable ecosystems. *Trends in Ecology & Evolution* 20:387–94.

Bottin, L., S. Le Cadre, A. Quilichini, P. Bardin, J. Moret, and N. Machon. 2007. Re-establishment trials in endangered plants: A review and the example of *Arenaria grandiflora*, a species on the brink of extinction in the Parisian region (France). *Ecoscience* 14:410–19.

Bouzat, J. L. 2010. Conservation genetics of population bottlenecks: The role of chance, selection, and history. *Conservation Genetics* 11:463–78.

Bowen, G. D., and C. Theodorou. 1979. Interactions between bacteria and ectomycorrhizal fungi. *Soil Biology and Biochemistry* 11:119–26.

Bowles, M., K. R. Bachtell, M. M. DeMauro, L. G. Sykora, and C. R. Bautista. 1988. Propagation techniques used in establishing a greenhouse population of *Astragalus tennesseensis* Gray. *Natural Areas Journal* 8:122.

Bowles, M., R. Flakne, K. McEachern, and N. Pavlovic. 1993. Recovery planning and reintroduction of the federally threatened pitcher's thistle (*Cirsium pitcheri*) in Illinois. *Natural Areas Journal* 13(3):164–76.

Bowles, M., and J. McBride. 1996. Pitcher's thistle (*Cirsium pitcheri*) reintroduction. In *Restoring Diversity: Strategies for Reintroduction of Endangered Plants*, edited by D. A. Falk, C. I. Millar, and M. Olwell, 423–31. Covelo, CA: Island Press.

Bowles, M., J. McBride, and R. F. Betz. 1998. Management and restoration ecology of the federal threatened mead's milkweed, *Asclepias meadii* (Asclepiadaceae). *Annals of the Missouri Botanical Garden* 85:110–25.

Bowles, M., J. McBride, and T. J. Bell. 2001. Restoration of the federally threatened Mead's milkweed (*Asclepias meadii*). *Ecological Restoration* 19:235–41.

Braham, R., C. Murray, and M. Boyer. 2006. Mitigating impacts to Michaux's sumac (*Rhus michauxii* Sarg.): A case study of transplanting an endangered shrub. *Castanea* 71(4):265–71.

Brauner, S. 1988. *Malheur wirelettuce* (Stephanomeria malheurensis) *Biology and Interactions with Cheatgrass: 1987 Study Results and Recommendations for a Recovery Plan.* Burns District, OR: Bureau of Land Management.

Brenes-Arguedas, T., M. Ríos, G. Rivas-Torres, C. Blundo, P. D. Coley, and T. A. Kursar. 2008. The effect of soil on the growth performance of tropical species with contrasting distributions. *Oikos* 117:1453–60.

Brewbaker, J. L. 1967. The distribution and phylogenetic significance of binucleate and trinucleate pollen grains in the angiosperms. *American Journal of Botany* 54:1069–83.

Brigham, C. A., and D. M. Thomson. 2003. Approaches to modeling population viability in plants: An overview. In *Population Viability in Plants: Conservation, Management, and Modeling of Rare Plants*, edited by C. A. Brigham and M. W. Schwartz, 145–72. Berlin: Springer Verlag.

Broadhurst, L. M., A. Lowe, D. J. Coates, S. A. Cunningham, M. McDonald, P. A. Vesk, and C. Yates. 2008. Seed supply for broadscale restoration: Maximizing evolutionary potential. *Evolutionary Applications* 1:587–97.

Brook, B. W., J. J. O'Grady, A. P. Chapman, M. A. Burgman, H. R. Akçakaya, and R. Franklin. 2000. Predictive accuracy of population viability analysis in conservation biology. *Nature* 404:385–87.

Brown, P. M. 2002. *Wild Orchids of Florida: With Reference to the Atlantic and Gulf Coastal Plains*. Gainesville: University of Florida Press.

Brown, A. H. D., and D. R. Marshall. 1995. A basic sampling strategy: Theory and practice. In *Collecting Plant Genetic Diversity: Technical Guidelines*, edited by L. Guarino, V. R. Rao, and R. Ried, 75–91. Wallingford, UK: CAB International.

Brown, B. J., R. J. Mitchell, and S. A. Graham. 2002. Competition for pollination between an invasive species (purple loosestrife) and a native congener. *Ecology* 83(8):2328–36.

Brown, J. H., and A. Kodric-Brown. 1977. Turnover rates in insular biogeography: Effects of immigration on extinction. *Ecology* 58:445–49.

Bruegmann, M. M., V. Caraway, P. Dunn, S. M. Gon III, R. Hobdy, J. D. Jacobi, K. Kawelo, et al. 2008. *An Integrated Plan for the Conservation of Hawai'i's Unique Plants and Their Ecosystems, Part 1: Hawaiian Plant Conservation Strategy*. Honolulu, HI: US Fish and Wildlife Service.

Brumback, W. E., and C. W. Fyler. 1996. Small whorled pogonia (*Isotria medeoloides*) transplant project. In *Restoring Diversity: Strategies for Reintroduction of Endangered Plants*, edited by D. A. Falk, C. I. Millar, and M. Olwell, 433–43. Washington, DC: Island Press.

Brumback, W. E., D. M. Weihrauch, and K. D. Kimball. 2004. Propagation and transplanting of an endangered alpine species Robbins' cinquefoil, *Potentilla robbinsiana* (Rosaceae). *Native Plants Journal* 5:91–97.

Brundrett, M., N. Bougher, B. Dell, T. Grove, and N. Malajczuk. 1996. *Working with Mycorrhizas in Forestry and Agriculture*. Canberra, A.C.T.: Australian Centre for International Agricultural Research.

Brundrett, M. C., A. Scade, A. L. Batty, K. W. Dixon, and K. Sivasithamparam. 2003. Development of in situ and ex situ seed baiting techniques to detect mycorrhizal fungi from terrestrial orchid habitats. *Mycological Research* 107(10):1210–20.

Brutti, C. B., E. J. Rubiio, B. E. Llorente, and N. M. Apostolo. 2002. Artichoke leaf morphology and surface features in different micropropagation stages. *Biologia Plantarum* 45:197–204.

Buckley, Y. M., S. Ramula, S. P. Blomberg, J. H. Burns, E. E. Crone, J. Ehrlén, T. M. Knight, J. Pichancourt, H. Quested, and G. M. Wardle. 2010. Causes and consequences of variation in plant population growth rate: A synthesis of matrix population models in a phylogenetic context. *Ecology Letters* 13(9):1182–97.

Bunn, E., and B. Dixon. 2008. Re-introduction of the endangered Bancroft's symonanthus in Western Australia. In *Global Re-introduction Perspectives: Re-introduction Case-Studies from around the World*, edited by P. S. Soorae, 225–28. Abu Dhabi, UAE: IUCN/SSC Re-introduction Specialist Group.

Butcher, D., and S. A. Marlowe. 1989. Asymbiotic germination of epiphytic and terrestrial orchids. In *Modern Methods in Orchid Conservation: The Role of Physiology, Ecology and Management*, edited by H. W. Pritchard, 31–38. Cambridge, England: Cambridge University Press.

Cairney, J. W. G. 1999. Intraspecific physiological variation: Implications for understanding functional diversity in ectomycorrhizal fungi. *Mycorrhiza* 9:125–35.

Callaway, R. M., and E. T. Aschehoug. 2000. Invasive plants versus their new and old neighbors: A mechanism for exotic invasion. *Science* 290:521–23.

Camacho, A. E. 2010. Assisted migration: Redefining nature and natural resource law under climate change. *Legal Studies Research Paper Series* 2009:37.

Carlquist, S. 1970. *Hawaii: A Natural History: Geology, Climate, Native Flora and Fauna above the Shoreline*. Garden City, NY: Natural History Press.

Case, T. J., and M. E. Gilpin. 1974. Interference competition and niche theory. *Proceedings of the National Academy of Sciences (USA)* 71:3073–77.

Castro, S. A., J. A. Figueroa, M. Muñoz-Schick, and F. M. Jaksic. 2005. Minimum resident time, biogeographical origin, and life cycle as determinants of the geographical extent of naturalized plants in continental Chile. *Diversity and Distributions* 11:183–91.

Caswell, H. 2001. *Matrix Population Models: Construction, Analysis and Interpretation*, 2nd ed. Sunderland, MA: Sinauer.

Caughley, G. 1994. Directions in conservation biology. *Journal of Animal Ecology* 63:215–44.

Cavallazzi, J. R. P., O. K. Filho, S. L. Stuermer, P. T. Rygiewicz, and M. M. de Mendonca. 2007. Screening and selecting arbuscular mycorrhizal fungi for inoculating micropropagated apple rootstocks in acid soils. *Plant Cell Tissue and Organ Culture* 90:117–29.

Center for Plant Conservation. 1991. Genetic sampling guidelines for conservation collections of endangered plants. In *Genetics and Conservation of Rare Plants*, edited by D. A. Falk and K. E. Holsinger, 225–38. New York: Oxford University Press.

Center for Plant Conservation. 1996. Guidelines for developing a rare plant reintroduction plan. In *Restoring Diversity*, edited by D. A. Falk, C. I. Millar, and M. Olwell, 453–90. Washington, DC: Island Press.

Center for Plant Conservation. 2009. *CPC International Reintroduction Registry*. Accessed September 1, 2009, http://www.centerforplantconservation.org/reintroduction /MN _ReintroductionEntrance.asp.

Charlesworth, D., and B. Charlesworth. 1987. Inbreeding depression and its evolutionary consequences. *Annual Review of Ecological Systems* 18:237–68.

Chen, B. L., and H. P. Nooteboom. 1993. Notes on Magnoliaceae: III: The Magnoliaceae of China. *Annals of the Missouri Botanical Garden* 80(4):999–1104.

Chen, F., Z. Xie, G. Xiong, Y. Liu, and H. Yang. 2005. Reintroduction and population reconstruction of an endangered plant *Myricaria laxiflora* in the Three Gorges Reservoir area, China. *Acta Ecologica Sinica* 25:1811–17.

Cieslak, E., G. Korbecka, and M. Ronikier. 2007. Genetic structure of the critically endangered endemic *Cochlearia polonica* (Brassicaceae): Efficiency of the last-chance transplantation. *Botanical Journal of the Linnean Society* 155:527–32.

Clausen, J., D. D. Keck, and W. M. Hiesey. 1940. *Experimental Studies on the Nature of Species. I. The Effect of Varied Environments on Western North American Plants.* Carnegie Institute of Washington Publication 520.

Clayton, S., and G. Myers. 2009. *Conservation Psychology: Understanding and Promoting Human Care for Nature.* Chichester, West Sussex, UK: Wiley-Blackwell.

Coates, D. J., and V. L. Hamley. 1999. Genetic divergence and the mating system in the endangered and geographically restricted species, *Lambertia orbifolia* Gardner (Proteaceae). *Heredity* 83:418–27.

Coates, D. J., J. Sampson, and C. Yates. 2007. Plant mating systems and assessing population persistence in fragmented landscapes. *Australian Journal of Botany* 55:239–49.

Cochran, M. E., and S. Ellner. 1992. Simple methods for calculating age-specific life history parameters from stage-structured models. *Ecological Monographs* 62:345–64.

Cochrane, A. 2004. Western Australia's ex situ program for threatened species: A model integrated strategy for conservation. In *Ex Situ Plant Conservation: Supporting Species Survival in the Wild*, edited by E. O. Guerrant Jr., K. Havens, and M. Maunder, 40–66. Washington, DC: Island Press.

Cochrane, A., and S. Barrett. 2009. The role of seed orchards in plant conservation. *Australian Plant Conservation: Journal of the Australian Network for Plant Conservation* 17:10–12.

Cochrane, A., L. Monks, and S. Juszkiewicz. 2000. Translocation trials for four threatened Western Australian plant taxa. *Danthonia* 9:7–9.

Colas, B., F. Kirchner, M. Riba, I. Olivieri, A. Mignot, E. Imbert, C. Beltrame, D. Carbonell, and H. Freville. 2008. Restoration demography: A 10-year demographic comparison between introduced and natural populations of endemic *Centaurea corymbosa* (Asteraceae). *Journal of Applied Ecology* 45:1468–76.

Collaboration for Environmental Evidence. 2009. Statistical Methods Group. Accessed February 1, 2011, http://www.environmentalevidence.org.

Collier. 2005a. IPCC DDC AR4 CSIRO-Mk3.0 1PCTTO2X run1. World Data Center for Climate. CERA-DB CSIRO_Mk3.0_1PCTTO2X_1 http://cera-www.dkrz.de /WDCC/ui/Compact.jsp?acronym=CSIRO_Mk3.0_1PCTTO2X_1.

Collier. 2005b. IPCC DDC AR4 CSIRO-Mk3.0 SRESA1B run1. World Data Center for Climate. CERA-DB "CSIRO_Mk3.0_SRESA1B_1" http://cera-www.dkrz.de/WDCC /ui/Compact.jsp?acronym=CSIRO_Mk3.0_SRESA1B_1.

Collier. 2005c. IPCC DDC AR4 CSIRO-Mk3.0 SRESB1 run1. World Data Center for Climate. CERA-DB "CSIRO_Mk3.0_SRESB1_1" http://cera-www.dkrz.de/WDCC /ui/Compact.jsp?acronym=CSIRO_Mk3.0_SRESB1_1.

Collinge, S. K., and C. Ray. 2009. Transient patterns in the assembly of vernal pool plant communities. *Ecology* 90:3313–23.

Collins, S. L., and R. E. Good. 1987. The seedling regeneration niche: Habitat structure of tree seedlings in an oak–pine forest. *Oikos* 48:89–98.

Colwell, R. K., and T. F. Rangel. 2009. Hutchinson's duality: The once and future niche. *Proceedings of the National Academy of Science* 106(Suppl 2):19651–58.

Cooke, B. 2008. Community-based monitoring: Exploring the involvement of Friends Groups in a terrestrial park management context. *Australasian Plant Conservation* 17:10–12.

Cordell, S., M. McClellan, Y. Carter, Y. Yarber, and L. J. Hadway. 2008. Towards restoration of Hawaiian tropical dry forests: The Kaupulehu Outplanting Programme. *Pacific Conservation Biology* 14:279–84.

Cornwell, W. K., and P. J. Grubb. 2003. Regional and local patterns in plant species richness with respect to resource availability. *Oikos* 100:417–28.

Coumbe, R., and S. Dopson. 1999. *Indigenous Plant Translocations in New Zealand: A Summary 1987–1999*. Wellington: New Zealand Department of Conservation.

Cowie, R. H. 1997. Catalog and bibliography of the nonindigenous nonmarine snails and slugs of the Hawaiian Islands. *Bishop Museum Occasional Papers* 50:1–66.

Cox, P. A., and T. Elmqvist. 2000. Pollinator extinction in the Pacific Islands. *Conservation Biology* 14:1237–39.

Cramer, J. R. 2008. Reviving the connection between children and nature through service learning restoration partnerships. *Native Plants* 9:278–86.

Crespi, M., and S. Gálvez. 2000. Molecular mechanisms in root nodule development. *Journal of Plant Growth Regulation* 19:155–66.

Crooks, J. A. 2005. Lag times and exotic species: The ecology and management of biological invasions in slow motion. *Ecoscience* 12:316–29.

Crutsinger, G. M., M. D. Collins, J. A. Fordyce, Z. Gompert, C. C. Nice, and N. J. Sanders. 2006. Plant genotypic diversity predicts community structure and governs an ecosystem process. *Science* 313:966–68.

Cully, A. 1996. Knowlton's cactus (*Pediocactus knowltonii*) reintroduction. In *Restoring Diversity: Strategies for Reintroduction of Endangered Plants*, edited by D. A. Falk, C. I. Millar, and M. Olwell, 403–10. Covelo, CA: Island Press.

Curnow, C., M. Griffiths, H. Mills, K. Sawyer, and P. Lewis. 2008. A toolkit for conservation on private land: Winning hearts and minds. *Australian Plant Conservation* 17:16–17.

Currin, R., and R. J. Meinke. 2008. *Malheur Wirelettuce* (Stephanomeria malheurensis) *Reintroduction and Seed Bulking: 2008 Recovery Efforts*. Report prepared for US Fish and Wildlife Service, Region 1, Portland, Oregon. Salem: Oregon Department of Agriculture.

Currin, R., R. J. Meinke, and A. Raven. 2007. *Malheur Wirelettuce* (Stephanomeria malheurensis) *Recovery Efforts: Reintroduction and Seed Bulking*. Prepared by Oregon Department of Agriculture and Berry Botanic Garden for US Fish and Wildlife Service, Region 1.

Daehler, C. C. 2001. Darwin's naturalization hypothesis revisited. *American Naturalist* 158:324–30.

Daehler, C. C., J. S. Denslow, S. Ansari, and H. Kuo. 2004. A risk assessment system for screening out invasive pest plants from Hawaii and other Pacific islands. *Conservation Biology* 18:360–68.

Dalrymple, S. E., and A. Broome. 2010. The importance of donor population identity and habitat type when creating new populations of small cow-wheat *Melampyrum sylvaticum* from seed in Perthshire, Scotland. *Conservation Evidence* 7:1–8.

Dalrymple, S. E., A. Broome, and P. Gallagher. 2008. Re-introduction of small cow-wheat into the Scottish Highlands, UK. In *Global Re-introduction Perspectives: Re-introduction Case-Studies from around the World*, edited by P. S. Soorae, 221–24. Abu Dhabi, UAE: IUCN/SSC Re-Introduction Specialist Group.

Dangremond, E. M., E. A. Pardini, and T. M. Knight. 2010. Apparent competition with an invasive plant hastens the extinction of an endangered lupine. *Ecology* 91:2261–71.

Danielsen, F., N. D. Burgess, and A. Balmford. 2005. Monitoring matters: Examining the potential of locally-based approaches. *Biodiversity Conservation* 14:2507–42.

Darwin, C. 1859. *The Origin of Species by Means of Natural Selection*. New York: Penguin.

Davidson, I., and C. Simkanin. 2008. Skeptical of assisted colonization. *Science* 322:1048–49.

Davis, A. S., R. D. Cousens, J. Hill, R. N. Mack, D. Simberloff, and S. Raghu. 2010. Screening bioenergy feedstock crops to mitigate invasion risk. *Frontiers in Ecology and the Environment* 8:533–39.

Davis, M. A. 2003. Biotic globalization: Does competition from introduced species threaten biodiversity? *BioScience* 53:481–89.

Davis, M. A., J. P. Grime, and K. Thompson. 2000. Fluctuating resources in plant communities: A general theory of invasibility. *Journal of Ecology* 88:528–34.

Davis, M. B. 1989. Lags in vegetation response to greenhouse warming. *Climate Change* 15:75–82.

Davis, M. B., and R. G. Shaw. 2001. Range shifts and adaptive responses to Quaternary climate change. *Science* 292:673–79.

Dehnen-Schmutz, K., J. Touza, C. Perrings, and M. Williamson. 2007a. A century of the ornamental plant trade and its impact on invasion success. *Diversity and Distribution* 13:527–34.

Dehnen-Schmutz, K., J. Touza, C. Perrings, and M. Williamson. 2007b. The horticultural trade and ornamental plant invasions in Britain. *Conservation Biology* 21(1):224–31.

Delcourt, H. R. 2002. *Forests in Peril*. Blacksburg, VA: McDonald & Woodward.

DeMauro, M. M. 1993. Relationship of breeding system to rarity in the lakeside daisy (*Hymenoxys acaulis* var. *glabra*). *Conservation Biology* 7:542–50.

DeMauro, M. M. 1994. Development and implementation of a recovery program for the federally threatened Lakeside daisy (*Hymenoxys acaulis* var. *glabra*). In *Restoration of Endangered Species: Conceptual Issues, Planning, and Implementation*, edited by M. L. Bowles and C. J. Whelan, 298–321. Cambridge, England: Cambridge University Press.

DerSimonian, R., and N. Laird. 1986. Meta-analysis in clinical trials. *Controlled Clinical Trials* 7:177–88.

Desprez-Loustau, M.-L., C. Robin, M. Buee, R. Courtecuisse, J. Garbaye, F. Suffert, I. Sache, and D. M. Rizzo. 2007. The fungal dimension of biological invasions. *Trends in Ecology & Evolution* 22:472–80.

Dickie, I. A., and R. G. FitzJohn. 2007. Using terminal restriction fragment length polymorphism (T-RFLP) to identify mycorrhizal fungi: A methods review. *Mycorrhiza* 17:259–70.

Dixon, B. 2004. The Corrigin grevillea (*Grevillea scapigera*): An update. *Australasian Plant Conservation* 13:14–15.

Dixon, B., and S. Krauss. 2001. Translocation of *Grevillea scapigera*: Is it working? *Danthonia* 10:2–3.

Dixon, B., and S. L. Krauss. 2008. Translocation of the Corrigin grevellia in south Western Australia. In *Global Re-introduction Perspectives: Re-introduction Case-Studies from around the World*, edited by P. S. Soorae, 229–34. Abu Dhabi, UAE: IUCN/SSC Reintroduction Specialist Group.

Dixon, K. W., S. P. Kell, R. L. Barrett, and P. J. Cribb, eds. 2003. *Orchid Conservation*. Borneo, Malaysia: Natural History Publications.

Dolan, R. W., D. L. Marr, and A. Schnabel. 2008. Capturing genetic variation during ecological restorations: An example from Kankakee Sands in Indiana. *Restoration Ecology* 16:386–96.

Dolcet-Sanjuan, R., E. Claveria, A. Camprubi, V. Estaun, and C. Calvet. 1996. Micropropagation of walnut trees (*Juglans regia* L.) and response to arbuscular mycorrhizal inoculation. *Agronomie* 16:639–45.

Dostálek, T., Z. Münzbergová, and I. Plačková. 2010. Genetic diversity and its effect on fitness in an endangered plant species, *Dracocephalum austriacum* L. *Conservation Genetics* 11:773–83.

Dougherty, D., and S. Reichard. 2004. Factors affecting the control of *Cytisus scoparius* and restoration of invaded sites. *Plant Protection Quarterly* 19:137–42.

Drake, D. R., and T. L. Hunt. 2009. Invasive rodents on islands: Integrating historical and contemporary ecology. *Biological Invasions* 11:1483–87.

Drayton, B., and R. B. Primack. 2000. Rates of success in the reintroduction by four methods of several perennial plant species in eastern Massachusetts. *Rhodora* 102:299–331.

Dumroese, R. K., T. Luna, and T. Landis, eds. 2009. *Nursery Manual for Native Plants: A Guide for Tribal Nurseries*. US Department of Agriculture Forest Service, Agriculture Handbook 730.

Duncan, R. P., and P. A. Williams. 2002. Darwin's naturalization hypothesis challenged. *Nature* 417:608–9.

Ecker, L. S. 1990. *Population Enhancement of a Rare Arizona Cactus*, Mamillaria thornberi *Orcutt (Cactaceae)*. Tempe: Arizona State University.

Eckert, C. G., S. Kalisz, M. A. Geber, R. Sargent, E. Elle, P. Cheptou, C. Goodwillie, et al. 2010. Plant mating systems in a changing world. *Trends in Ecology & Evolution* 25:35–43.

Eckstein, R. L., and A. Otte. 2005. Effects of cleistogamy and pollen source on seed production and offspring performance in three endangered violets. *Basic and Applied Ecology* 6:339–50.

Edmands, S. 2007. Between a rock and a hard place: Evaluating the relative risks of inbreeding and outbreeding for conservation and management. *Molecular Ecology* 16:463–75.

Ehleringer, J. R., S. L. Phillips, W. S. F. Schuster, and D. R. Sandquist. 1991. Differential utilization of summer rains by desert plants. *Oecologia* 88:430–34.

Ehrlén, J. 1995. Demography of the perennial herb *Lathyrus vernus*, II. Herbivory and population dynamics. *Journal of Ecology* 83:297–308.

Elith, J., and C. H. Graham. 2009. Do they? How do they? Why do they differ? On finding reasons for differing performances of species distribution models. *Ecography* 32(1):66–77.

Elith, J., C. Graham, R. Anderson, M. Dudik, S. Ferrier, A. Guisan, R. Hijmans, et al. 2006. Novel methods improve prediction of species' distributions from occurrence data. *Ecography* 29:129–51.

Elith, J., and J. R. Leathwick. 2009. Species distribution models: Ecological explanation and prediction across space and time. *Annual Review of Ecology, Evolution and Systematics* 40:677–97.

Ellstrand, N. C., and D. R. Elam. 1993. Population genetic consequences of small population size: Implications for plant conservation. *Annual Review of Ecology and Systematics* 24:217–42.

Elton, C. 1927. *Animal Ecology*. London: Sedgwick and Jackson.

Elton, C. S. 1958. *The Ecology of Invasions by Animals and Plants*. Chicago: University of Chicago Press.

Elzinga, C. L., D. W. Salzer, and D. W. Willoughby. 1998. *Measuring and Monitoring Plant Populations*. Denver, CO: Bureau of Land Management.

ESRI. 2009. ERSI ArcMap 9.3. Redlands, CA: ESRI.

Estrada-Luna, A. A., and F. T. Davies Jr. 2003. Arbuscular mycorrhizal fungi influence water relations, gas exchange, abscisic acid and growth of micropropagated chile ancho pepper (*Capsicum annuum*) plantlets during acclimatization and post-acclimatization. *Journal of Plant Physiology* 160:1073–83.

Estrada-Luna, A. A., F. T. Davies Jr., and J. N. Egilla. 2000. Mycorrhizal fungi enhancement of growth and gas exchange of micropropagated guava plantlets (*Psidium guajava* L.) during ex vitro acclimatization and plant establishment. *Mycorrhiza* 10:1–8.

Etterson, J. R., and R. G. Shaw. 2001. Constraint to adaptive evolution in response to global warming. *Science* 294:151–54.

Evans, K., and M. R. Guariguata. 2008. *Participatory Monitoring in Tropical Forest Management: A Review of Tools, Concepts and Lessons Learned*. Bogor, Indonesia: Center for International Forestry Research.

Falk, D. A. 1990. Integrated strategies for conserving plant genetic diversity. *Annals of the Missouri Botanical Garden* 77:38–47.

Falk, D. A., and K. E. Holsinger. 1991. *Genetics and Conservation of Rare Plants*. New York: Oxford University Press.

Falk, D. A., E. E. Knapp, and E. O. Guerrant. 2001. *An Introduction to Restoration Genetics*. Plant Conservation Alliance, Bureau of Land Management, US Department of Interior, US Environmental Protection Authority. http://www.ser.org/pdf/SER -restoration-genetics.pdf.

Falk, D. A., C. I. Millar, and M. Olwell. 1996. *Restoring Diversity: Strategies for Reintroduction of Endangered Plants*. Washington, DC: Island Press.

Falk, D. A., C. M. Richards, A. M. Montalvo, and E. E. Knapp. 2006. Population and ecological genetics in restoration ecology. In *Foundations of Restoration Ecology*, edited by D. A. Falk, M. A. Palmer, and J. B. Zedler, 14–41. Washington, DC: Island Press.

Fant, J. B., R. M. Holmstrom, E. Sirkin, J. R. Etterson, and S. Masi. 2008. Genetic structure of threatened native populations and propagules used for restoration in a clonal species, American beachgrass (*Ammophila breviligulata* Fern.). *Restoration Ecology* 16:594–603.

Fay, M. F. 1992. Conservation of rare and endangered plants using in vitro methods. *In Vitro Cellular and Developmental Biology* 28:1–4.

Fazey, I., and J. Fischer. 2009. Assisted colonization is a techno-fix. *Trends in Ecology & Evolution* 24:475.

Fenster, C. B., and M. R. Dudash. 1994. Genetic considerations for plant population restoration and conservation. In *Restoration of Endangered Species*, edited by M. L. Bowles and C. J. Whelan, 43–62. Cambridge, England: Cambridge University Press.

Fernandez, O. A., and M. M. Caldwell. 1975. Phenology and dynamics of root growth of three cool semi-desert shrubs under field conditions. *Journal of Ecology* 63(2):703–14.

Ferreira, J., and S. Smith. 1987. Methods for increasing native populations of *Erysimum menziessii*. In *Conservation and Management of Rare and Endangered Plants*, edited by T. S. Elias and J. Nelson, 507–11. Sacramento: California Native Plant Society.

Fiedler, P. L., and R. D. Laven. 1996. Selecting reintroduction sites. In *Restoring Diversity: Strategies for Reintroduction of Endangered Plants*, edited by D. A. Falk, C. I. Millar, and M. Olwell, 157–170. Washington, DC: Island Press.

Figlar, R. B., and H. P. Nooteboom. 2004. Notes on Magnoliaceae 4. *Blumea* 49(1):87–100.

Fila, G., F. W. Badeck, S. Meyer, Z. Cerovic, and J. Ghashghaie. 2006. Relationships between leaf conductance to CO_2 diffusion and photosynthesis in micropropagated grapevine plants, before and after ex vitro acclimatization. *Journal of Experimental Botany* 57:2687–95.

Fischer, J., and D. B. Lindenmayer. 2000. An assessment of the published results of animal relocations. *Biological Conservation* 96:1–11.

Fischer, M., M. Hock, and M. Paschke. 2003. Low genetic variation reduces crosscompatibility and offspring fitness in populations of a narrow endemic plant with a self-incompatibility system. *Conservation Genetics* 4:325–36.

Fisher, A. 2002. *Radical Ecopsychology: Psychology in the Service of Life*. Albany, NY: SUNY Press.

Fisher, J. B., and K. Jayachandran. 2002. Arbuscular mycorrhizal fungi enhance seedling growth in two endangered plant species from South Florida. *International Journal of Plant Sciences* 163:559.

Fisher, J. B., and K. Jayachandran. 2006. *Effects of Arbuscular Mycorrhizal Fungi on Seedling Growth and Transplant Survival of Two Endangered Plants from South Florida.* Unpublished report.

FLEPPC. 2009. FLEPPC 2009 List of Invasive Plant Species, Fall 2009. Accessed November 1, 2010, http://www.fleppc.org/list/List-WW-F09-final.pdf.

Fordham, M. C., R. S. Harrison-Murray, L. Knight, and C. M. Clay. 2001. Decline in stomatal response to leaf water deficit in *Corylus maxima* cuttings. *Tree Physiology* 21:489–96.

Forsyth, D. M., and R. P. Duncan. 2001. Propagule size and the relative success of exotic ungulate and bird introductions to New Zealand. *American Naturalist* 157:583–95.

Foster-Smith, J., and S. M. Evans. 2003. The value of marine ecological data collected by volunteers. *Biological Conservation* 113:199–213.

Frank, A. B. 1894. Die Bedeutung det Mykorrhiza-pilze fur die gemeine Kiefer. *Forstwissenschaeftliches* 16:1852–90.

Frankham, R., J. D. Ballou, and D. A. Briscoe. 2002. *Introduction to Conservation Genetics.* Cambridge, England: Cambridge University Press.

Frey-Klett, P., J. Garbaye, and M. Tarkka. 2007. The mycorrhizal helper bacteria revisited. *New Phytologist* 176:22–36.

Friar, E. A., T. Ladoux, E. H. Roalson, and R. H. Robichaux. 2000. Microsatellite analysis of a population crash and bottleneck in the Mauna Kea silversword, *Argyroxiphium sandwicense* ssp. *sandwicense* (Asteraceae), and its implications for reintroduction. *Molecular Ecology* 9:2027–34.

Fried, J. S., J. K. Gilless, W. J. Riley, T. J. Moody, C. Simon de Blas, K. Hayhoe, M. Moritz, et al. 2008. Predicting the effect of climate change on wildfire behavior and initial attack success. *Climatic Change* 87(Suppl 1):S251–64.

Fuller, R., K. Irvine, P. Devine-Wright, P. Warren, and K. Gaston. 2007. Psychological benefits of greenspace increase with biodiversity. *Biology Letters* 3:390–94.

Gangaprasad, A. N., W. S. Decruse, S. Seeni, and S. Menon. 1999. Micropropagation and restoration of the endangered Malabar daffodil orchid *Ipsea malabarica*. *Lindleyana* 14:38–46.

Gangaprasad, A., S. W. Decruse, S. Seeni, and G. M. Nair. 2005. Micropropagation and ecorestoration of *Decalepis arayalpathra* (Joseph & Chandra.) Venter: An endemic and endangered ethnomedicinal plant of Western Ghats. *Indian Journal of Biotechnology* 4:265–70.

Gange, A., and H. West. 1993. Interactions between foliar-feeding insects and VA mycorrhizas. *Bulletin of the British Ecological Society* 24:72–76.

Gann, G. D., K. A. Bradley, and S. W. Woodmansee. 2002. *Rare Plants of South Florida: Their History, Conservation, and Restoration.* Miami, FL: The Institute for Regional Conservation.

Garbaye, J., and R. Duponnois. 1992. Specificity and function of mycorrhization helper bacteria (MHB) associated with the *Pseudotsuga menziesii–Laccaria laccata* symbiosis. *Symbiosis* 14:335–44.

García, M. B., and J. Ehrlén 2002. Reproductive effort and herbivory timing in a perennial herb: Fitness components at the individual and population levels. *American Journal of Botany* 89:1295–1302.

Gardner, R. O., and J. W. Early. 1996. The naturalisation of banyan figs (*Ficus* spp. Moraceae) and their pollinating wasps (Hymenoptera: Agaonidae) in New Zealand. *New Zealand Journal of Botany* 34:103–10.

Garfin, G., M. A. Crimmins, and K. Jacobs. 2007. Drought, climate variability and implication for water supply and management. In *Arizona Water Policy: Management Innovation in an Urbanizing, Arid Region. Resources for the Future*, edited by B. G. Colby and K. Jacobs, 61–78. Washington, DC: RFF Press.

Gaston, K. J. 1994. *Rarity*. London: Springer.

Gaston, K. J., and W. E. Kunin. 1997. Rare–common differences: An overview. In *The Biology of Rarity*, edited by W. E. Kunin and K. J. Gaston, 12–29. London: Chapman and Hall.

Gentry, H. S. 1982. *Agaves of Continental North America*. Tucson: University of Arizona.

George, S., W. J. Snape, and M. Senatore. 1998. *State Endangered Species Acts: Past, Present and Future*. Washington, DC: Defenders of Wildlife.

Gianinazzi, S., and M. Vosatka. 2004. Inoculum of arbuscular mycorrhizal fungi for production systems, science meets business. *Canadian Journal of Botany* 82:1264–71.

Gilbert, N. 2010. Threats to the world's plants assessed. *Nature*. doi:10.1038/news.2010.499.

Gilfedder, L., J. B. Kirkpatrick, and S. Wells. 1997. The endangered Tunbridge buttercup (*Ranunculus prasinus*): Ecology, conservation status and introduction to the Township Lagoon Nature Reserve, Tasmania. *Australian Journal of Ecology* 22:347–51.

Gillespie, I. G., and E. B. Allen. 2008. Restoring the rare forb *Erodium macrophyllum* to exotic grassland in southern California. *Restoration Ecology* 5:65–72.

Gilly, C., R. Rohr, and A. Chamel. 1997. Ultrastructure and radiolabelling of leaf cuticles from ivy (*Hedera helix* L.) plants in vitro and during ex vitro acclimatization. *Annals of Botany* 80:139–45.

Giorgi, F., and R. Francisco. 2000. Uncertainties in regional climate change prediction: A regional analysis of ensemble simulations with the HADCM2 coupled AOGCM. *Climate Dynamics* 16:69–182.

Glitzenstein, J. S., D. R. Streng, D. D. Wade, and J. Brubaker. 2001. Starting new populations of longleaf pine ground-layer plants in the outer coastal plain of South Carolina, USA. *Natural Areas Journal* 21:89–110.

Godefroid, S., C. Piazz, G. Rossi, S. Buord, A.-D. Stevens, R. Aguraiuja, C. Cowell, et al. 2011. How successful are plant species reintroductions? *Biological Conservation* 144:672–82.

Gomez-Aparicio, L., J. M. Gomez, and R. Zamora. 2005. Microhabitats shift rank in suitability for seedling establishment depending on habitat type and climate. *Journal of Ecology* 93:1194–1202.

Gomez-Aparicio, L., R. Zamora, J. M. Gomez, J. A. Hodar, J. Castro, and E. Baraza. 2004. Applying plant facilitation to forest restoration: A meta-analysis of the use of shrubs as nurse plants. *Ecological Applications* 14:1128–38.

Gon, S. M. III, K. Poiani, T. Menard, S. Tom, M. Fox, and M. White. 2006. *An Ecoregional Assessment of Biodiversity Conservation for the Hawaiian High Islands.* A web-mediated publication by The Nature Conservancy of Hawai'i. Accessed November 18, 2010, http://www.hawaiiecoregionplan.info.

Goodman, J. L., S. J. Wright, and J. Maschinski. 2007. *Assessing Impacts on Populations and Taking Conservation Steps for the Endangered Key Tree Cactus* (Pilosocereus robinii): *Sept 2007.* Vero Beach: US Fish and Wildlife Service, South Florida Ecological Services Office.

Gordon, D. R. 1994. Translocation of species into conservation areas: A key for natural resource managers. *Natural Areas Journal* 14:31–37.

Gordon, D. R. 1996a. Apalachicola rosemary (*Conradina glabra*) reintroduction. In *Restoring Diversity: Strategies for Reintroduction of Endangered Plants*, edited by D. A. Falk, C. I. Millar, and M. Olwell, 417–22. Washington, DC: Island Press.

Gordon, D. R. 1996b. Experimental translocation of the endangered shrub Apalachicola rosemary (*Conradina glabra*) to the Apalachicola Bluffs and Ravines Preserve, Florida. *Biological Conservation* 77:19–26.

Gordon, D. R. 1998. Effects of invasive, non-indigenous plant species on ecosystem processes: Lessons learned from Florida. *Ecological Applications* 8:975–89.

Gordon, D., and C. Gantz. 2008. Screening new plant introductions for potential invasiveness: A test of impacts for the United States. *Conservation Letters* 1:227–35.

Gordon, D. R., B. Mitterdorfer, P. Pheloung, S. Ansari, C. Buddenhagen, C. Chimera, C. Daehler, et al. 2010. Guidance for addressing the Australian weed risk assessment questions. *Plant Protection Quarterly* 25(2):56–74.

Gordon, D. R., D. A. Onderdonk, A. M. Fox, and R. K. Stocker. 2008a. Consistent accuracy of the Australian weed risk assessment system across varied geographies. *Diversity and Distribution* 14:234–42.

Gordon, D. R., D. A. Onderdonk, A. M. Fox, R. K. Stocker, and C. Gantz. 2008b. Predicting invasive plants in Florida using the Australian weed risk assessment. *Invasive Plant Science and Management* 1:176–95.

Grace, J., F. Berninger, and L. Nagy. 2002. Impacts of climate change on the tree line. *Annals of Botany* 90:537–44.

Graham, C. H., S. Ferrier, F. Huettman, C. Moritz, and A. T. Peterson. 2004. New developments in museum-based informatics and application in biodiversity analysis. *Trends in Ecology & Evolution* 16(9):497–503.

Grange, O., H. Bartschi, and G. Gay. 1997. Effect of the ectomycorrhizal fungus *Hebeloma cylindrosporum* on in vitro rooting of micropropagated cuttings of arbuscular mycorrhiza-forming *Prunus avium* and *Prunus cerasus*. *Trees* 12:49–56.

Grinnell, J. 1917. The niche-relationships of the California thrasher. *Auk* 34:427–33.

Groff, P. A., and D. R. Kaplan. 1988. The relation of root systems to shoot systems in vascular plants. *The Botanical Review* 54:387–422.

Grøndahl, E., and B. K. Ehlers. 2008. Local adaptation to biotic factors: Reciprocal transplants of four species associated with aromatic *Thymus pulegioides* and *T. serpyllum*. *Journal of Ecology* 96:981–92.

Groom, M. J. 1998. Allee effects limit population viability of an annual plant. *American Naturalist* 151:487–96.

Grubb, P. J. 1977. The maintenance of species-richness in plant communities: The importance of the regeneration niche. *Biological Reviews* 52:107–45.

Guerrant, E. O. Jr. 1996a. Designing populations: Demographic, genetic, and horticultural dimensions. In *Restoring Diversity: Ecological Restoration and Endangered Plants*, edited by D. Falk, C. Millar, and P. Olwell, 171–207. New York: Island Press.

Guerrant, E. O. Jr. 1996b. Reintroduction of *Stephanomeria malheurensis*, a case study. In *Restoring Diversity: Strategies for Reintroduction of Endangered Species*, edited by D. A. Falk, C. I. Millar, and M. Olwell, 399–402. Covelo, CA: Island Press.

Guerrant, E. O. Jr. 1996c. Western lily, *Lilium occidentale* (Liliaceace). *Kalmiopsis* 6:16–18.

Guerrant, E. O. Jr. 2001. Experimental reintroduction of the endangered western lily (*Lilium occidentale*). In *Conservation of Washington's Native Plants and Ecosystems*, edited by S. H. Reichard, P. W. Dunwiddie, J. G. Gamon, A. R. Kruckeberg, and D. L. Salstrom, 201–11. Seattle: Washington Native Plant Society.

Guerrant, E. O. Jr., and P. L. Fiedler. 2004. Accounting for Sample Decline During Ex Situ Storage and Reintroduction. In *Ex Situ Plant Conservation: Supporting Species Survival in the Wild*, edited by E. O. Guerrant Jr., K. Havens, and M. Maunder, 365–85. Covelo, CA: Island Press.

Guerrant, E. O. Jr., P. L. Fiedler, K. Havens, and M. Maunder. 2004a. Revised genetic sampling guidelines for conservation collections of rare and endangered plants. In *Ex Situ Plant Conservation: Supporting Species Survival in the Wild*, edited by E. O. Guerrant Jr., K. Havens, and M. Maunder, 419–38. Washington, DC: Island Press.

Guerrant, E. O. Jr., K. Havens, and M. Maunder, eds. 2004b. *Ex Situ Plant Conservation: Supporting Species Survival in the Wild*. Washington, DC: Island Press.

Guerrant, E. O. Jr., and T. N. Kaye. 2007. Reintroduction of rare and endangered plants: Common factors, questions, and common approaches. *Australian Journal of Botany* 55:362–70.

Guerrant, E. O. Jr., and B. M. Pavlik. 1998. Reintroduction of rare plants: Genetics, demography and the role of ex situ conservation methods. In *Conservation Biology for the Coming Decade*, edited by P. L. Fiedler and P. Kareiva, 80–108. New York: Chapman and Hall.

Guisan, A., O. Broennimann, R. Engler, M. Vust, N. G. Yoccoz, A. Lehmann, and N. E. Zimmerman. 2006. Using niche-based models to improve the sampling of rare species. *Conservation Biology* 20:501–11.

Guisan, A., and W. Thuiller. 2005. Predicting species distribution: Offering more than simple habitat models. *Ecology Letters* 8:993–1009.

Gunderson, L. H. 2000. Ecological resilience—in theory and application. *Annual Review of Ecology and Systematics* 31:425–39.

Gurevitch, J., and D. K. Padilla. 2004. Are invasive species a major cause of extinctions? *Trends in Ecology & Evolution* 19:470–74.

Gustafson, D. J., D. J. Gibson, and D. L. Nickrent. 2002. Genetic diversity and competitive abilities of *Dalea purpurea* (Fabaceae) from remnant and restored grasslands. *International Journal of Plant Sciences* 163:979–90.

Gustafson, D. J., D. J. Gibson, and D. L. Nickrent. 2004a. Competitive relationships of *Andropogon gerardii* (big bluestem) from remnant and restored native populations and select cultivated varieties. *Functional Ecology* 18:451–57.

Gustafson, D. J., D. J. Gibson, and D. L. Nickrent. 2004b. Conservation genetics of two co-dominant grass species in an endangered grassland ecosystem. *Journal of Applied Ecology* 41:389–97.

Gutiarrez-Miceli, F. A., T. Ayora-Talavera, M. Abud-Archila, M. Salvador-Figueroa, L. Adriano-Anaya, M. L. Arias Hernandez, and L. Dendooven. 2008. Acclimatization of micropropagated orchid *Guarianthe skinnerii* inoculated with *Trichoderma harzianum*. *Asian Journal of Plant Sciences* 7:327–30.

Hackney, E. E., and J. B. McGraw. 2001. Experimental demonstration of an Allee effect in American ginseng. *Conservation Biology* 15:129–36.

Hadfield, M. S., and B. S. Mountain. 1980. A field study of a vanishing species, *Achatinella mustelina* (Gastropoda, Pulmonata), in the Waianae Mountains of Oahu. *Pacific Science* 34:345–58.

Hagmann, M. 2001. The world in 2050: More crowded, urban and aged. *Bulletin of the World Health Organization* 79(5):482–83. doi:10.1590/S0042-96862001000500020. Accessed July 29, 2010, http://www.scielosp.org/scielo.php?script=sci_arttext&pid=S0042-96862001000500020&lng=en.

Hall, I. R., and A. Kelson. 1981. An improved technique for the production of endomycorrhizal infested soil pellets. *New Zealand Journal of Agricultural Research* 24:221–22.

Hamrick, J. L., and M. J. W. Godt, 1996. Conservation genetics of endemic plant species. In *Conservation Genetics: Case Histories from Nature*, edited by J. C. Avise, and J. L. Hamrick, 281–304. New York: Chapman & Hall.

Hanks, L. M., J. G. Millar, T. D. Paine, and C. D. Campbell. 2000. Classical biological control of the Australian weevil *Gonipterus scutellatus* (Coleoptera: Curculionidae) in California. *Environmental Entomology* 29:369–75.

Hanley, J. A., and B. J. McNeil. 1982. The meaning and use of the area under a receiver operating characteristic (ROC) curve. *Radiology* 29:773–85.

Hannah, L. 2003. Regional biodiversity impact assessments for climate change: A guide for protected area managers. In *Buying Time: A User's Manual for Building Resistance and Resilience to Climate Change in Natural Systems*, edited by L. J. Hansen, J. L. Biringer, and J. R. Hoffman, 235–44. Berlin: World Wildlife Fund Climate Change Program.

Hanski, I. 1999. *Metapopulation Ecology*. New York: Oxford University Press.

Hanski, I., and O. Ovaskainen. 2000. The metapopulation capacity of a fragmented landscape. *Nature* 404:755–58.

Hansson, L. 2003. Why ecology fails at application: Should we consider variability more than regularity? *Oikos* 100:624–27.

Hanzawa, F. M., and S. Kalisz. 1993. The relationship between age, size, and reproduction in *Trillium grandiflorum* (Liliaceae). *American Journal of Botany* 80:405–10.

Harper, J. L. 1977. *Population Biology of Plants*. New York: Academic Press.

Harper, K. T. 1979. Some reproductive and life history characteristics of rare plants and the implications for their management. *Great Basin Naturalist Memoirs* 3:129–37.

Hartl, D. L. 2000. *A Primer of Population Genetics*, 3rd ed. Sunderland, MA: Sinauer.

Havens, K. 1998. The genetics of plant restoration. *Restoration & Management Notes* 16:68–72.

Hawaii Conservation Alliance. 2005. *Controlling Ungulate Populations in Native Ecosystems in Hawaii*. Position Paper. Accessed February 1, 2009, http://hawaiiconservation.org/library/documents/ungulates.pdf.

Hayden, W. J. 1987. The identity of the genus *Neowawraea*. *Brittonia* 39:268–77.

Hayman, D. S. 1986. Mycorrhizae of nitrogen-fixing legumes. *MIRCEN Journal* 2:121–45.

Hayworth, D., M. Bowles, B. Schaal, and K. Williamson. 2001. Clonal population structure of the federal threatened Mead's milkweed, as determined by RAPD analysis, and its conservation implications. In *Proceedings of the Seventh North American Prairie Conference*, edited by N. P. Bernstein and L. J. Ostrander, 182–90. Mason City: North Iowa Area Community College.

HDOA. 2009a. *List of Active Special Local Need Registrations by SLN Number*. Accessed February 1, 2009, http://hawaii.gov/hdoa/pi/pest/SLN.pdf.

HDOA. 2009b. *List of Currently Licensed Pesticides by EPA Registration Number*. Accessed February 1, 2009, http://hawaii.gov/hdoa/pi/pest/liclist_numeric.pdf.

Helenurm, K. 1998. Outplanting and differential source population success in *Lupinus guadalupensis*. *Conservation Biology* 12:118–27.

Hellman, J. J., J. McLachlan, D. Sax, and M. Schwartz. 2008. Managed Relocation Working Group. Accessed March 1, 2008, http://www.nd.edu/~hellmann/MRWorkingGroup/Managed_relocation.html.

Hereford, J. 2009. A quantitative survey of local adaptation and fitness trade-offs. *The American Naturalist* 173:579–83.

Hierro, J. L., J. L. Maron, and R. M. Callaway. 2005. A biogeographical approach to plant invasions: The importance of studying exotics in their introduced and native range. *Journal of Ecology* 93:5–15.

Higgins, J. P. T., and S. G. Thompson. 2004. Controlling the risk of spurious findings from meta-regression. *Statistics in Medicine* 23:1663–82.

Hijmans, R. J., S. E. Cameron, J. L. Parra, P. G. Jones, and A. Jarvis. 2005. Very high resolution interpolated climate surfaces for global land areas. *International Journal of Climatology* 25:1965–78.

Hines, J. M., H. R. Hungerford, and A. N. Tomera. 1987. Analysis and synthesis of research on responsible environmental behavior: A meta-analysis. *Journal of Environmental Education* 18:1–8.

Hobbs, R. J., and D. A. Norton. 1996. Towards a conceptual framework for restoration ecology. *Restoration Ecology* 4:93–110.

Hobbs, R. J., and D. A. Norton. 2004. Ecological filters, thresholds, and gradients in resistance to ecosystem assembly. In *Assembly Rules and Restoration Ecology: Bridging the Gap between Theory and Practice*, edited by V. M. Temperton, R. J. Hobbs, T. Nuttle, and S. Halle, 72–95. Washington, DC: Island Press.

Hoegh-Guldberg, O., L. Hughes, S. McIntyre, D. B. Lindemayer, C. Parmesan, H. P. Possingham, and C. D. Thomas. 2008. Assisted colonization and rapid climate change. *Science* 321:345–46.

Hogbin, P. M., and R. Peakall. 1999. Evaluation of the contribution of genetic research to the management of the endangered plant *Zieria prostrata*. *Conservation Biology* 13:514–22.

Holl, K. D., and G. F. Hayes. 2006. Challenges to introducing and managing disturbance regimes for *Holocarpha macradenia*, an endangered annual grassland forb. *Conservation Biology* 20:1121–31.

Holling, C. S. 1973. Resilience and the stability of ecological systems. *Annual Review of Ecology and Systematics* 4:1–23.

Holling, C. S. 1996. Surprise for science, resilience for ecosystems, and incentives for people. *Ecological Applications* 6:733–35.

Honjo, M., S. Ueno, Y. Tsumura, T. Handa, I. Washitani, and R. Ohsawa. 2008. Tracing the origins of stocks of the endangered species *Primula sieboldii* using nuclear microsatellites and chloroplast DNA. *Conservation Genetics* 9:1139–47.

Horvitz, C. C., and D. W. Schemske. 1995. Spatiotemporal variation in demographic transitions of a tropical understory herb: Projection matrix analysis. *Ecological Monographs* 65:155–92.

Hufford, K., and S. J. Mazer. 2003. Plant ecotypes: Genetic differentiation in the age of ecological restoration. *Trends in Ecology & Evolution* 18:147–55.

Hunt, T. L., and C. P. Lipo. 2007. Chronology, deforestation, and collapse: Evidence vs. faith in Rapa Nui prehistory. *Rapa Nui Journal* 21:85–97.

Hunter, M. L. 2007. Climate change and moving species: Furthering the debate on assisted colonization. *Conservation Biology* 21:1356–58.

Husband, B. C., and D. W. Schemske. 1996. Evolution of the magnitude and timing of inbreeding depression in plants. *Evolution* 50:54–70.

Huston, M. A. 1999. Local processes and regional patterns: Appropriate scales for understanding variation in the diversity of plants and animals. *Oikos* 86:393–401.

Hutchinson, G. E. 1957. Concluding remarks. *Cold Spring Harbor Symposium Quantitative Biology* 22:415–27.

Huynh, T. T., C. B. McLean, and A. C. Lawrie. 2002. Seed germination and propagation of *Archnorchis formosa*. *International Plant Propagators Society Proceedings* 52:161–66.

Igic, I. B., and J. R. Kohn. 2006. The distribution of plant mating systems: Study bias against obligately outcrossing species. *Evolution* 60(5):1098–1103.

IPCC. 2007. *Climate Change 2007: Synthesis Report. Contribution of Working Groups I, II and III to the Fourth Assessment Report of the Intergovernmental Panel on Climate Change*. Geneva, Switzerland: IPCC.

Iriondo, J. M., M. J. Alber, and A. Escudero. 2003. Structural equation modeling: An alternative for assessing causal relationships in threatened plant populations. *Biological Conservation* 113:367–77.

IUCN. 1998a. *Guidelines for Reintroductions*. Prepared by IUCN/SSC Re-introduction Specialist Group. Gland, Switzerland and Cambridge, UK: IUCN.

IUCN. 1998b. Species Survival Commission Reintroduction Specialist Group Members Database. Accessed February 1, 2011, http://www.iucnsscrsg.org/rsg_database.php.

IUCN. 2002. *Global Strategy for Plant Conservation*. Accessed February 8, 2011, http://iucn.org/about/work/programmes/species/our_work/plants/what_we_do_plants/the_global_strategy_for_plant_conservation/index.cfm and http://intranet.iucn.org/web files/doc/SSC/SSCwebsite/Plants/global_strategy.pdf.

IUCN. 2010. *IUCN Red List of Threatened Species*. Version 2010.3. Accessed October 5, 2010, http://www.iucnredlist.org.

IUCN Species Survival Commission. 2004. *2004 Red List of Threatened Species. A Global Species Assessment*, edited by J. E. M. Baillie, C. Hilton-Taylor, and S. N. Stuart. Cambridge, MA: IUCN Publications Services Unit.

IUCN/SSC Re-introduction Specialist Group. 2007. RSG Members. Accessed February 1, 2010, http://www.iucnsscrsg.org/rsg_database.php.

Jackson, S. T., and R. J. Hobbs. 2009. Ecological restoration in the light of ecological history. *Science* 325:567–69.

Jackson, S. T., and J. T. Overpeck. 2000. Responses of plant populations and communities to environmental changes of the late Quaternary. *Paleobiology* 26:194–220.

Jäderlund, L., V. Arthurson, U. Granhall, and J. K. Jansson. 2008. Specific interactions between arbuscular mycorrhizal fungi and plant growth-promoting bacteria as revealed by different combinations. *FEMS Microbiology Letters* 287:174–80.

James, S. A. 2009. *Population Genetics of* Delissea waianaeensis *Lammers (Campanulaceae) and Parental Identification of a Potential New Founder Individual*. Final Report. Prepared for the Oahu Army Natural Resource Program, Schofield Barracks, Hawaii.

Janes, J. K. 2009. *Techniques for Tasmanian Native Orchid Germination*. Nature Conservation Report 09/1. Tasmania: Department of Primary Industries and Water.

Janssen, G. K., and P. S. Williamson. 1996. Encouraging conservation of endangered plants on private lands: A case study of Johnston's frankenia (*Frankenia johnstonii*), an endangered South Texas subshrub. In *Southwestern Rare and Endangered Plants*, edited by J. Maschinski and L. Holter, 1–7. Flagstaff, AZ: US Department of Agriculture, Forest Service, Rocky Mountain Research Station.

Janzen, D. H. 1985. The natural history of mutualisms. In *The Biology of Mutualism*, edited by D. H. Boucher, 40–99. New York: Oxford University Press.

Jarvi, S. I., C. T. Atkinson, and R. C. Fleischer. 2001. Immunogenetics and resistance to avian malaria in Hawaiian honeycreepers (Drepanidinae). *Studies in Avian Biology* 22:254–63.

Jasper, A., E. M. Freitas, E. L. Musskopf, and J. Bruxel. 2005. Methodology of the preservation of Bromeliaceae, Cactaceae and Orchidaceae in the Forqueta Fall Power Plant, Sao Jose do Herval. *Pesquisas Botanica* 56:265–83.

Joe, S. M., and C. C. Daehler. 2008. Invasive slugs as under-appreciated obstacles to rare plant restoration: Evidence from the Hawaiian Islands. *Biological Invasions* 10:245–55.

Jogar, U., and M. Moora. 2008. Reintroduction of a rare plant (*Gladiolus imbricatus*) population to a river floodplain: How important is meadow management? *Restoration Ecology* 16:382–85.

Johnson, M. T. J., M. J. Lajeunesse, and A. A. Agrawal. 2006. Additive and interactive effects of plant genotypic diversity on arthropod communities and plant fitness. *Ecology Letters* 9:24–34.

Johnson, N. C., J. H. Graham, and F. A. Smith. 1997. Functioning of mycorrhizal associations along the mutualism–parasitism continuum. *New Phytologist* 135:575–85.

Jones, C. G., and F. T. Last. 1991. Ectomycorrhizae and trees: Implications for aboveground herbivory. In *Microbial Mediation of Plant–Herbivore Interactions*, edited by P. Barbosa, V. A. Krischik, and C. G. Jones, 65–103. New York: Wiley.

Jones, T. A., and D. A. Johnson. 1998. Integrating genetic concepts into planning rangeland seedings. *Journal of Range Management* 51:594–606.

Joshi, J., and K. Vrieling. 2005. The enemy release and EICA hypothesis revisited: Incorporating the fundamental difference between specialist and generalist herbivores. *Ecology Letters* 8:704–14.

Judd, W. S., C. S. Campbell, E. A. Kellogg, P. F. Stevens, and M. J. Donoghue. 2008. *Plant Systematics: A Phylogenetic Approach*, 3rd ed. Sunderland, MA: Sinauer.

Jump, A. S., and J. Peñuelas. 2005. Running to stand still: Adaptation and the response of plants to rapid climate change. *Ecology Letters* 8:1010–20.

Jumpponen, A. 2001. Dark septate endophytes: Are they mycorrhizal? *Mycorrhiza* 11:207–11.

Jumpponen, A., and J. M. Trappe. 1998. Dark septate endophytes: A review of facultative biotrophic root-colonizing fungi. *New Phytology* 140:295–310.

Jusaitis, M. 1991. Endangered *Phebalium* (Rutaceae) species return to South Australia. *Re-introduction News* 4.

Jusaitis, M. 1996. Experimental translocations of endangered *Phebalium* spp. (Rutaceae) in South Australia: An update. *Re-introduction News* 7–8.

Jusaitis, M. 1997. Experimental translocations: Implications for the recovery of endangered plants. In *Conservation into the 21st Century* (Proceedings of the 4th International Botanical Gardens Conservation Congress), edited by D. H. Touchell, K. W. Dixon, A. S. George, and A. T. Wills, 181–96. Perth, Western Australia: Kings Park & Botanic Gardens.

Jusaitis, M. 2005. Translocation trials confirm specific factors affecting the establishment of three endangered plant species. *Ecological Management and Restoration* 6:61–67.

Jusaitis, M., and L. Polomka. 2008. Weeds and propagule type influence translocation success in the endangered Whibley wattle, *Acacia whibleyana* (Leguminosae: Mimosoideae). *Ecological Management and Restoration* 9:72–76.

Jusaitis, M., L. Polomka, and B. Sorensen. 2004. Habitat specificity, seed germination and experimental translocation of the endangered herb *Brachycome muelleri* (Asteraceae). *Biological Conservation* 116:251–66.

Jusaitis, M., and B. Sorensen. 2007. Successful augmentation of an *Acacia whibleyana* (Whibley wattle) population by translocation. *Australasian Plant Conservation* 16:23–24.

Jusaitis, M., and J. Val. 1997. Herbivore grazing: An important consideration in plant translocations. *Re-introduction News* 11–12.

Kahn, P. H. Jr. 2002. Children's affiliations with nature: Structure, development, and the problem of environmental generational amnesia. In *Children and Nature: Psychological, Sociocultural, and Evolutionary Investigations*, edited by P. H. Kahn Jr. and S. R. Kellert, 93–116. Cambridge, MA: MIT Press.

Kapoor, R., D. Sharma, and A. K. Bhatnagar. 2008. Arbuscular mycorrhizae in micropropagation systems and their potential applications. *Scientia Horticulturae* 116:227–39.

Kardol, P., N. J. Cornips, M. M. L. van Kempen, J. M. T. Bakx-Schotman, and W. H. van der Putten. 2007. Microbe-mediated plant–soil feedback causes historical contingency effects in plant community assembly. *Ecological Monographs* 77(2):147–62.

Karl, T. R., and K. E. Trenberth. 2003. Modern global climate change. *Science* 302:1719–23.

Kartsonas, E., and M. Papafotiou. 2007. Mother plant age and seasonal influence on in vitro propagation of *Quercus euboica* Pap., an endemic, rare and endangered oak species of Greece. *Plant Cell Tissue and Organ Culture* 90:111–16.

Kaufman, B., and L. Mehrhoff. 2009. *Large Scale Restoration as an Adaptive Strategy to Climate Change*. Honolulu: Hawaii Conservation Conference.

Kaye, T. 2001. *Propogation and Population Re-establishment for Tall Bugbane (Cimicifuga elata) on the Salem District, BLM*. Salem, OR: Salem District, Bureau of Land Management and Institute for Applied Ecology.

Kaye, T. 2008. Vital steps toward success of endangered plant reintroductions. *Native Plants* 9:313–22.

Kaye, T. N., and J. R. Cramer. 2003. Direct seeding or transplanting: The cost of restoring populations of Kincaid's lupine. *Ecological Restoration* 21:224–25.

Kaye, T., and D. A. Pyke. 2003. The effect of stochastic technique on estimates of population viability from transition matrix models. *Ecology* 84:1464–76.

Keane, R. M., and M. J. Crawley. 2002. Exotic plant invasions and the enemy release hypothesis. *Trends in Ecology & Evolution* 17:164–70.

Kearns, C. A., D. W. Inouye, and N. M. Waser. 1998. Endangered mutualisms: The conservation of plant–pollinator interactions. *Annual Review of Ecology and Systematics* 29:83–112.

Keddy, P. 2001. *Competition*. Dordrecht, The Netherlands: Kluwer.

Keel, B. G. 2005. Assisted migration. In *Oxford Dictionary of Ecology*, edited by M. Allaby, 36. Oxford, England: Oxford University Press.

Keel, B. G. 2007. *Assisted Migration as a Conservation Strategy for Rapid Climate Change: Investigating Extended Photoperiod and Mycobiont Distributions for Habe - naria repens Nuttall (Orchidaceae) as a Case Study*. PhD dissertation, Antioch University New England, Keene, NH.

Kephart, S. R. 2004. Inbreeding and reintroduction: Progeny success in rare *Silene* populations of varied density. *Conservation Genetics* 5:49–61.

Kephart, S., and C. Paladino. 1997. Demographic change and microhabitat variability in a grassland endemic, *Silene douglasii* var. *oraria* (Caryophyllaceae). *American Journal of Botany* 84:179–89.

Kessell, S. L. 1927. Soil organisms. The dependence of certain pine species on a biological soil factor. *Empire Forestry Journal* 6:70–74.

Khade, S. W., and B. F. Rodrigues. 2008. Effects of arbuscular mycorrhizal fungi on micropropagated banana. *Journal of Economic and Taxonomic Botany* 32:510–15.

Khan, P. S. S. V., H. S. Devi, R. K. Kishor, and B. N. Rao. 2009. Micropropagation and some acclimatization characteristics of *Centella asiatica* (Linn.) Urban. *Indian Journal of Plant Physiology* 14:353–59.

Kimmins, J. P. 1996. *Forest Ecology: A Foundation for Sustainable Management.* 2nd ed. Upper Saddle River, NJ: Prentice Hall.

Kirchner, F., A. Robert, and B. Colas. 2006. Modelling the dynamics of introduced populations in the narrow-endemic *Centaurea corymbosa*: A demo-genetic integration. *Journal of Applied Ecology* 43:1011–21.

Knapp, E. E., and K. J. Rice. 1994. Starting from seed: Genetic issues in using native grasses for restoration. *Restoration and Management Notes* 12:40–45.

Knight, T. M. 2003a. Effects of herbivory and its timing across populations of *Trillium grandiflorum* (Liliaceae). *American Journal of Botany* 90:1207–14.

Knight, T. M. 2003b. Floral density, pollen limitation and reproductive success in *Trillium grandiflorum*. *Oecologia* 137:557–63.

Knight, T. M. 2004. The effects of herbivory and pollen limitation on a declining population of *Trillium grandiflorum*. *Ecological Applications* 14:915–28.

Knight, T. M., H. Caswell, and S. Kalisz. 2009. Population growth rate of a common understory herb decreases non-linearly across a gradient of deer herbivory. *Forest Ecology and Management* 257:1095–1103.

Kohn, D., and P. Lusby. 2004. Translocation of twinflower (*Linnaea borealis* L.) in the Scottish Borders. *Botanical Journal of Scotland* 56(1):25–37.

Kramer, A. T., and K. Havens. 2009. Plant conservation genetics in a changing world. *Trends in Plant Science* 14:599–607.

Krauss, S. L., B. Dixon, and K. Dixon. 2002. Rapid genetic decline in a translocated population of the endangered plant *Grevillea scapigera*. *Conservation Biology* 16(4):986–94.

Krauss, S. L., L. Hermanutz, S. D. Hopper, and D. J. Coates. 2007. Population-size effects on seeds and seedlings from fragmented eucalypt populations: Implications for seed sourcing for ecological restoration. *Australian Journal of Botany* 55(3):390–99.

Krauss, S. L., and J. M. Koch. 2004. Rapid genetic delineation of provenance for plant community restoration. *Journal of Applied Ecology* 41:1162–73.

Kucharczyk, M., and E. Teske. 1996. Active protection of extremely 1433 small populations of plants: *Primula vulgaris* Hudson. *Bulletin of the Polish Academy of Sciences Biological Sciences* 44:121–25.

Kuussaari, M., R. Bommarco, R. K. Heikkinen, A. Helm, J. Krauss, R. Lindborg, E. Öckinger, et al. 2009. Extinction debt: A challenge for biodiversity conservation. *Trends in Ecology & Evolution* 24:564–71.

Lambers, H., F. S. Chapin III, and T. L. Pons. 1998. *Plant Physiological Ecology.* New York: Springer.

Lammers, T. G. 2005. Revision of *Delissea* (Campanulaceae–Lobelioideae). *Systematic Botany Monographs* 73:1–75.

Lande, R. 1988 Genetics and demography in biological conservation. *Science* 241:1455–60.

Lankau, R. A., V. Nuzzo, G. Spyreas, and A. S. Davis. 2009. Evolutionary limits ameliorate the negative impact of an invasive plant. *Proceedings of the National Academy of Sciences* 106:15362–67.

Larkin, D., G. Vivian-Smith, and J. B. Zedler. 2006. Topographic heterogeneity theory and ecological restoration. In *Foundations of Restoration Ecology*, edited by D. A. Falk, M. A. Palmer, and J. B. Zedler, 142–64, Washington, DC: Island Press.

Latta, R. G. 2008. Conservation genetics as applied evolution: From genetic pattern to evolutionary process. *Evolutionary Applications* 1:84–94.

Lawrence, B. A., and T. N. Kaye. 2009. Reintroduction of *Castilleja levisecta*: Effects of ecological similarity, source population genetics, and habitat quality. *Restoration Ecology*. doi:10.1111/j.1526–100X.2009.00549.x.

Lawrence, M. J., D. F. Marshall, and P. Davies. 1995a. Genetics of genetic conservation. I. Sample size when collecting germplasm. *Euphytica* 84:89–99.

Lawrence, M. J., D. F. Marshall, and P. Davies. 1995b. Genetics of genetic conservation. II. Sample size when collecting seed of cross-pollinating species and the information that can be obtained from the evaluation of material held in gene banks. *Euphytica* 84:101–7.

Ledig, F. T. 1996. *Pinus torreyana* at the Torrey Pines State Reserve, California. In *Restoring Diversity: Strategies for Reintroduction of Endangered Plants*, edited by D. A. Falk, C. I. Millar, and M. Olwell, 265–71. Washington, DC: Island Press.

Lee, C. H., and K. S. Kim. 2009. Acclimatization of *Dendranthema zawadskii* Tzvelev by using in vitro paclobutrazol pretreatment. *Horticulture Environment and Biotechnology* 50:154–59.

Leonard, Y. 2006a. Reintroduction of perennial knawel *Scleranthus perennis prostratus* to sheep-grazed grassheath at West Stow, Suffolk, England. *Conservation Evidence* 3:15–16.

Leonard, Y. 2006b. Soil disturbance & seedling transplanting as a method of reintroduction of perennial knawel *Scleranthus perennis prostratus* at Icklingham, Suffolk, England. *Conservation Evidence* 3:17–18.

Lesica, P. 1995. The demography of *Astragalus scaphoides* and effects of herbivory on population growth. *Great Basin National Park* 55:142–50.

Lesica, P., and F. W. Allendorf. 1999. Ecological genetics and the restoration of plant communities: Mix or match? *Restoration Ecology* 7:42–50.

Lesica, P., and B. McCune. 2004. Decline of arctic–alpine plants at the southern margin of their range following a decade of climate warming. *Journal of Vegetation Science* 15:679–90.

Levine, J., and C. D'Antonio. 1999. Elton revisited: A review of evidence linking diversity and invisibility. *Oikos* 87:15–26.

Lewis, C. 2009. Reversing teenagers' disconnect from nature. *Plant Science Bulletin* 55:113–17.

Li, Y.-Y., X.-Y. Chen, X. Zhang, T.-Y. Wu, H.-P. Lu, and Y.-W. Cai. 2005. Genetic differences between wild and artificial populations of *Metasequoia glyptostroboides*: Implications for species recovery. *Conservation Biology* 19:224–31.

Li, Z., and M. Kafatos. 2000. Interannual variability of vegetation in the United States and its relation to El Niño/Southern Oscillation. *Remote Sensing of Environment* 71:239–47.

Lin, G., S. L. Phillips, and J. R. Ehleringer. 1996. Monsoonal precipitation responses of shrubs in a cold desert community on the Colorado Plateau. *Oecologia* 106:8–17.

Lindborg, R., and J. Ehrlén. 2002. Evaluating the extinction risk of a perennial herb: Demographic data versus historical records. *Conservation Biology* 16:683–90.

Linhart, Y. B. 1995. Restoration, revegetation, and the importance of genetic and evolutionary perspectives. In *Proceedings: Wildland Shrub and Arid Land Restoration Symposium*, edited by B. A. Roundy, E. McArthur, J. S. Haley, and D. K. Mann, 271–83. Las Vegas, NV: US Department of Agriculture, Forest Service, Intermountain Research Station.

Linhart, Y. B., and M. C. Grant. 1996. Evolutionary significance of local genetic differentiation in plants. *Annual Review of Ecology and Systematics* 27:237–77.

Liu, C., P. M. Berry, T. P. Dawson, and R. G. Pearson. 2005. Selecting thresholds of occurrence in the prediction of species distributions. *Ecography* 28:385–93.

Liu, G. H., J. Zhou, D. S. Huang, and W. Li. 2004. Spatial and temporal dynamics of a restored population of *Oryza rufipogon* in Huli Marsh, South China. *Restoration Ecology* 12:456–63.

Liu, H., C.-L. Feng, Y.-B. Luo, B.-S. Chen, Z.-S. Wang, and H.-Y. Gu. 2010. Potential challenges of climate change to orchid conservation in a wild orchid hotspot in southwestern China. *Botanical Review* 76:174–92.

Liu, H., and P. Stiling. 2006. Testing the enemy release hypothesis: A review and meta-analysis. *Biological Invasions* 8:1535–45.

Liu, H., P. Stiling, and R. W. Pemberton. 2007. Does enemy release matter? Evidence from a comparison of insect damage level among invasive, non-invasive, and native congeners. *Biological Invasions* 9:773–81.

Liu, H., P. Stiling, R. W. Pemberton, and J. Pena. 2006. Insect herbivore faunal diversity among invasive, non-invasive and native *Eugenia* species: Implications for the enemy release hypothesis. *Florida Entomology* 89:475–84.

Liu, M.-H., X.-Y. Chen, X. Zhang, and D.-W. Shen. 2008. A population genetic evaluation of ecological restoration with the case study on *Cyclobalanopsis myrsinaefolia* (Fagaceae). *Plant Ecology* 197:31–41.

Loarie, S. R., B. E. Carter, K. Hayhoe, S. McMahon, R. Moe, C. A. Knight, and D. D. Ackerly. 2008. Climate change and the future of California's endemic flora. *PLoS ONE* 3(6):e2502. doi:10.1371/journal.pone.0002502.

Lobo, J. M., A. Jimenez-Valverde, and R. Real. 2008. AUC: A misleading measure of the performance of predictive distribution models. *Global Ecology and Biogeography* 17:145–51.

Lockwood, D. R., G. M. Volk, and C. M. Richards. 2007. Wild plant sampling strategies: The roles of ecology and evolution. *Plant Breeding Reviews* 29:286–314.

Lockwood, J. L., D. Simberloff, M. L. McKinney, and B. Von Holle. 2001. How many, and which, plants will invade natural areas? *Biological Invasions* 3:1–6.

Lofflin, D. L., and S. R. Kephart. 2005. Outbreeding, seedling establishment, and maladaptation in natural and reintroduced populations of rare and common *Silene douglasii* (Caryophyllaceae). *American Journal of Botany* 92:1691–1700.

Long, S. R. 1996. Rhizobium symbiosis: Nod factors in perspective. *The Plant Cell* 8:1885–98.

Louda, S. M. 1982. Distribution ecology: Variation in plant recruitment over a gradient in relation to insect seed predation. *Ecological Monographs* 52:25–41.

Louro, R. P., L. J. M. Santiago, A. V. dos Santos, and R. D. Machado. 2003. Ultrastructure of *Eucalyptus grandis x E. urophylla* plants cultivated ex vitro in greenhouse and field conditions. *Trees* 17:11–22.

Lowe, A. J., D. H. Boshier, D. Ward, C. F. E. Bacles, and C. Navarro. 2005. Genetic resource loss following habitat fragmentation and degradation: Reconciling predicted theory with empirical evidence. *Heredity* 95:255–73.

Ludington, S., B. C. Moring, R. J. Miller, P. A. Stone, A. A. Bookstrom, D. R. Bedford, J. G. Evans, et al. 2007. Preliminary integrated geologic map database for the US western states: California, Nevada, Arizona, Washington, Oregon, Idaho and Utah. Version 1.3. Accessed November 18, 2010, http://pubs.usgs.gov/of/2005/1305.

Lugo, A. E. 2009. The emerging era of novel tropical forests. *Biotropica* 41:589–91.

Luo, Y., S. Wan, D. Hui, and L. L. Wallace. 2001. Acclimatization of soil respiration to warming in a tall grass prairie. *Nature* 413:622–25.

Lusby, P., S. Lindsay, and A. F. Dyer. 2002. Principles, practice and problems of conserving the rare British fern *Woodsia ilvensis* (L.) R.Br. *Fern Gazette* 16:350–55.

Lynch, D., and D. Weihrauch. 2002. Endangered and threatened wildlife and plants; removal of *Potentilla robbinsiana* (Robbins' cinquefoil) from the Federal List of Endangered and Threatened Plants. *Federal Register* 67(166):54968–75.

MacArthur, R. H. 1972. *Geographical Ecology: Patterns in the Distribution of Species.* Princeton, NJ: Princeton University Press.

MacDonald, G. M., and R. A. Case. 2005. Variations in the Pacific Decadal Oscillation over the past millennium. *Geophysical Research Letters* 32:L08703.

Macel, M., C. S. Lawson, S. R. Mortimer, M. Šmilauerova, A. Bischoff, L. Crémieux, J. Doležal, et al. 2007. Climate vs. soil factors in local adaptation of two common plant species. *Ecology* 88(2):424–33.

Mack, R. N., D. Simberloff, W. M. Lonsdale, H. Evans, M. Clout, and F. A. Bazzaz. 2000. Biotic invasions: Causes, epidemiology, global consequences and control. *Ecological Applications* 10:689–710.

Malecki, R. A., B. Blossey, S. D. Hight, D. Schroeder, L. T. Kok, and J. R. Coulson. 1993. Biological control of purple loosestrife. *BioScience* 43:680–86.

Managed Relocation Working Group. 2008. *Managed Relocation.* Accessed November 1, 2008, http://www.nd.edu/~hellmann/MRWorkingGroup/Managed_relocation.html.

Mardon, D. K. 2003. Conserving montane willow scrub on Ben Lawers NNR. *Botanical Journal of Scotland* 55:189–203.

Maron, J. L., and E. Crone. 2006. Herbivory: Effects on plant abundance, distribution and population growth. *Proceedings of the Royal Society B* 273:2575–84.

Marsico, T. D., and J. J. Hellmann. 2009. Dispersal limitation inferred from an experimental translocation of *Lomatium* (Apiaceae) species outside their geographic ranges. *Oikos* 118:1783–92.

Martin, K. P. 2003. Clonal propagation, encapsulation and reintroduction of *Ipsea malabarica* (Reichb. f.) J.D. Hook, an endangered orchid. *In Vitro Cellular and Developmental Biology: Plant* 39:322–26.

Martinez, T. A., and S. L. McMullin. 2004. Factors affecting decisions to volunteer in nongovernmental organizations. *Environment and Behavior* 36:112–26.

Martinez-Meyer, E., A. T. Peterson, J. I. Servin, and L. F. Kiff. 2006. Ecological niche modeling and prioritizing areas for species reintroductions. *Oryx* 40:411–18.

Martins, A., A. Casimiro, and M. S. Pais. 1997. Influence of mycorrhization of physiological parameters of micropropagated *Castanea sativa* Mill. plants. *Mycorrhiza* 7:161–65.

Maschinski, J. 2001. Impacts of ungulate herbivores on a rare willow at the southern edge of its range. *Biological Conservation* 101:119–30.

Maschinski, J. 2006. Implications of population dynamic and metapopulation theory for restoration. In *Foundations of Restoration Ecology*, edited by D. A. Falk, M. A. Palmer, and J. B. Zedler, 59–87. Washington, DC: Island Press.

Maschinski, J., J. E. Baggs, and C. F. Sacchi. 2004a. Seedling recruitment and survival of an endangered limestone endemic in its natural habitat and experimental reintroduction sites. *American Journal of Botany* 91:689–98.

Maschinski, J., and J. Duquesnel. 2007. Successful reintroductions of the endangered long-lived Sargent's cherry palm, *Pseudophoenix sargentii*, in the Florida Keys. *Biological Conservation* 134:122–29.

Maschinski, J., and J. L. Goodman. 2008. *Assessment of Population Status and Causes of Decline for* Pilosocereus robinii *(Lem.) Byles & G. D. Rowley in the Florida Keys*. Report to US Fish and Wildlife Service, South Florida Ecological Services Office, Vero Beach.

Maschinski, J., J. Goodman, S. J. Wright, D. Walters, J. Possley, and C. Lewis. 2009. The Connect to Protect Network: Botanic gardens working to restore habitats and conserve rare species. *BGJournal* 6:6–9.

Maschinski, J., B. Schaffer, S. J. Wright, K. S. Wendelberger, and J. Roncal. 2007. *Ongoing Efforts to Reintroduce and Study Two Endangered Plant Species, Beach Jacquemon-*

tia and Crenulate Lead-Plant. Report to US Fish and Wildlife Service, South Florida Ecological Services Office, Vero Beach, FL.

Maschinski, J., K. S. Wendelberger, S. J. Wright, J. Possley, D. Walters, J. Roncal, and J. Fisher. 2006. *Conservation of South Florida Endangered and Threatened Flora: 2005–2006 Program at Fairchild Tropical Garden*. Final Report Contract #009706. Gainesville: Florida Department of Agriculture and Consumer Services, Division of Plant Industry.

Maschinski, J., K. Wendelberger, S. J. Wright, H. Thornton, A. Frances, J. Possley, and J. Fisher. 2004b. *Conservation of South Florida Endangered and Threatened Flora: 2004 Program at Fairchild Tropical Garden*. Gainesville: Florida Department of Agriculture and Consumer Services, Division of Plant Industry.

Maschinski, J., M. S. Ross, H. Liu, J. O'Brien, E. J. von Wettberg, and K. E. Haskins. 2011. Sinking ships: Conservation options for endemic taxa threatened by sea level rise. *Climatic Change* 107:147–67.

Maschinski, J., and S. Rutman. 1993. The price of waiting may be too high: *Astragalus cremnophylax* var. *cremnophylax* at Grand Canyon National Park. In *Proceedings of the Southwestern Rare and Endangered Plant Conference*, edited by R. Sivinski and K. Lightfoot. Santa Fe: New Mexico Forestry and Resources Conservation Division, Santa Fe.

Maschinski, J., and S. J. Wright. 2006. Using ecological theory to plan restorations of the endangered beach *Jacquemontia* in fragmented habitats. *Journal for Nature Conservation* 14:180–89.

Maschinski, J., S. J. Wright, J. Possley, D. Powell, L. Krueger, V. Pence, and J. Pascarella. 2010. *Conservation of South Florida Endangered and Threatened Flora: 2009–2010 Program at Fairchild Tropical Garden*. Final Report Contract #014880. Florida Department of Agriculture and Consumer Services, Division of Plant Industry, Gainesville.

Maschinski, J., S. J. Wright, H. Thornton, J. Fisher, J. Possley, B. Pascarella, C. Lane, E. Pinto-Torres, and S. Carrara. 2003. *Restoration of Jacquemontia reclinata to the South Florida Ecosystem*. Final Report to the US Fish and Wildlife Service for Grant Agreement 1448-40181-99-G-173. Vero Beach, FL.

Maschinski, J., S. J. Wright, K. Wendelberger, J. Possley, and J. Fisher. 2005. *Conservation of South Florida Endangered and Threatened Flora: 2004–2005 Program at Fairchild Tropical Botanic Garden. Final Report*. Gainesville: Florida Department of Agriculture and Consumer Services, Division of Plant Industry.

Maschinski, J., S. J. Wright, K. S. Wendelberger, H. E. B. Thornton, and A. Muir. 2003. *Conservation of South Florida Endangered and Threatened Flora: Final report*. Gainesville: Florida Department of Agriculture and Consumer Services. Contract #007182.

Matthies, D., I. Brauer, W. Maibom, and T. Tscharntke. 2004. Population size and the risk of local extinction: Empirical evidence from rare plants. *Oikos* 105:481–88.

Matyas, C., and C. W. Yeatman. 1992. Effect of geographical transfer on growth and survival of jackpine *Pinus banksiana* Lamb. populations. *Silvae Genetica* 41:370–76.

Maunder, M. 1992. Plant reintroduction: An overview. *Biodiversity and Conservation* 1:51–61.

Maunder, M., A. Culham, B. Alden, G. Zizka, C. Orliac, W. Lobin, A. Bordeu, J. M. Ramirez, and S. Glissman-Gough. 2000. Conservation of the Toromiro tree: Case study in the management of a plant extinct in the wild. *Conservation Biology* 14(5):1341–50.

May, D. E., P. J. Webber, and T. A. May. 1982. Success of transplanted alpine tundra plants on Niwot Ridge, Colorado. *Journal of Applied Ecology* 19:965–76.

McClain, W. E., and J. E. Ebinger. 2008. Reintroduction of lakeside daisy (*Tetraneuris herbacea* Greene, Asteraceae) at Manito Prairie Nature Preserve, Tazewell County, Illinois. *Transactions of the Illinois State Academy of Science* 10:79–85.

McDonald, A. W., and C. R. Lambrick. 2006. Apium repens *Creeping Marshwort Species Recovery Programme 1995–2005*. Peterborough, England: English Nature.

McDonald, C. B. 1996. The regulatory and policy context. In *Restoring Diversity: Ecological Restoration and Endangered Plants*, edited by D. Falk, C. Millar, and P. Olwell, 171–207. New York: Island Press.

McDonald, R. J. 2005. Reproductive ecology and re-establishment of *Argusia argentea* on Ashmore Reef. *Beagle* S1:153–62.

McDougall, K. L., and J. W. Morgan. 2005. Establishment of native grassland vegetation at Organ Pipes National Park near Melbourne, Victoria: Vegetation changes from 1989 to 2003. *Ecological Management and Restoration* 6(1):34–42.

McEachern, A. K. 1992. *Disturbance dynamics of pitcher's thistle* (Cirsium pitcheri) *in Great Lakes sand dune landscapes*. Ph.D. dissertation, University of Wisconsin, Madison.

McEachern, A. K., M. Bowles, and N. Pavlovic. 1994. A metapopulation approach to pitcher's thistle (*Cirsium pitcheri*) recovery in southern Lake Michigan dunes. In *Restoraton of Endangered Species*, edited by M. Bowles and C. J. Whelan, 194–218. Cambridge, England: Cambridge University Press.

McGlaughlin, M., K. Karoly, and T. Kaye. 2002. Genetic variation and its relationship to population size in reintroduced populations of pink sand verbena, *Abronia umbellata* subsp. *breviflora* (Nyctaginaceae). *Conservation Genetics* 3:411–20.

McHaffie, H. S. 2005. Re-introduction of a rare fern: Oblong woodsia at four sites in the UK. *Re-introduction News* 24:48–50.

McHaffie, H. 2006. A reintroduction programme for *Woodsia ilvensis* (L.) R. Br. in Britain. *Botanical Journal of Scotland* 58:75–80.

McKay, J. K., J. G. Bishop, J.-Z. Lin, J. H. Richards, A. Sala, and T. Mitchell-Olds. 2001. Local adaptation across a climatic gradient despite small effective population size in the rare sapphire rockcress. *Proceedings of the Royal Society of London B Series* 268:1715–21.

McKay, J. K., C. E. Christian, S. Harrison, and K. J. Rice. 2005. "How local is local?" A review of practical and conceptual issues in the genetics of restoration. *Restoration Ecology* 13:432–40.

McKenzie, D., C. Miller, and D. A. Falk. 2010. *The Landscape Ecology of Fire*. Ecological Studies Series. New York: Springer.

McLachlan, J. S., J. J. Hellmann, and M. W. Schwartz. 2007. A framework for debate of assisted migration in an era of climate change. *Conservation Biology* 21:297–302.

McLaughlin, J. F., J. J. Hellmann, C. L. Boggs, and P. R. Ehrlich. 2002. Climate change hastens population extinctions. *Proceedings of the National Academy of Sciences* 99(9):6070–74.

Meehan, A. J., and R. J. West. 2002. Experimental transplanting of *Posidonia australis* seagrass in Port Hacking, Australia, to assess the feasibility of restoration. *Marine Pollution Bulletin* 44:25–31.

Meeks, J. C. 1998. Symbiosis between nitrogen-fixing cyanobacteria and plants. *BioScience* 48(4):266–76.

Mehrhoff, L. A. 1996. FOCUS: Reintroducing endangered Hawaiian plants. In *Restoring Diversity: Strategies for Reintroduction of Endangered Plants*, edited by D. A. Falk, C. I. Millar, and M. Olwell, 101–20. Washington, DC: Island Press.

Melbourne, B. A., H. V. Cornell, K. F. Davies, C. J. Dugaw, S. Elmendorf, A. L. Freestone, A. Hastings, et al. 2007. Invasion in a heterogeneous world: Resistance, coexistence or hostile takeover? *Ecology Letters* 10:77–94.

Menges, E. S. 1991. The application of minimum viable population theory to plants. In *Genetics and Conservation of Rare Plants*, edited by D. A. Falk and K. E. Holsinger, 45–61. New York: Oxford University Press.

Menges, E. S. 2000. Population viability analysis in plants: Challenges and opportunities. *Trends in Ecology & Evolution* 15:51–56.

Menges, E. S. 2008. Restoration demography and genetics of plants: When is a translocation successful? *Australian Journal of Botany* 56:187–96.

Menges, E. S., and D. R. Gordon. 1996. Three levels of monitoring intensity for rare plant species. *Natural Areas Journal* 16:227–37.

Menges, E. S., E. O. Guerrant Jr., and S. Hamzé. 2004. Effects of seed collection on the extinction risk of perennial plants. In *Ex Situ Plant Conservation: Supporting Species Survival in the Wild*, edited by E. O. Guerrant Jr., K. Havens, and M. Maunder, 305–24. Covelo, CA: Island Press.

Menges, E. S., and P. F. Quintana-Ascencio. 2004. Population viability with fire in *Eryngium cuneifolium*: Deciphering a decade of demographic data. *Ecological Monographs* 74:79–99.

Midgley, G. F., L. Hannah, D. Millar, W. Thuiller, and A. Booth. 2003. Developing regional and species-level assessment of climate change impacts on biodiversity in the Cape Floristic Region. *Biological Conservation* 112:87–97.

Millar, C. I., and W. J. Libby. 1989. Restoration: Disneyland or a native ecosystem? A question of genetics. *Restoration and Management Notes* 7(1):18–23.

Millar, C. I., N. L. Stephenson, and S. L. Stephens. 2007. Climate change and forests of the future: Managing in the face of uncertainty. *Ecological Applications* 17:2145–51.

Miller, J. R. 2005. Biodiversity conservation and the extinction of experience. *Trends in Ecology & Evolution* 20:430–34.

Mills, D., Y. Zhou, and A. Benzioni. 2009. Effect of substrate medium composition, irradiance and ventilation on jojoba plantlets at the rooting stage of micropropagation. *Scientia Horticulturae* 121:113–18.

Minteer, B. A., and J. P. Collins. 2010. Move it or lose it? The ecological ethics of relocating species under climate change. *Ecological Applications* 20(7):1801–04.

Misic, D. M., N. A. Ghalawenji, D. V. Grubisic, and R. M. Konjevic. 2005. Micropropagation and reintroduction of *Nepeta rtanjensis*, an endemic and critically endangered perennial of Serbia. *Phyton (Horn)* 45:9–20.

Mistretta, O. 1994. Genetics of species re-introductions: Applications of genetic analysis. *Biodiversity and Conservation* 3:184–90.

Mistretta, O., and S. D. White. 2001. Introducing two federally listed carbonate-endemic plants onto a disturbed site in the San Bernardino Mountains, California. In *Southwestern Rare and Endangered Plants: Proceedings of the Third Conference, September 25–28*, edited by J. Maschinski and L. Holter, 20–26. Fort Collins, CO: US Department of Agriculture, Forest Service, Rocky Mountain Research Station.

Mohan, J. E., L. H. Ziska, W. H. Schlesinger, R. B. Thomas, R. C. Sicher, K. George, and J. S. Clark. 2006. Biomass and toxicity responses of poison ivy (*Toxicodendron radicans*) to elevated atmospheric CO_2. *Proceedings of the National Academy of Science* 103:9086–89.

Molofsky, J., and C. K. Augspurger. 1992. The effect of leaf litter on early seedling establishment in a tropical forest. *Ecology* 73:68–77.

Monks, L. 2002. Assessing translocation success. *Danthonia* 11(2):2–3.

Monks, L. 2009. Experimental approaches in threatened plant translocations: How failures can lead to success. *Australasian Plant Conservation* 17(3):8–10.

Monks, L., and D. Coates. 2002. The translocation of two critically endangered *Acacia* species. *Conservation Science Western Australia* 4(3):54–61.

Montalvo, A. M., and N. C. Ellstrand. 2000. Transplantation of the subshrub *Lotus scoparius*: Testing the home site advantage hypothesis. *Conservation Biology* 14:1034–45.

Moody, M., and R. Mack. 1988. Controlling the spread of plant invasions: The importance of nascent foci. *Journal of Applied Ecology* 25:1009–21.

Moora, M., N. Öpik, R. Sen, and M. Zobel. 2004. Native arbuscular mycorrhizal fungal communities differentially influence the seedling performance of rare and common *Pulsatilla* species. *Functional Ecology* 18:554–62.

Morgan, J. W. 1999. Have tubestock plantings successfully established populations of rare grassland species into reintroduction sites in western Victoria? *Biological Conservation* 89:235–43.

Morgan, J. W. 2000. Reproductive success in reestablished versus natural populations of a threatened grassland daisy (*Rutidosis leptorrhynchoides*). *Conservation Biology* 14:780–85.

Morin, X., C. Augspurger, and I. Chuine. 2007. Process-based modeling for tree species' distributions. What limits temperate tree species' range boundaries? *Ecology* 88:2280–91.

Morin, X., and M. J. Lechowicz. 2008. Contemporary perspectives on the niche that can improve models of species range shifts under climate change. *Biology Letters* 4:573–76.

Moritz, K. N., and D. C. Bickerton. 2010. Recovery plan for the Peep Hill hop-bush *Dodonaea subglandulifera* 2010. In A. G. D. o. t. E. Report to the Recovery Planning and Implementation Section, Water, Heritage and the Arts, Canberra.

Morris, W. F., P. L. Bloch, B. R. Hudgens, L. C. Moyle, and J. R. Stinchcombe. 2002. Population viability analysis in endangered species recovery plans: Past use and future improvements. *Ecological Applications* 12:708–12.

Morris, W. F., and D. F. Doak. 2002. *Quantitative Conservation Biology.* Sunderland, MA: Sinauer.

Morte, M. A., G. Diaz, and M. Honrubia. 1996. Effect of arbuscular mycorrhizal inoculation on micropropagated *Tetraclinis articulata* growth and survival. *Agronomie* 16:633–37.

Motley, T. J., and G. D. Carr. 1998. Artificial hybridization in the Hawaiian endemic genus *Labordia* (Loganiaceae). *American Journal of Botany* 85:654–60.

Mottl, L. M., C. M. Mabry, and D. R. Farrar. 2006. Seven-year survival of perennial herbaceous transplants in temperate woodland restoration. *Restoration Ecology* 14(3):330–38.

Moulin, L., A. Munive, B. Dreyfus, and C. Boivin-Masson. 2001. Nodulation of legumes by members of the ª-subclass of Proteobacteria. *Nature* 411:948–50.

MR Working Group. 2010. Managed Relocation Working Group. University of Notre Dame. http://www.nd.edu/~hellmann/MRWorkingGroup/Managed_relocation.html.

Mueller, J. M., and J. J. Hellman. 2008. An assessment of invasion risk from assisted migration. *Conservation Biology* 22:562–67.

Mulvaney, M. 2001. The effect of introduction pressure on the naturalization of ornamental woody plants in south-eastern Australia. In *Weed Risk Assessment*, edited by R. H. Groves, F. D. Panetta, and J. G. Virtue, 186–93. Melbourne, Australia: CSIRO Publishing.

Mummey, D. L., and M. C. Rillig. 2007. Evaluation of LSU rRNA-gene PCR primers for the study of arbuscular mycorrhizal fungal communities via terminal restriction fragment length polymorphism analysis. *Journal of Microbiological Methods* 70:200–204.

Münzbergová, Z., M. Milden, J. Ehrlen, and T. Herben. 2005. Population viability and reintroduction strategies: A spatially explicit landscape-level approach. *Ecological Applications* 15:1377–86.

Murashige, T., and F. Skoog. 1962. A revised medium for rapid growth and bioassays with tobacco tissue cultures. *Physiologia Plantarum* 15:473–97.

Mustart, P., J. Juritz, C. Makua, S. W. VanderMerwe, and N. Wessels. 1995. Restoration of the Clanwilliam cedar *Widdringtonia cedarbergensis*: The importance of monitoring seedlings planted in the Cedarberg, South Africa. *Biological Conservation* 72:73–76.

Nadel, H., J. H. Frank, and R. J. Knight. 1992. Escapees and accomplices: The naturalization of exotic *Ficus* and their associated faunas in Florida. *Florida Naturalist* 75:29–38.

Nakicenovic, N., J. Alcamo, G. Davis, B. de Vries, J. Fenhann, S. Gaffin, K. Gregory, et al. 2000. *Special Report on Emissions Scenarios: A Special Report of Working Group III of the Intergovernmental Panel on Climate Change.* Cambridge, UK: Cambridge University Press. Accessed from http://www.grida.no/climate/ipcc/emission/index.htm.

Nelson, J. T., B. L. Woodworth, S. G. Fancy, G. D. Lindsey, and E. J. Tweed. 2002. Effectiveness of rodent control and monitoring techniques for a montane rainforest. *Wildlife Society Bulletin* 30:82–92.

Newman, C., C. D. Buesching, and D. W. Macdonald. 2003. Validating mammal monitoring methods and assessing the performance of volunteers in wildlife conservation — "*Sed quis custodiet ipsos custodies?*" *Biological Conservation* 113:189–97.

Newman, D., and D. Pilson. 1997. Increased probability of extinction due to decreased genetic effective populations size: Experimental populations of *Clarkia pulchella.* *Evolution* 51:354–62.

Nijjer, S., W. E. Rogers, and E. Siemann. 2007. Negative plant–soil feedbacks may limit persistence of an invasive tree due to rapid accumulation of soil pathogens. *Proceedings of the Royal Society B* 274:2621–27.

Nobuoka, T., T. Nishimoto, and T. Kimie. 2005. Wing and light promote graft-take and growth of grafted tomato seedlings. *Journal of the Japanese Society for Horticultural Science* 74:170–75.

North Carolina Department of Agriculture and Consumer Services. 2010. *Plant Conservation Scientific Committee.* Accessed August 10, 2010, http://www.ncagr.gov/plant industry/plant/plantconserve/scicom.htm.

North Carolina Plant Conservation Program Scientific Committee. 2005. *Rare Plant Reintroduction Guidelines.* Accessed August 10, 2011, http://ncbg.unc.edu/uploads /files/RarePlantReintroductionGuidelines.pdf.

Noss, R. F. 2001. Beyond Kyoto: Forest management in a time of rapid climate change. *Conservation Biology* 15(3):578–90.

Nozawa. 2005a. IPCC DDC AR4 CCSR-MIROC3.2_(med-res) 1PCTTO2X run1. World Data Center for Climate. CERA-DB MIROC3.2_mr_1PCTTO2X_1. Accessed November 1, 2010, http://cera-www.dkrz.de/WDCC/ui/Compact.jsp?acronym =MIROC3.2_mr_1PCTTO2X_1.

Nozawa. 2005b. IPCC DDC AR4 CCSR-MIROC3.2_(med-res) 1PCTTO4X run1. World Data Center for Climate. CERA-DB MIROC3.2_mr_1PCTTO4X_1. Accessed November 1, 2010, http://cera-www.dkrz.de/WDCC/ui/Compact.jsp?acronym =MIROC3.2_mr_1PCTTO4X_1.

Nozawa. 2005c. IPCC DDC AR4 CCSR-MIROC3.2_(med-res) SRESA1B run1. World Data Center for Climate. CERA-DB MIROC3.2_mr_SRESA1B_1. Accessed November 1, 2010, http://cera-www.dkrz.de/WDCC/ui/Compact.jsp?acronym=MIROC3.2 _mr_SRESA1B_1.

Nozawa. 2005d. IPCC DDC AR4 CCSR-MIROC3.2_(med-res) SRESA2 run1. World Data Center for Climate. CERA-DB MIROC3.2_mr_SRESA2_1. Accessed November 1, 2010, http://cera-www.dkrz.de/WDCC/ui/Compact.jsp?acronym=MIROC3.2 _mr_SRESA2_1.

Nozawa. 2005e. IPCC DDC AR4 CCSR-MIROC3.2_(med-res) SRESB1 run1. World Data Center for Climate. CERA-DB MIROC3.2_mr_SRESB1_1. Accessed November 1, 2010, http://cera-www.dkrz.de/WDCC/ui/Compact.jsp?acronym=MIROC3.2 _mr_SRESB1_1.

Nybom, H. 2004. Comparison of different nuclear DNA markers for estimating intraspecific genetic diversity in plants. *Molecular Ecology* 13:1143–55.

Nybom, H., and I. V. Bartish. 2000. Effects of life history traits and sampling strategies on genetic diversity estimates obtained with RAPD markers in plants. *Perspectives in Plant Ecology, Evolution and Systematics* 3:93–114.

Obee, E. M., and R. J. Cartica. 1997. Propagation and reintroduction of the endangered hemiparasite *Schwalbea americana* (Scrophulariaceae). *Rhodora* 99:134–47.

Ogura-Tsujita, Y., and T. Yukawa. 2008. High mycorrhizal specificity in a widespread mycoheterotrophic plant, *Eulophia zollingeri* (Orchidaceae). *American Journal of Botany* 95:93–97.

Ohmann, J. L., and T. A. Spies. 1998. Regional gradient analysis and spatial patterns of woody plant communities of Oregon forests. *Ecological Monographs* 68(2):151–82.

Okon, Y., and C. A. Labandera-Gonzalez. 1994. Agronomic application of *Azospirillum*: An evaluation of 20 years worldwide field inoculation. *Soil Biology and Biochemistry* 26(12):1591–1601.

Olli, E., G. Grendstad, and D. Wolleback. 2001. Correlates of environmental behaviors: Bringing back social context. *Environment and Behavior* 33:181–208.

Olwell, M., A. Cully, and P. Knight. 1990. The establishment of a new population of *Pediocactus knowltonii*: Third year assessment. In *Ecosystem Management: Rare Species and Significant Habitats*, edited by R. S. Mitchell, C. J. Sheviak, and D. J. Leopold, 189–93. *New York State Museum Bulletin* 471.

Olwell, M., A. Cully, P. Knight, and S. Brack. 1987. *Pediocactus knowtonii* recovery efforts. In *Conservation and Management of Rare and Endangered Plants*, edited by T. S. Elias and J. Nelson, 519–22. Sacramento: California Native Plant Society.

Paau, A. S. 1989. Improvement of *Rhizobium* inoculants. *Applied and Environmental Microbiology* 55:862–65.

Packard, S. 1991. Broadcasting seed restores prairie fringed orchid, other small-seeded forbs (Illinois). *Restoration and Management Notes* 9:121–22.

Packer, A., and K. Clay. 2000. Soil pathogens and spatial patterns of seedling mortality in a temperate tree. *Nature* 404:278–81.

Padilla, F. M., and F. I. Pugnaire. 2006. The role of nurse plants in the restoration of degraded environments. *Frontiers in Ecology and the Environment* 4:196–202.

Padilla, I. M. G., E. Carmona, N. Westendorp, and C. L. Encina. 2006. Micropropagation and effects of mycorrhiza and soil bacteria on acclimatization and development of lucumo (*Pouteria lucuma* R. and Pav.) var. *la molina*. *In Vitro Cellular and Developmental Biology: Plant* 42:193–96.

Panwar, J., and J. C. Tarafdar. 2006. Distribution of three endangered medicinal plant species and their colonization with arbuscular mycorrhizal fungi. *Journal of Arid Environments* 65:337–50.

Panwar, J., and A. Vyas. 2002. AM fungi: A biological approach towards conservation of endangered plants in Thar Desert, India. *Current Science* 82(5):576–78.

Parenti, R., and E. O. Guerrant Jr. 1990. Down but not out: Reintroduction of the extirpated Malheur wirelettuce, *Stephanomeria malheurensis*. *Endangered Species Update* 8:62–63.

Parker, K. A. 2008. Translocations: Providing outcomes for wildlife, resource managers, scientists, and the human community. *Restoration Ecology* 16:204–9.

Parmesan, C. 1996. Climate change and species range. *Nature* 382:765–66.

Parmesan, C. 2006. Ecological and evolutionary responses to recent climate change. *Annual Review of Ecology, Evolution, and Systematics* 37:637–69.

Parolo, G., G. Rossi, and A. Ferrarini. 2008. Toward improved species niche modeling: *Arnica montana* in the Alps as a case study. *Journal of Applied Ecology* 45:1410–18.

Parsons, L. S., and J. B. Zedler. 1997. Factors affecting reestablishment of an endangered annual plant at a California salt marsh. *Ecological Applications* 7:253–67.

Pavlik, B. M. 1991. *Reintroduction of* Amsinckia grandiflora *to Three Sites across Its Historic Range*. Sacramento: Endangered Plant Program, California Department of Fish and Game.

Pavlik, B. M. 1996. Defining and measuring success. In *Restoring Diversity, Strategies for Reintroduction of Endangered Plants*, edited by D. A. Falk, C. I. Millar, and M. Olwell, 127–55. Covelo, CA: Island Press.

Pavlik, B. M., and E. K. Espeland. 1998. Demography of natural and reintroduced populations of *Acanthomintha duttonii*, an endangered serpentinite annual in northern California. *Madrono* 45:31–39.

Pavlik, B. M., D. L. Nickrent, and A. M. Howald. 1993. The recovery of an endangered plant. I. Creating a new population of *Amsinckia grandiflora*. *Conservation Biology* 7:510–26.

Pearce, J., and D. Lindenmayer. 1998. Bioclimatic analysis to enhance reintroduction biology of the endangered helmeted honeyeater (*Lichenostomus melanops cassidix*) in southeastern Australia. *Restoration Ecology* 6(3):238–43.

Pearman, D. A., and K. Walker. 2004. Rare plant introductions in the UK: Creative conservation or wildflower gardening? *British Wildlife* 15:174–82.

Pearson, R. G., C. J. Raxworthy, M. Nakamura, and A. T. Peterson. 2007. Predicting species distributions from small numbers of occurrence records: A test case using cryptic geckos in Madagascar. *Journal of Biogeography* 34:102–17.

Pelabon, C., M. L. Carlson, T. F. Hansen, and W. S. Armbruster. 2005. Effects of crossing distance on offspring fitness and developmental stability in *Dalechampia scandens* (Euphorbiaceae). *American Journal of Botany* 92:842–51.

Pemberton, R. W., and H. Liu. 2008a. The naturalization of the oil collecting bee *Centris nitida* (Hymenoptera, Apidae, Centrini), a potential pollinator of selected native ornamental, and invasive plants in Florida. *Florida Entomologist* 91:101–9.

Pemberton, R. W., and H. Liu. 2008b. A naturalized orchid bee pollinates resin reward flowers in southern Florida: Novel and known mutualism. *Biotropica* 40(6):714–18.

Pemberton, R. W., and H. Liu. 2008c. Potential of invasive and native solitary specialist bee pollinators to help restore the rare cowhorn orchid (*Cyrtopodium punctatum*) in Florida. *Biological Conservation* 7:1758–64.

Pemberton, R. W., and H. Liu. 2009. Marketing time predicts naturalization of horticultural plants. *Ecology* 90:69–80.

Pence, V., S. Murray, L. Whitman, D. Cloward, H. Barnes, and R. Van Buren. 2008. Supplementation of the autumn buttercup population in Utah, USA, using in vitro propagated plants. In *Global Re-introduction Perspectives: Re-introduction Case-Studies from around the Globe*, edited by P. S. Soorae, 239–43. Abu Dhabi, UAE: IUCN/SSC Reintroduction Specialist Group.

Pennisi, E. 2010. Tending the global garden. *Science* 329:1274–77.

Perez, H., A. B. Shiels, H. M. Zaleski, and D. R. Drake. 2008. Germination after simulated rat damage in seeds of two endemic Hawaiian palm species. *Journal of Tropical Ecology* 24:555–58.

Peterson, A. T., M. Papes, and D. A. Kluza. 2003. Predicting the potential invasive distribution of four alien plant species in North America. *Weed Science* 51:863–68.

Pheloung, P., P. A. Williams, and S. R. Halloy. 1999. A weed risk assessment model for use as a biosecurity tool evaluating plant introductions. *Journal of Environmental Management* 57:239–51.

Phillips, S. J., R. P. Anderson, and R. E. Schapire. 2006. Maximum entropy modeling of species geographic distributions. *Ecological Modelling* 190:231–59.

Phillips, S. J., M. Dudik, J. Elith, C. Graham, A. Lehmann, J. Leathwick, and S. Ferrier. 2009. Sample selection bias and presence-only distribution models: Implication for background and pseudo-absence data. *Ecological Applications* 19(1):181–97.

Pickett, S. T. A. 1980. Non-equilibrium co-existence of plants. *Bulletin Torrey Botanic Club* 107:238–48.

Pigott, C. D. 1988. The reintroduction of *Cirsium tuberosum* L. All. in Cambridgeshire UK. *Watsonia* 17:149–52.

Pimentel, D., R. Zuniga, and D. Morrison. 2005. Update on the environmental and economic costs associated with alien-invasive species in the United States. *Ecological Economics* 52:273–88.

Pinto-Torres, E., and S. Koptur. 2009. Hanging by a coastal strand: Breeding system of a federally endangered morning-glory of the south-eastern Florida coast, *Jacquemontia reclinata*. *Annals of Botany* 104:1301–11.

Pipoly, J. I., J. Maschinski, J. B. Pascarella, S. J. Wright, and J. Fisher. 2006. *Demography of Coastal Dunes Vines: Endangered* Jacquemontia reclinata, *Endangered* Okenia hypogaea, *and Threatened* Cyperus pedunculatus, *from South Florida*. Final report to the Florida Fish and Wildlife Conservation Commission, Tallahassee, FL.

Pitman, N. C. A., and P. M. Jorgensen. 2002. Estimating the size of the world's threatened flora. *Nature* 298:989.

Polhill, R. M., and P. H. Raven, eds. 1981. *Advances in Legume Systematics*, Parts 1 and 2. Richmond, UK: Royal Botanic Gardens, Kew.

Porley, R. 2005. *Translocation of* Carex vulpina, *Murcott Meadows SSSI, Oxfordshire*. Unpublished report to English Nature.

Pospíšilová, J., D. Haisel, H. Synkova, and P. Batkova-Spoustova. 2009. Improvement of ex vitro transfer of tobacco plantlets by addition of abscisic acid to the last subculture. *Biologia Plantarum* 53:617–24.

Pospíšilová, J., I. Tichá, P. Kadleček, D. Haisel, and Š. Plzáková. 1999. Acclimation of micropropagated plants to ex vitro conditions. *Biologia Plantarum* 42(4):481–97.

Possley, J., K. Hines, J. Maschinski, J. G. Dozier, and C. Rodriguez. 2007. Multiple agencies and volunteers unite to reintroduce goatsfoot passionflower to rockland hammocks of Miami, Florida. *Native Plants Journal* 8:252–58.

Possley, J., and J. Maschinski. 2009. *Year 6 Report: Biological Monitoring for Plant Conservation in Miami–Dade County Natural Areas.* Prepared by Fairchild Tropical Botanic Garden for Miami–Dade County Resolution #R-808-07.

Possley, J., and J. Maschinski. 2010. *Year 7 Report: Biological Monitoring for Plant Conservation in Miami–Dade County Natural Areas.* Miami–Dade County Resolution #R-808-07.

Possley, J., J. Maschinski, C. Rodriguez, and J. Dozier. 2009. Alternatives for reintroducing a rare ecotone species: Manually thinned forest edge versus restored habitat remnant. *Restoration Ecology* 17:668–77.

Power, P. J. 1995. Reintroduction of Texas wildrice (*Zizania texana*) in Spring Lake: Some important environmental and biotic considerations. In *Southwest Rare and Endangered Plants. Proc. 2nd Conf., Sept. 11–14, Flagstaff, AZ. RM-GTR-283,* edited by J. Maschinski et al., 179–86. Ft. Collins, CO: USDA Forest Service.

Pratt, H. D. 1994. Avifaunal change in the Hawaiian islands, 1893–1993. *Studies in Avian Biology* 15:103–18.

Pratt, H. D. 2005. *The Hawaiian Honeycreepers.* Oxford, UK: Oxford University Press.

Preece, J. E., and T. P. West. 2006. Greenhouse growth and acclimatization of encapsulated *Hibiscus mosheutos* nodal segments. *Plant Cell Tissue and Organ Culture* 87:127–38.

Price, J. P., S. M. Gon, J. D. Jacobi, and D. Matsuwaki. 2007. *Mapping Plant Species Ranges in the Hawaiian Islands: Developing a Methodology and Associated GIS Layers.* Accessed November 1, 2010, http://www.uhh.hawaii.edu/hcsu/documents/Priceetal008pdfFinal.pdf.

Primack, R. B. 1996. Lessons from ecological theory: Dispersal, establishment, and population structure. In *Restoring Diversity, Strategies for Reintroduction of Endangered Plants,* edited by D. A. Falk, C. I. Millar, and M. Olwell, 209–33. Washington, DC: Island Press.

Primack, R. B. 2006. *Essentials of Conservation Biology,* 4th ed. Sunderland, MA: Sinauer.

Primack, R. B., and S. L. Miao. 1992. Dispersal can limit local plant distribution. *Conservation Biology* 6:513–19.

Pringle, A., J. D. Bever, M. Gardes, J. L. Parrent, M. C. Rillig, and J. N. Klironomos. 2009. Mycorrhizal symbioses and plant invasions. *Annual Review of Ecology and Systematics* 40:699–715.

Pulliam, H. R. 1988. Sources, sinks and population regulation. *American Naturalist* 132:652–61.

Pulliam, H. R. 2000. On the relationship between niche and distribution. *Ecology Letters* 3:349–61.

Pyšek, P., M. Křivanek, and V. Jarošik. 2009. Planting intensity, residence time, and species traits determine invasive success of alien woody species. *Ecology* 90:2734–44.

Pyšek, P., and D. M. Richardson. 2007. Traits associated with invasiveness in alien plants: Where do we stand? In *Biological Invasions, Ecological Studies*, Vol. 193, edited by W. Nentwig, 97–125. Berlin: Springer-Verlag.

Quinn, R. M., J. H. Lawton, B. C. Eversham, and S. N. Wood. 1994. The biogeography of scarce vascular plants in Britain with respect to habitat preference, dispersal ability and reproductive biology. *Biological Conservation* 70:149–57.

Raabová, J., Z. Münzbergová, and M. Fischer. 2007. Ecological rather than geographic or genetic distance affects local adaptation of the rare perennial herb, *Aster amellus*. *Biological Conservation* 139:348–57.

Rabin, L. B., and R. S. Pacovsky. 1985. Reduced larva growth of two lepidoptera (Noctuidae) on excised leaves of soybean infected with a mycorrhizal fungus. *Journal of Economic Entomology* 78:1358–63.

Rai, A. N. 1990. *Handbook of Symbiotic Cyanobacteria*. Boca Raton, FL: CRC Press.

Ramp, J. M., S. K. Collinge, and T. A. Ranker. 2006. Restoration genetics of the vernal pool endemic *Lasthenia conjugens* (Asteraceae). *Conservation Genetics* 7:631–49.

Ramsay, M. M., and K. W. Dixon. 2003. Propagation science, recovery and translocation of terrestrial orchids. In *Orchid Conservation*, edited by K. W. Dixon, S. P. Kell, R. L. Barrett, and P. J. Cribb, 259–88. Borneo, Malaysia: Natural History Publication.

Ramula, S., T. M. Knight, J. H. Burns, and Y. M. Buckley. 2008. General guidelines for invasive plant management based on comparative demography of invasive and native plant populations. *Journal of Applied Ecology* 45:1124–33.

Rasmussen, H. N. 1995. *Terrestrial orchids from seed to mycotrophic plant*. Cambridge, UK: Cambridge University Press.

Rasmussen, H. N., and D. Whigham. 1993. Seed ecology of dust seeds in situ: A new study technique and its application in terrestrial orchids. *American Journal of Botany* 80:1374–78.

Raunkiaer, C. 1934. *The Life Forms of Plants and Statistical Plant Geography, Being the Collected Papers of C. Raunkiaer*. Oxford, England: Oxford University Press. (Reprinted 1978, edited by F. N. Egerton, in the Ayer Co. Pub. History of Ecology series.)

Rawson, H. M. 1992. Plant responses to temperature under conditions of elevated CO_2. *Australian Journal of Botany* 40:473–90.

Redecker, D. 2000. Specific PCR primers to identify arbuscular mycorrhizal fungi within colonized roots. *Mycorrhiza* 10:73–80.

Redman, R. S., K. B. Sheehan, R. G. Stout, R. J. Rodriguez, and J. M. Henson. 2002. Thermotolerance generated by plant/fungal symbiosis. *Science* 298:1581.

Reed, D. H., and R. Frankham. 2003. Correlation between fitness and genetic diversity. *Conservation Biology* 17:230–37.

Reichard, S. 1997. Preventing the introduction of invasive plants. In *Assessment and Management of Plant Invasions*, edited by J. Luken and J. Thieret, 215–27. New York: Springer-Verlag.

Reichard, S. H., and C. W. Hamilton. 1997. Predicting invasions of woody plants introduced into North America. *Conservation Biology* 11:193–203.

Reichard, S. H., and P. White. 2001. Horticulture as a pathway of invasive plant introductions in the United States. *BioScience* 51:103–13.

Reichard, S., and P. White. 2003. Invasion biology: An emerging field of study. *Annals of the Missouri Botanical Garden* 90:64–66.

Reilly, S. R. 1997. *Iron (Ferric) Phosphate Pesticide Fact Sheet*. Washington, DC: US Environmental Protection Agency Office of Prevention, Environmental Protection Pesticides Agency and Toxic Substances (7501C).

Rejmánek, M., and D. M. Richardson. 1996. What attributes make some plant species more invasive? *Ecology* 77:655–61.

Renison, D., A. M. Cingolani, R. Suarez, E. Menoyo, C. Coutsiers, A. Sobral, and I. Hensen. 2005. The restoration of degraded mountain woodlands: Effects of seed provenance and microsite characteristics on *Polylepis australis* seedling survival and growth in central Argentina. *Restoration Ecology* 13:129–37.

Requena, N., I. Jimenez, M. Toro, and J. M. Barea. 1997. Interactions between plant-growth–promoting rhizobacteria (PGPR), arbuscular mycorrhizal fungi and *Rhizobium* spp. in the rhizosphere of *Anthyllis cystoides*, a model legume for revegetation in mediterranean semi-arid ecosystems. *New Phytologist* 136:667–77.

Reusch, T. B. H., A. Ehlers, A. Hammerli, and B. Worm. 2005. Ecosystem recovery after climatic extremes enhanced by genotypic diversity. *Proceedings of the National Academy of Sciences* 102:2826–31.

Ricciardi, A., and S. K. Atkinson. 2004. Distinctiveness magnifies the impact of biological invaders in aquatic ecosystems. *Ecology Letters* 7:781–84.

Ricciardi, A., and D. Simberloff. 2009a. Assisted colonization is not a viable conservation strategy. *Trends in Ecology & Evolution* 24:248–53.

Ricciardi, A., and D. Simberloff. 2009b. Assisted colonization: Good intentions and dubious risk assessment. *Trends in Ecology & Evolution* 24:476–77.

Rice, K. J., and N. C. Emery. 2003. Managing microevolution: Restoration in the face of global change. *Frontiers in Ecology and the Environment* 1:469–78.

Rich, T. C. G., C. Gibson, and M. Marsden. 1999. Re-establishment of the extinct native plant *Filago gallica* L. (Asteraceae), narrow-leaved cudweed, in Britain. *Biological Conservation* 91:1–8.

Rich, T. C. G., C. R. Lambrick, C. Kitchen, and M. A. R. Kitchen. 1998. Conserving Britain's biodiversity. I: *Thlaspi perfoliatum* L. (Brassicaceae), Cotswold Pennycress. *Biodiversity and Conservation* 7:915–26.

Richardson, D. M., N. Allsopp, C. M. D'Antonio, S. J. Milton, and M. Rejmanek. 2000. Plant invasions: The role of mutualisms. *Biological Review* 75:65–93.

Richardson, D. M., J. J. Hellmann, J. S. McLachlan, D. F. Sax, M. W. Schwartz, P. Gonzalez, E. J. Brennan, et al. 2009. Multidimensional evaluation of managed relocation. *Proceedings of the National Academy of Sciences* 106:9721–24.

Richardson, D. M., and S. I. Higgins. 1998. Pines as invaders in the Southern Hemisphere. In *Ecology and Biogeography of* Pinus, edited by D. M. Richardson, 450–73. Cambridge, UK: Cambridge University Press.

Ricketts, T. H., E. Dinerstein, D. M. Olson, and C. Loucks. 1999. *Terrestrial Ecoregions of North America: A Conservation Assessment.* Washington, DC: Island Press.

Rieseberg, L. H., and S. M. Swensen. 1996. Conservation genetics of endangered island plants. In *Conservation Genetics: Case Histories from Nature,* edited by J. C. Avise and J. L. Hamrick, 305–34. New York: Chapman & Hall.

Riffle, J. W., and R. W. Tinus. 1982. Ectomycorrhizal characteristics, growth, and survival of artificially inoculated ponderosa and Scots pine in a greenhouse and plantation. *Forest Science* 28:646–60.

Rimer, R. L., and K. A. McCue. 2005. Restoration of *Helenium virginicum* Blake, a threatened plant of the Ozark Highlands. *Natural Areas Journal* 25:86–90.

Ritland, K. 2002. Extension of models for the estimation of mating systems using n independent loci. *Heredity* 88:367–68.

Robichaux, R. H., E. A. Friar, and D. W. Mount. 1997. Molecular genetic consequences of a population bottleneck associated with reintroduction of the Mauna Kea silversword (*Argyroxiphium sandwicense* ssp. *sandwicense* [Asteraceae]). *Conservation Biology* 11:1140–46.

Rodriguez-Romero, A. S., M. S. P. Guerra, and M. C. Jaizme-Vega. 2005. Effect of arbuscular mycorrhizal fungi and rhizobacteria on banana growth and nutrition. *Agronomy for Sustainable Development* 25:395–99.

Rogers, D. L., and A. M. Montalvo. 2004. *Genetically Appropriate Choices for Plant Materials to Maintain Biological Diversity.* Lakewood, CO: USDA Forest Service, Rocky Mountain Region.

Roncal, J., J. Maschinski, B. Schaffer, S. M. Gutierrez, and D. Walters. In press. Testing appropriate habitat outside of historic range: The case of *Amorpha herbacea* var. *crenulata* (Fabaceae). *Journal for Nature Conservation.*

Rosbrook, P. A. 1990. Effect of inoculum type and placement on nodulation and growth of *Casuarina cunninghamiana* seedlings. *Forest Ecology and Management* 36:135–47.

Rosenzweig, M. L. 2001. The four questions: What does the introduction of exotic species do to diversity? *Evolutionary Ecology Research* 3:361–67.

Rout, T. M., C. E. Hauser, and H. P. Possingham. 2007. Minimise long-term loss or maximise short-term gain? Optimal translocation strategies for threatened species. *Ecological Modelling* 201:67–74.

Roy, J. 1990. In search of the characteristics of plant invaders. In *Biological Invasions in Europe and the Mediterranean Basin,* edited by A. diCastri, J. Hansen, and M. Debussche, 335–52. Dordrecht, The Netherlands: Kluwer Academic Press.

Rubluo, A., V. Chave, and A. Martinez. 1989. In vitro seed germination and reintroduction of *Bletia urbana* (Orchidaceae) in its natural habitat. *Lindleyana* 4:68–73.

Rudgers, J. A., and K. Clay. 2008. An invasive plant–fungal mutualism reduces arthropod diversity. *Ecology Letters* 11:831–40.

Ruhren, S., and S. N. Handel. 2003. Herbivory constrains survival, reproduction, and mutualisms when restoring nine temperate forest herbs. *Journal Torrey Botanic Society* 130:34–42.

Russell, S. K., and E. W. Schupp. 1998. Effects of microhabitat patchiness on patterns of seed dispersal and seed predation of *Cercocarpus ledifolius* (Rosaceae). *Oikos* 81:434–43.

Safford, H. D., J. J. Hellmann, J. McLachlan, D. F. Sax, and M. W. Schwartz. 2009. Managed Relocation of Species: Noah's Ark or Pandora's Box? *Eos Transactions American Geophysical Union* 90:15.

Sainz-Ollero, H., and J. E. Hernandez-Bermejo. 1979. Experimental reintroductions of endangered plant-species in their natural habitats in Spain. *Biological Conservation* 16:195–206.

Sakai, A. K., F. W. Allendorf, J. S. Holt, D. M. Lodge, J. Molofsky, K. A. With, S. Baughmann, et al. 2001. The population biology of invasive species. *Annual Review of Ecology and Systematics* 32:305–32.

Sakai, A. K., W. L. Wagner, D. M. Ferguson, and D. R. Herbst. 1995a. Biogeographical and ecological correlates of dioecy in the Hawaiian flora. *Ecology* 76:2530–43.

Sakai, A. K., W. L. Wagner, D. M. Ferguson, and D. R. Herbst. 1995b. Origins of dioecy in the Hawaiian flora. *Ecology* 76:2517–29.

Sampson, J. F., D. J. Coates, and S. J. Van Leeuwen. 1996. Mating system variation in animal-pollinated rare and endangered plant populations in Western Australia. In *Gondwanan Heritage: Past, Present and Future of the Western Australian Biota*, edited by S. D. Hopper, J. A. Chappill, M. S. Harvey, and A. S. George, 187–95. Chipping Norton, Australia: Surrey Beatty & Sons.

Sanders, S., and J. B. McGraw. 2005. Population differentiation of a threatened plant: Variation in response to local environment and implications for restoration. *Journal of the Torrey Botanical Society* 132:561–72.

SAS Institute. 2001. SAS proprietary software release, version 8.0. Cary, NC: SAS Institute Inc.

Satterthwaite, W. H., K. D. Holl, G. F. Hayes, and A. L. Barber. 2007. Seed banks in plant conservation: Case study of Santa Cruz tarplant restoration. *Biological Conservation* 135:57–66.

Sauer, J. D. 1988. *Plant Migration: The Dynamics of Geographic Patterning in Seed Plant Species*. Los Angeles: University of California Press.

Sax, D. F., K. F. Smith, and A. R. Thompson. 2009. Managed relocation: A nuanced evaluation is needed. *Trends in Ecology & Evolution* 24:472–73.

Schaal, B. A., and L. G. Leverich. 2004. Population genetic issues in ex situ plant conservation. In *Ex Situ Plant Conservation: Supporting Species Survival in the Wild*, edited by E. O. Guerrant Jr., K. Havens, and M. Maunder, 267–85. Washington, DC: Island Press.

Schaal, B. A., and W. J. Leverich. 2005. Conservation genetics: Theory and practice. *Annals of the Missouri Botanical Gardens* 92:1–11.

Schemske, D. W., B. C. Husband, M. H. Ruckelshaus, C. Goodwillie, I. M. Parker, and J. B. Bishop. 1994. Evaluating approaches to the conservation of rare and endangered plants. *Ecology* 75:584–606.

Schlaepfer, M. A., W. D. Helenbrook, K. B. Searing, and K. T. Shoemaker. 2009. Assisted colonization: Evaluating contrasting management actions (and values) in the face of uncertainty. *Trends in Ecology & Evolution* 24:471–72.

Schultz, P. W. 2000. Empathizing with nature: The effects of perspective taking on concern for environmental issues. *Journal of Social Issues* 56:391–406.

Schupp, E. W. 1995. Seed–seedling conflicts, habitat choice, and patterns of plant recruitment. *American Journal of Botany* 82:399–409.

Schwartz, M. W. 2003. Assessing population viability in long-lived plants. In *Population Viability in Plants: Conservation, Management, and Modeling of Rare Plants*, edited by C. A. Brigham and M. W. Schwartz, 239–66. Berlin: Springer-Verlag.

Schwartz, M. W. 2005. Conservationists should not move *Torreya taxifolia*. *Wild Earth* Winter 2005:73–79.

Schwartz, M. W., J. J. Hellmann, and J. S. McLachlan. 2009. The precautionary principle in managed relocation is misguided advice. *Trends in Ecology & Evolution* 24:474.

Schwartz, M. W., J. D. Hoeksema, C. A. Gehring, N. C. Johnson, J. N. Klironomos, L. K. Abbott, and A. Pringle. 2006. The promise and the potential consequences of the global transport of mycorrhizal fungal inoculum. *Ecology Letters* 9:601–16.

Schwinning, S., J. Belnap, D. R. Bowling, and J. R. Ehleringer. 2008. Sensitivity of the Colorado Plateau to change: Climate, ecosystems, and society. *Ecology and Society* 13:37–57.

Schwinning, S., B. I. Starr, and J. R. Ehleringer. 2003. Dominant cold desert plants do not partition warm season precipitation by event size. *Oecologia* 136:252–60.

Scott, J. M., S. Mountainspring, F. L. Ramsey, and C. B. Kepler. 1986. *Forest Bird Communities of the Hawaiian Islands: Their Dynamics, Ecology, and Conservation*. Lawrence, KS: Allen Press and Cooper Ornithological Society (Studies in Avian Biology 9).

Secretariat of the Convention on Biological Diversity. 2010. *Global Biodiversity Outlook 3*. Montréal: Convention on Biological Diversity.

Seddon, P. J. 2010. From reintroduction to assisted colonization: Moving along the conservation translocation spectrum. *Restoration Ecology* 18(6):796–802.

Seddon, P. J., D. P. Armstrong, and R. F. Maloney. 2007. Developing the science of reintroduction biology. *Conservation Biology* 21:303–12.

Seddon, P. J., D. P. Armstrong, P. Soorae, F. Launay, S. Walker, C. R. Ruiz-Miranda, S. Molur, H. Koldewey, and D. G. Kleiman. 2009. The risks of assisted colonization. *Conservation Biology* 23:788–89.

Severns, P. M. 2003. Propagation of a long-lived and threatened prairie plant, *Lupinus sulphureus* ssp. *kincaidii*. *Restoration Ecology* 11:334–42.

Sharma, D., R. Kapoor, A. K. Bhatnagar. 2008. Arbuscular mycorrhizal (AM) technology for the conservation of *Curcurligo orchioides* Gaertn.: An endangered medicinal herb. *World Journal of Microbiology and Biotechnology* 24:395–400.

Sharma, J., L. W. Zettler, J. W. van Sambeek, M. R. Ellersieck, and C. J. Starbuck. 2003. Symbiotic seed germination and mycorrhizae of federally threatened *Platanthera praeclara* (Orchidaceae). *American Midland Naturalist* 149:104–20.

Sheridan, P. M., and N. Penick. 2000. Highway rights-of-way as rare plant restoration habitat in coastal Virginia. In *Seventh International Symposium on Environmental Concerns in Rights-of-Way-Management*, 185–91. Amsterdam: Elsevier.

Shim, S.-W., E.-J. Hahn, and K.-Y. Paek. 2003. In vitro and ex vitro growth of grapevine rootstock "5BB" as influenced by number of air exchanges and the presence or absence of sucrose in culture media. *Plant Cell Tissue and Organ Culture* 75:57–62.

Shirey, P. D., and G. A. Lamberti. 2010. Assisted migration under the U.S. Endangered Species Act. *Conservation Letters* 3:1–8.

Shirey, P. D., and G. A. Lamberti. 2011. Regulate trade in rare plants. *Nature* 469:465–67.

Siano, A. B., C. K. Kim, M. Y. Chung, J. S. Park, K. B. Lim, and J.-D. Chung. 2007. Effect of several culture conditions, sucrose and CO_2 concentration on growth of chrysanthemum propagules. *Horticulture Environment and Biotechnology* 48:332–36.

Siemann, E., W. E. Rogers, and S. J. Dewalt. 2006. Rapid adaptation of insect herbivores to an invasive plant. *Proceedings of the Royal Society B* 273:2763–69.

Silvertown, J. M. 1982. *Introduction to Plant Population Ecology*. London: Longman.

Silvertown, J., M. Dodd, D. Gowing, C. Lawson, and K. McConway. 2006. Phylogeny and the hierarchical organization of plant diversity. *Ecology* 87:S39–49.

Silvertown, J., M. Franco, and E. Menges. 1996. Interpretation of elasticity matrices as an aid to the management of plant populations for conservation. *Conservation Biology* 10:591–97.

Silvertown, J., M. Franco, I. Pisanty, and A. Mendoza. 1993. Comparative plant demography: Relative importance of life-cycle components to the finite rate of increase in woody and herbaceous perennials. *Journal of Ecology* 81:465–76.

Simard, S. W., D. A. Perry, M. D. Jones, D. D. Myrold, D. M. Durall, and R. Molina. 1997. Net transfer of carbon between ectomycorrhizal tree species in the field. *Nature* 388:579–82.

Simberloff, D. 2009. Invasions of plant communities: More of the same, something very different, or both? *American Midland Naturalist* 163:220–33.

Simonich, M. T., and M. D. Morgan. 1990. Researchers successful in transplanting dwarf lake iris ramets (Wisconsin). *Restoration and Management Notes (Ecological Restoration)* 8:131–32.

Sinclair, A., and P. M. Catling. 2004. Restoration of *Hydrastis canadensis*: Experimental test of a disturbance hypothesis after two growing seasons. *Restoration Ecology* 12:184–89.

Sinclair, E., and R. J. Hobbs. 2009. Sample size effects on estimates of population genetic structure: Implications for ecological restoration. *Restoration Ecology* 17:837–44.

Sivinski, R. 2008. *Knowlton's cactus* (Pediocactus knowltonii). Santa Fe: New Mexico Forestry Division.

Sivinski, R., and P. Tonne. 2008. *Holy Ghost Ipomopsis Recovery Plan*. Albuquerque, NM: US Fish and Wildlife Service.

Slobodkin, L. B. 2001. The good, the bad and the reified. *Evolutionary Ecology Research* 3:1–13.

Smith, S., A. A. Sher, and T. A. I. Grant. 2007. Genetic diversity in restoration materials and the impacts of seed collection in Colorado's restoration plant production industry. *Restoration Ecology* 15:369–74.

Smith, S. E., and D. J. Read. 1997. *Mycorrhizal Symbiosis*, 2nd ed. San Diego, CA: Academic Press.

Smith, T. 1998. Reintroduction of running buffalo clover (*Trifolium stoloniferum*) in Missouri. *Missouriensis* 19:12–19.

Smith, T. E. 1999. *Geocarpon* successfully relocated at a western Missouri site. *Ecological Restoration* 17:91–92.

Smith, T. E. 2003. Observations on the experimental planting of *Lindera melissifolia* (Water) Blume in southeastern Missouri after ten years. *Castanea* 68:75–80.

Smith, Z. F., E. A. James, M. J. McDonnell, and C. B. McLean. 2009. Planting conditions improve translocation success of the endangered terrestrial orchid *Diuris fragrantissima* (Orchidaceae). *Australian Journal of Botany* 57:200–209.

Smulders, M. J. M., J. van der Schoot, R. H. E. M. Geerts, A. G. Antonisse-de Jong, H. Korevaar, A. van der Werf, and B. Vosman. 2000. Genetic diversity and the reintroduction of meadow species. *Plant Biology* 2:447–54.

Soberon, J. 2007. Grinnellian and Eltonian niches and geographic distributions of species. *Ecology Letters* 10:1115–23.

Society for Ecological Restoration. 2009. Global Restoration Network website. http://www.globalrestorationnetwork.org.

Society for Ecological Restoration Science & Policy Working Group (SER). 2002. *The SER Primer on Ecological Restoration.* Accessed November 1, 2010, http://www.ser.org.

South, A. 1992. *Terrestrial Slugs, Biology, Ecology and Control.* New York: Chapman & Hall.

Srivastava, D. 1999. Using local–regional richness plots to test for species saturation: Pitfalls and potential. *Journal of Animal Ecology* 68:1–16.

Stadler, J., A. Trefflich, S. Klotz, and R. Brandl. 2000. Exotic plant species invade diversity hotspots: The alien flora of northwestern Kenya. *Ecography* 23:169–76.

StataCorp, USA. 2011. Stata version 11.0.

Steenhoudt, O., and J. Vanderleyden. 2000. *Azospirillum*, a free-living nitrogen-fixing bacterium closely associated with grasses: Genetic, biochemical and ecological aspects. *FEMS Microbiology Reviews* 24:487–506.

Stiling, P., A. Rossi, and D. Gordon. 2000. The difficulties of single factor thinking in restoration: Replanting a rare cactus in the Florida Keys. *Biological Conservation* 94:327–33.

Stockwell, D. R. B., and A. T. Peterson. 2002. Effects of sample size on accuracy of species distribution models. *Ecological Modelling* 248:1–13.

Stockwell, C. A., P. Andrew, A. P. Hendry, and M. T. Kinnison. 2003. Contemporary evolution meets conservation biology. *Trends in Ecology & Evolution* 18:94–101.

Stoeser, D. B., G. N. Green, L. C. Morath, W. D. Heran, A. B. Wilson, D. W. Moore, and B. S. Van Gosen. 2007. *Preliminary Integrated Geologic Map Databases for the US: Central States: Montana, Wyoming, Colorado, New Mexico, North Dakota, South Dakota, Nebraska, Kansas, Oklahoma, Texas, Iowa, Missouri, Arkansas and Louisiana.* Version 1.2. Accessed November 1, 2010, http://pubs.usgs.gov/of/2005/1351.

Stohlgren, T. J., D. Binkely, G. W. Chong, M. A. Kalkhan, L. D. Schell, K. A. Bull, Y. Otsuki, et al. 1999. Exotic plant species invade hot spots of native plant diversity. *Ecological Monographs* 69:25–46.

Stone, C. P., and J. M. Scott. 1984. *Hawai'i's Terrestrial Ecosystems: Preservation and Management*. Honolulu: University of Hawaii Cooperative National Park Resources Studies Unit.

Stone, R. 2010. Home, home outside the range? *Science* 329:1592–94.

Strauss, S. Y., C. O. Webb, and N. Salamin. 2006. Exotic taxa less related to native species are more invasive. *Proceedings of the National Academy of Science* 103:5841–45.

Suding, K. N., and K. L. Gross. 2006. The dynamic nature of ecological systems: Multiple states and restoration trajectories. In *Foundations of Restoration Ecology*, edited by D. A. Falk, M. A. Palmer, and J. B. Zedler, 190–209. Washington, DC: Island Press.

Sugii, N., and C. Lamoureux. 2004. Tissue culture as a conservation method: An empirical view from Hawaii. In *Ex Situ Plant Conservation: Supporting Species Survival in the Wild*, edited by E. O. Guerrant Jr., K. Havens, and M. Maunder, 189–205. Washington, DC: Island Press.

Sutter, R. D. 1996. Monitoring. In *Restoring Diversity, Strategies for Reintroduction of Endangered Plants*, edited by D. A. Falk, C. I. Millar, and M. Olwell, 235–64. Covelo, CA: Island Press.

Svenning, J.-C., C. Flojgaard, N. Morueta-Holme, J. Lenoir, S. Normand, and F. Skov. 2009. Big moving day for biodiversity? A macroecological assessment of the scope for assisted colonization as a conservation strategy under global warming. *Earth and Environmental Science* 8:1–12.

Sylvia, D. M. 1989. Nursery inoculation of sea oats with vesicular–arbuscular mycorrhizal fungi and outplanting performance on Florida beaches. *Journal of Coastal Research* 5(4):747–54.

Taylor, S. I., and F. Levy. 2002. Responses to soils and a test for preadaptation to serpentine in *Phacelia dubia* (Hydrophyllaceae). *New Phytologist* 155:437–47.

Tecic, D. L., J. L. McBride, M. L. Bowles, and D. L. Nickrent. 1998. Genetic variability in the federal threatened Mead's milkweed, *Asclepias meadii* Torrey (Asclepiadaceae), as determined by allozyme electrophoresis. *Annals of the Missouri Botanical Garden* 85:97–109.

Temperton, V. M., R. J. Hobbs, T. Nuttle, and S. Halle, eds. 2004. *Assembly Rules and Restoration Ecology: Bridging the Gap between Theory and Practice*. Washington, DC: Island Press.

Thomas, C. D., A. Cameron, R. E. Green, M. Bakkenes, L. J. Beaumont, Y. C. Collingham, B. F. N. Erasmus, et al. 2004. Extinction risk from climate change. *Nature* 427:145–48.

Thompson, J., A. E. Lugo, and J. Thomlinson. 2007. Land use history, hurricane disturbance, and the fate of introduced species in a subtropical wet forest in Puerto Rico. *Plant Ecology* 192:289–301.

Thomson, A. M., and W. H. Parker. 2008. Boreal forest provenance tests used to predict optimal growth and response to climate change. 1. Jack pine. *Canadian Journal of Forest Research* 38:157–70.

Thomson, D. 2005. Measuring the effects of invasive species on the demography of a rare endemic plant. *Biological Invasions* 7:615–24.

Thomson, J. D., and S. C. H. Barret. 1981. Selection for outcrossing, sexual selection, and the evolution of dioecy in plants. *American Naturalist* 118:443–49.

Thornton, H. E. B., J. Roncal, C. E. Lewis, J. Maschinski, and J. Francisco-Ortega. 2008. Conservation genetics of *Jacquemontia reclinata* (Convolvulaceae), an endangered species from southern Florida: Implications for restoration management. *Biotropica* 40:507–14.

Thornton, H. E. B., and S. J. Wright. 2003. Site-specific recommendations for management of *Jacquemontia reclinata* populations. In *Restoration of* Jacquemontia reclinata *to the South Florida ecosystem*, edited by J. Maschinski, S. J. Wright, and H. E. B. Thornton, 97–104. Final Report to the US Fish and Wildlife Service, Vero Beach, FL.

Thorpe, A. S., R. T. Massatti, and T. N. Kaye. 2008a. Abronia umbellata *var.* breviflora *on the Oregon Coast: Reintroduction and Population Monitoring*. Corvallis, OR: Institute for Applied Ecology, USDA Forest Service, Siuslaw National Forest, USDI Bureau of Land Management, Coos Bay District, and Oregon Department of Parks and Recreation.

Thorpe, A., R. T. Massatti, and T. Kaye. 2008b. *Reintroduction of* Lupinus sulphureus *ssp.* kincaidii *(Kincaid's lupine) to Dragonfly Bend in the West Eugene Wetlands 2008 Report*. Corvallis, OR: Institute for Applied Ecology.

Thuiller, W., S. Lavorel, M. B. Araujo, M. T. Sykes, and I. C. Prentice. 2005. Climate change threats to plant diversity in Europe. *Proceedings of the National Academy of Science* 102(23):8245–50.

Tilman, D. 1988. *Plant Strategies and the Dynamics and Structure of Plant Communities*. Princeton, NJ: Princeton University Press.

Tilman, D. 2004. Niche tradeoffs, neutrality, and community structure: A stochastic theory of resource competition, invasion, and community assembly. *Proceedings of the National Academy of Science* 101:10854–61.

Tilman, D., and C. Lehman. 2001. Human-caused environmental change: Impacts on plant diversity and evolution. *Proceedings of the National Academy of Sciences of the United States of America* 98:5433–40.

Tilman, D., D. Wedin, and J. A. Knops. 1996. Productivity and sustainability influenced by biodiversity in grassland ecosystems. *Nature* 379:718–20.

Timm, O., and H. F. Diaz. 2009. Synoptic–statistical approach to regional downscaling of IPCC 21st century climate projections: Seasonal rainfall over the Hawaiian Islands. *American Meteorological Society* 22:4261–80.

Titus, J. H., and P. H. Titus. 2008. Assessing the reintroduction potential of the endangered Huachua (water umble) in southeastern Arizona. *Ecological Restoration* 26:312–20.

Tobin, M. E., R. T. Sugihara, and A. E. Koehler. 1997. Bait placement and acceptance by rats in macadamia orchards. *Crop Protection* 16:507–10.

Towns, D. R., I. A. E. Atkinson, and C. H. Daugherty. 2006. Have the harmful effects of introduced rats on islands been exaggerated? *Biological Invasions* 8:863–91.

Traill, L. W., C. J. A. Bradshaw, and B. W. Brook. 2007. Minimum viable population size:

A meta-analysis of 30 years of published estimates. *Biological Conservation* 139:159–66.

Travis, S. E., and P. Sheridan. 2006. Genetic structure of natural and restored shoalgrass *Halodule wrightii* populations in the NW Gulf of Mexico. *Marine Ecology Progress Series* 322:117–27.

Tremblay, R. L. 2008. Ecological correlates and short-term effects of relocation of a rare epiphytic orchid after Hurricane Georges. *Endangered Species Research* 5:83–90.

Turner, M. G. 2010. Disturbance and landscape dynamics in a changing world. *Ecology* 91(10):2833–49.

Udvardy, M. 1975. *A Classification of the Biogeographical Provinces of the World. Man and Biosphere Programme.* IUCN Occasional Paper No. 18. Morges, Switzerland.

United Nations Environment Program. 2009. *Climate Change Science Compendium.* Accessed October 1, 2010, http://www.unep.org/compendium2009.

Urgenson, L., S. Reichard, and C. Halpern. 2009. Community and ecosystem consequences of giant knotweed (*Polygonum sachalinense*) invasion into riparian forests of western Washington, USA. *Biological Conservation* 142:1536–41.

US Army Garrison Hawaii and Makua Implementation Team. 2003. *Final Draft: Makua Implementation Plan.* Prepared for the US Army Garrison, Schofield Barracks, HI.

US Fish and Wildlife Service. 1999. *South Florida Multi-Species Recovery Plan.* Atlanta, GA: Southeast Region, US Fish and Wildlife Service.

US Fish and Wildlife Service. 2005. Cyanea superba *(Haha) 5-Year Review Summary and Evaluation.* Honolulu: US Fish and Wildlife Service.

US Fish and Wildlife Service. 2008. *Amendment of the Biological Opinion of the U.S. Fish and Wildlife Service for Military Training at Makua Military Reservation* (1-2-2005-F-356). Honolulu, HI: US Fish and Wildlife Service.

US Fish and Wildlife Service. 2009. *Endangered Species Program.* Accessed October 1, 2009, http://www.fws.gov/endangered/.

Valentine, D. H. 1977. The pollination of introduced species, with special reference to the British Isles and the genus *Impatiens.* In *The Pollination of Flowers by Insects,* edited by A. J. Richards, 117–23. New York: Academic Press.

Vallee, L., T. Hogbin, L. Monks, B. Makinson, M. Matthes, and M. Rossetto. 2004. *Guidelines for the Translocation of Threatened Plants in Australia,* 2nd ed. Australian Network for Plant Conservation, Canberra. Accessed November 1, 2010, http://www.anbg.gov.au/anpc/books.html#Translocation.

van Andel, J. 1998. Intraspecific variability in the context of ecological restoration projects. *Perspectives in Plant Ecology, Evolution and Systematics* 1:221–27.

Van Groenendael, J. M., N. J. Ouborg, and R. J. J. Hendriks. 1998. Criteria for the introduction of plant species. *Acta Botanica Neerlandica* 47:3–13.

van Riper, C. III, and J. M. Scott. 2001. Limiting factors affecting Hawaiian native birds. In *Evolution, Ecology, Conservation, and Management of Hawaiian Birds: A Vanishing Avifauna,* edited by J. M. Scott, S. Conant, and C. van Riper III, 220–33. Berkeley: University of California Press.

Varma, A., S. Verma, Sudha, N. Sahay, B. Bütehorn, and P. Franken. 1999. *Piriformospora indica*, a cultivable plant-growth-promoting root endophyte. *Applied and Environmental Microbiology* 65:2741–44.

Veblen, T. T. 1992. Regeneration dynamics. In *Plant Succession: Theory and Prediction*, edited by D. C. Glenn-Lewin, R. K. Peet, and T. T. Veblen, 152–87. London: Chapman & Hall.

Vellinga, E. C., B. E. Wolfe, and A. Pringle. 2009. Global patterns of ectomycorrhizal introductions. *New Phytologist* 181:960–73.

Verma, S., A. Varma, K. H. Rexer, G. Kost, A. Sarbhoy, P. Bisen, B. Bütehorn, and P. Franken. 1998. *Piriformospora indica*, gen. et sp. nov., a new root-colonizing fungus. *Mycologia* 95:896–903.

Vitt, P., K. Havens, and O. Hoegh-Guldberg. 2009. Assisted migration: Part of an integrated conservation strategy. *Trends in Ecology & Evolution* 24:473–74.

Vitt, P., K. Havens, A. T. Kramer, D. Sollenberger, and E. Yates. 2010. Assisted migration of plants: Changes in latitudes, changes in attitudes. *Biological Conservation* 143:18–27.

Von Holle, B., and D. Simberloff. 2005. Ecological resistance to biological invasion overwhelmed by propagule pressure. *Ecology* 86:3212–18.

Wacholder, S. 1986. Binomial regression in GLIM: Estimating risk ratios and risk differences. *American Journal of Epidemiology* 123:174–84.

Wagner, W. L., D. R. Herbst, and S. H. Sohmer. 1999. *Manual of the Flowering Plants of Hawai'i*, revised ed. Honolulu: University of Hawai'i Press/Bishop Museum Press.

Wagner, W. L., S. G. Weller, and A. Sakai. 2005. Monograph of *Schiedea* (Caryophyllaceae–Alsininuideae). *Systematic Botany Monographs* 72.

Wall, L. G. 2000. The actinorhizal symbiosis. *Journal of Plant Growth Regulation* 19:167–82.

Walter, M. 2005. Transplanting and sowing seeds of common cow-wheat *Melampyrum pratense* to increase its distribution at Blean Woods RSPB Reserve, Kent, England. *Conservation Evidence* 2:41–42.

Walther, G.-R., E. Post, P. Convey, A. Menzel, C. Parmesan, T. J. C. Beebee, J. Fromenti, O. Hoegh-Guldberg, and F. Bairlein. 2002. Ecological responses to recent climate change. *Nature* 416:389–95.

Wang, H., S. Parent, A. Gosselin, and Y. Desjardins. 1993. Study of vesicular–arbuscular mycorrhizal peat-based substrates on symbiosis establishment and growth of three micropropagated species. *Journal of the American Society for Horticultural Science* 118:896–901.

Warcup, J. H. 1971. Specificity of mycorrhizal association in some Australian terrestrial orchids. *New Phytologist* 70:41–46.

Warcup, J. H. 1973. Symbiotic germination of some Australian terrestrial orchids. *New Phytologist* 72:387–92.

Warren, C. R. 2007. Perspectives on the "alien" versus "native" species debate: A critique of concepts, language and practice. *Progress in Human Geography* 31:427–46.

Waser, N. M., and M. V. Price. 1994. Crossing-distance effects in *Delphinium nelsonii*: Outbreeding and inbreeding depression in progeny fitness. *Evolution* 48:842–52.

Weekley, C. W. 2004. Experimental introduction of Florida *Ziziphus* at Lake Wales Ridge National Wildlife Refuge. *Saving Our Research* 4–5. Lake Wales Ridge National Wildlife Refuge newsletter.

Weekley, C. W., T. L. Kubisiak, and T. M. Race. 2002. Genetic impoverishment and cross-incompatibility in remnant genotypes of *Ziziphus celata* (Rhamnaceae), a rare shrub endemic to the Lake Wales Ridge, Florida. *Biodiversity and Conservation* 11:2027–46.

Weekley, C. W., and E. S. Menges. 2008. Experimental introductions of Florida ziziphus on Florida's Lake Wales Ridge, USA. In *Global Re-introduction Perspectives: Re-introduction Case-Studies from around the Globe*, edited by P. S. Soorae, 256–60. Abu Dhabi, UAE: IUCN/SSC Re-introduction Specialist Group.

Weekley, C. W., and T. M. Race. 1999. *10-Year Status Report on* Ziziphus celata, *a Federally Listed Endemic Plant of the Lake Wales Ridge, Florida*. Report prepared for the US Fish and Wildlife Service.

Weekley, C. W., and T. M. Race. 2001. The breeding system of *Ziziphus celata* Judd and D.W. Hall (Rhamnaceae), a rare endemic plant of the Lake Wales Ridge, Florida, USA: Implications for recovery. *Biological Conservation* 100:207–13.

Weekley, C., T. Race, and D. Hardin. 1999. Saving Florida ziziphus: Recovery of a rare Lake Wales Ridge endemic. *The Palmetto* 19(2):9–10, 20.

Weinbaum, B. S., M. F. Allen, and E. B. Allen. 1996. Survival of arbuscular mycorrhizal fungi following reciprocal transplanting across the Great Basin, USA. *Ecological Applications* 6(4):1365–72.

Wendelberger, K. S., M. Q. N. Fellows, and J. Maschinski. 2008. Rescue and restoration: Experimental translocation of *Amorpha herbacea* Walter var. *crenulata* (Rybd.) Isley into a novel urban habitat. *Restoration Ecology* 16:542–52.

Wendelberger, K. S., and J. Maschinski. 2006. *Microhabitat Preference and Demography of* Tephrosia angustissima *var.* corallicola *in an Experimental Introduction*. Gainesville, FL: FDACS.

Wendelberger, K. S., and J. Maschinski. 2009. Linking GIS, observational and experimental studies to determine optimal seedling microsites of an endangered plant in a subtropical urban fire-adapted ecosystem. *Restoration Ecology* 17:845–53.

Wheeler, B. M. 2001. *Starfruit* Damasonium alisma *Project in 2000: Part One: Starfruit in 2000; Part Two: Survey of Wetland Plants and Aquatic Macroinvertebrates in Five Starfruit Ponds with Notes on Management*. Plantlife Report no. 167. London: Plantlife.

White, P. S., and A. Jentsch. 2004. Disturbance, succession and community assembly in terrestrial plant communities. In *Assembly Rules and Restoration Ecology*, edited by V. M. Temperton, R. J. Hobbs, T. Nuttle, and S. Halle, 342–66. Washington, DC: Island Press.

White, P. S., and J. L. Walker. 1997. Approximating nature's variation: Selecting and using reference information in restoration ecology. *Restoration Ecology* 5(4):338–49.

Whittaker, R. H. 1975. The design and stability of plant communities. In *Unifying Concepts in Ecology*, edited by W. H. van Dobben and R. H. Lowe-McConnell, 169–74. The Hague, The Netherlands: Junk.

Wiens, J. A., D. Stralberg, D. Jongsomjit, C. A. Howell, and M. A. Snyder. 2009. Niches, models, and climate change: Assessing the assumptions and uncertainties. *Proceedings of the National Academy of Sciences* 106:19729–36.

Wilby, R. L., S. P. Charles, E. Zorita, B. Timbal, P. Whetton, and L. O. Mearns. 2004. Guidelines for use of climate scenarios developed from statistical downscaling methods. In *Supporting Material of Procedures for the Preparation, Review, Acceptance, Adoption, Approval and Publication of IPCC Reports*, edited by S. Solomon, D. Qin, M. Manning, Z. Chen, M. Marquis, K. B. Averyt, M. Tignor, and H. L. Miller, 1–27. Cambridge, England: Cambridge University Press.

Wilby, R. L., and L. W. Perry. 2006. Climate change, biodiversity and the urban environment: A critical review based on London, UK. *Progress in Physical Geography* 30:73–98.

Wilcove, D. S., D. Rothstein, J. Dubrow, A. Phillips, and E. Losos. 1998. Quantifying threats to imperiled species in the United States. *BioScience* 48:607–15.

Wilkinson, D. M. 2001. Is local provenance important in habitat creation? *Journal of Applied Ecology* 38:1371–73.

Williams, J. W., S. T. Jackson, and J. E. Kutzbac. 2007. Projected distributions of novel and disappearing climates by 2100 AD. *Proceedings of the National Academy of Sciences USA* 104:5738.

Wiliams, K. J. H., and J. Cary. 2002. Landscape preferences, ecological quality, and biodiversity protection. *Environmental Behavior* 34:257–74.

Williams, S. L., and C. A. Davis. 1996. Population genetic analyses of transplanted eelgrass (*Zostera marina*) beds reveal reduced genetic diversity in southern California. *Restoration Ecology* 4:163–80.

Williams, S. L., and R. J. Orth. 1998. Genetic diversity and structure of natural and transplanted eelgrass populations in the Chesapeake and Chincoteague bays. *Estuaries* 21:118–28.

Willis, C. G., B. Ruhfel, R. B. Primack, A. J. Miller-Rushing, and C. C. Davis. 2008. Phylogenetic patterns of species loss in Thoreau's woods are driven by climate change. *Proceedings of the National Academy of Sciences* 105:17029–33.

Wiser, S. K., R. K. Peet, and P. S. White. 1998. Prediction of rare plant occurrence: A southern Appalachian example. *Ecological Applications* 8:669–80.

Witmer, G., and J. D. Eisemann. 2007. Rodenticide use in rodent management in the United States: An overview. In *Proceedings of the 12th Wildlife Damage Management Conference, Corpus Christi, Texas*, edited by D. L. Nolte, W. M. Arjo, and D. Stalman, 160–67. Washington, DC: The Wildlife Damage Management Working Group, the Wildlife Society.

Wolfe, A. D., and A. Liston. 1998. Contributions of PCR-based methods to plant systematics and evolutionary biology. In *Molecular Systematics of Plants II: DNA Sequencing*, edited by D. E. Soltis, P. S. Soltis, and J. J. Doyle, 43–86. Boston: Kluwer.

Woodward, F. I. 1987. *Climate and Plant Distribution*. Cambridge Studies in Ecology. Cambridge, UK: Cambridge University Press.

Wright, S. J. 2003a. Attributes of wild *J. reclinata* plants and associated habitats/microenvironments. In *Restoration of* Jacquemontia reclinata *to the South Florida Ecosystem*. Final Report to the US Fish and Wildlife Service for Grant Agreement 1448-40181-99-G-173, edited by J. Maschinski, S. J. Wright, and H. Thornton, 97–100. Vero Beach, FL: Fairchild Tropical Botanic Garden.

Wright, S. J. 2003b. Effects of environmental gradients within the coastal dune on survivorship of outplantings. In *Restoration of* Jacquemontia reclinata *to the South Florida Ecosystem*. Final Report to the US Fish and Wildlife Service for Grant Agreement 1448-40181-99-G-173, edited by J. Maschinski, S. J. Wright, and H. Thornton, 184–88. Vero Beach, FL: Fairchild Tropical Botanic Garden.

Wright, S. J., and M. W. Fidelibus. 2004. Shade limited root mass and carbohydrate reserves of the federally endangered beach clustervine (*Jacquemontia reclinata*) grown in containers. *Native Plants Journal* 5:27–32.

Wright, S. J., and H. Thornton. 2003. Identification of restoration sites for *Jacquemontia reclinata*. In *Restoration of* Jacquemontia reclinata *to the South Florida Ecosystem*. Final Report to the US Fish and Wildlife Service for Grant Agreement 1448-40181-99-G-173, edited by J. Maschinski, S. J. Wright, and H. Thornton, 100–104. Vero Beach, FL: Fairchild Tropical Botanic Garden.

Wu, Z. Y., P. H. Raven, and D. Y. Hong, eds. 2007. *Flora of China*. Vol. 12. *Hippocastanaceae through Theaceae*. Beijing: Science Press; St. Louis: Missouri Botanical Garden Press.

Yandell, B. S. 1997. *Practical Data Analysis for Designed Experiments*. London: Chapman and Hall.

Yates, C. J., and L. M. Broadhurst. 2002. Assessing limitations on population growth in two critically endangered *Acacia* taxa. *Biological Conservation* 108:13–26.

Yates, C. J., D. J. Coates, C. Elliott, and M. Byrne. 2007a. Composition of the pollinator community, pollination and the mating system for a shrub in fragments of species rich kwongan in south-west Western Australia. *Biodiversity and Conservation* 16:1379–95.

Yates, C. J., C. Elliot, M. Byrne, D. J. Coates, and R. Fairman. 2007b. Seed production, germinability and seedling growth for a shrub in fragments of species rich kwongon in south-west Australia. *Biological Conservation* 136:306–14.

Yates, C. J., R. J. Hobbs, and R. W. Bell. 1994. Landscape-scale disturbances and regeneration in semi-arid woodlands of southwestern Australia. *Pacific Conservation Biology* 1:214–21.

Young, A. G., A. H. D. Brown, B. G. Murray, P. H. Thrall, and C. Miller. 2000. Genetic erosion, restricted mating and reduced viability in fragmented populations of the endangered grassland herb *Rutiodsis leptorrhynchoides*. In *Genetics, Demography and Viability of Fragmented Populations*, edited by A. G. Young and G. M. Clarke, 335–59. Cambridge, England: Cambridge University Press.

Young, A. G., and B. G. Murray. 2000. Genetic bottlenecks and dysgenic gene flow into re-established populations of the grassland daisy, *Rutidosis leptorrhynchoides*. *Australian Journal of Botany* 48:409–16.

Young, C. C., L. W. Morrison, M. I. Kelrick, and M. D. DeBacker. 2008. Monitoring *Lesquerella filiformis* Rollins (Missouri bladderpod): Application and evaluation of a grid-based survey approach. *Natural Areas Journal* 28:370–78.

Young, T. P., D. A. Petersen, and J. J. Clary. 2005. The ecology of restoration: Historical links, emerging issues and unexplored realms. *Ecology Letters* 8:662–73.

Zettler, L. W., S. L. Stewart, M. L. Bowles, and K. A. Jacobs. 2001. Mycorrhizal fungi and cold-assisted symbiotic germination of the federally threatened eastern prairie fringed orchid, *Platanthera leucophaea* (Nuttall) Lindley. *American Midland Naturalist* 145:168–75.

Zubek, S., K. Turnau, M. Tsimilli-Michael, and R. J. Strasser. 2009. Response of endangered plant species to inoculation with arbuscular mycorrhizal fungi and soil bacteria. *Mycorrhiza* 19:113–23.

Matthew A. Albrecht is an assistant curator for conservation biology at the Missouri Botanical Garden, St. Louis, Missouri. He has studied the population biology of medicinal forest herbs and the restoration of eastern oak forest ecosystems. His current research focuses on the demography, reintroduction, and autecology of rare and endangered plants.

Esther Banks is a research assistant at the James Hutton Institute, UK.

Timothy Bell is a professor of botany at Chicago State University, Chicago, Illinois, and Research Associate at the Morton Arboretum, Lisle, Illinois. His research interests include population viability analysis and reintroduction ecology of endangered plants, including Mead's milkweed, pitcher's thistle, eastern prairie fringed orchid, and leafy prairie clover.

Marlin L. Bowles is a plant conservation biologist at the Morton Arboretum, Lisle, Illinois, and a conservation officer for the Center for Plant Conservation. His research on restoration ecology of endangered plants has included authorship or coauthorship of more than ten journal articles or book chapters on the US-listed pitcher's thistle, Mead's milkweed, and eastern prairie fringed orchid, as well as membership on the federal recovery teams for these species. He has also coedited the book *Restoration of Endangered Species*, published by Cambridge University Press.

David Coates is a senior principal research scientist and program leader for the Flora Conservation and Herbarium Program and responsible for coordinating and facilitating research in flora conservation and plant systematics in the Department of Environment and Conservation Western Australia. He is a member of the Western Australian Threatened Species Scientific Committee. His research interests include conservation genetics, plant mating systems, and reintroduction of threatened plants.

Sarah E. Dalrymple is a research officer at the Centre for Evidence-Based Conservation, in the School of the Environment at Bangor University, Bangor, UK. Her research on plant conservation includes reintroductions, habitat restoration, and population dynamics of threatened species, with an emphasis on linking research with conservation practice. Sarah is a member of the IUCN Re-introduction Specialist Group and recently joined colleagues from the Re-introduction Specialist Group and the IUCN Invasive Species Specialist Group to form the Task Force on Moving Plants and Animals for Conservation Purposes. This group aims to revise and expand the international guidelines on reintroductions and related interventions.

Donald A. Falk is associate professor in the School of Natural Resources, University of Arizona, Tucson. He holds joint appointments in the Laboratory of Tree-Ring Research and the Institute for Earth and Society at the University of Arizona. His research focuses on fire history, fire ecology, fire climatology, and restoration ecology. His publications include numerous journal articles and four books. He serves on the editorial board for the Island Press–SER series *Science and Practice of Restoration Ecology*. Dr. Falk was the first executive director of the Society for Ecological Restoration and cofounder and executive director of the Center for Plant Conservation.

Edward O. Guerrant Jr. worked from 1989 to 2011 as the conservation director and seedbank curator at the Berry Botanic Garden, which is a founding participating institution of the Center for Plant Conservation. He oversaw the operation of the garden's seedbank for rare and endangered plants and conducted research into seed germination and reintroduction projects, usually in cooperation with federal and state land management agencies. He, the seedbank, and conservation program have moved to the Department of Environmental Science and Management at Portland State University, Oregon, where the work will continue.

Susan Ching Harbin is a botanist with the Hawaii Plant Extinction Prevention Program, which works to protect Hawaii's rarest plants from extinction in cooperation with state, federal, and private conservation partners.

Kristin E. Haskins is the research scientist at The Arboretum at Flagstaff and holds an adjunct faculty position in the Department of Biology at Northern Arizona University, Flagstaff, Arizona. Her work focuses on rare plant conservation of Colorado Plateau species, including the US-listed sentry milkvetch *Astragalus cremnophylax* var. *cremnophylax*. Dr. Haskins specializes in belowground plant and soil microbe interactions, with an emphasis on mycorrhizal ecology.

Chad Husby is collections manager and botanist at the Montgomery Botanical Center in Miami, Florida. His work focuses on botanical, horticultural, and statistical research and ex situ conservation of tropical plants in botanical gardens. He has pursued fieldwork in South America, the Caribbean, New Caledonia, and New Zealand.

Stephanie M. Joe is a research specialist with the Oahu Army Natural Resource Program. She completed her MSc in ecology, evolution, and conservation biology at the University of Hawaii at Manoa, where she focused on native plant response to alien slug predation. Her current work focuses on improving management of invasive arthropods and molluscs.

H. Kapua Kawelo is a botanist with the Oahu Army Natural Resource Program, working to stabilize more than fifty endangered Hawaiian plant taxa. She also works with native ecosystems, rare snails, birds, and arthropods.

Brian G. Keel is a conservation biologist specializing in orchids. Research interests include the effects of climate change on orchids and their pollinators, climate-driven orchid migration, and the application of assisted migration. He is working on a project to reclaim a woodlot from a species of nonnative honeysuckle in preparation for the reintroduction of native orchids to the site.

Matthew J. Keir is a botanist currently managing plant conservation tasks including field surveys, research, genetic conservation, and reintroductions of rare plant species for the Oahu Army Natural Resource Program.

Kathryn Kennedy is executive director of the Center for Plant Conservation. The center helps build expertise and capacity for strong, community-based plant conservation programs in US institutions. Dr. Kennedy received her MS degree from New Mexico State University in plant ecology and a PhD in plant systematics from the University of Texas at Austin. Past work included positions with the US Fish and Wildlife Service, the Texas Parks and Wildlife Department, and the

Lady Bird Johnson Wildflower Center. She has served as chair of the Conservation Committee for the Botanical Society of America and on the IUCN Plant Conservation Committee and the International Advisory Council of Botanic Gardens Conservation International.

Tiffany M. Knight is an associate professor of biology and director of the Environmental Studies Program at Washington University in St. Louis and a research associate at both the Missouri Botanical Garden and Chicago Botanic Garden. Her research focuses on understanding threats to rare plant populations using demographic data collection, experiments, and modeling and applying results to on-the-ground conservation and restoration.

Crystal Krause is a research ecologist with the US Geological Survey, studying how climate change may affect the Papahanaumokuakea Marine National Monument of the northwestern Hawaiian Islands. Her research interests include species distribution modeling with future climate projections, dispersal abilities, and identifying extinction risk.

Mike Kunz is the conservation ecologist at the North Carolina Botanical Garden, Chapel Hill. His background is in population and community ecology of invasive plant species. Currently, he conducts research on the reintroduction and demographics of rare plants. He also manages the imperiled plant seedbank, monitors rare plant populations, and performs natural area management. He serves on the board of the Friends of Plant Conservation.

Caroline Lewis earned a BS in biology, a BEd in secondary science and environmental science education, and an MS in educational leadership. For 22 years she worked as a science teacher, dean, academic counselor, and high school principal in Trinidad, New York, and Miami. Formerly director of education at Fairchild Tropical Botanic Garden, she currently serves as an education strategist, founder, and CEO of the CLEO Institute, working to build capacity, promote innovation, and inspire leadership in individual educators and educational institutions.

Xiaoya Li is the project manager for Fauna and Flora International, China Programme. Li received an MSc in landscape ecology from Beijing Forestry University & Research Center for Eco-Environmental Sciences, Chinese Academy of Sciences, and a BSc in Soil and Water Conservation at Beijing Forestry University. Li coordinates magnolia conservation projects in Yunnan, focusing on reinforcement, monitoring, management, and recovery of *Manglietiastrum sinicum* and its habitat. Li also developed the community participatory awareness-raising campaign for the project. Li has also been involved with conifer conservation projects in south China and the FFI Karst Biodiversity Conservation Programme in the limestone area of Southwest Guangxi/EU–China Biodiversity Programme.

Hong Liu is an assistant professor in the Department of Earth and Environment at Florida International University and a research ecologist at the Center for Tropical Plant Conservation, Fairchild Tropical Botanic Garden in Miami, Florida. Her research interests include invasive species biology, conservation ecology, plant mating systems, population viability analysis, and plant–animal interactions. Her current research addresses important environmental issues such as predicting horticultural plant naturalization and the ecological consequences of specialized invasive pollinator and invasive mutualism. She currently leads several conservation and restoration research projects on orchids in tropical and subtropical southwestern China.

Joyce Maschinski is the conservation ecologist leading the south Florida conservation program at Fairchild Tropical Botanic Garden, Miami and is adjunct faculty at Northern Arizona University, Florida International University, and University of Miami. For two decades her research has centered on understanding factors that limit reproduction, growth, and expansion of rare plant populations, the impact of human activities, and potential management solutions for their conservation. She and her colleagues at Fairchild and at The Arboretum at Flagstaff have conducted more than sixty-five rare plant reintroductions of seventeen species. She has organized several rare plant task forces and conferences on endangered plants in Arizona and Florida and has made presentations to international audiences at scientific conferences. She has published widely about plant reintroductions and threats to rare plants, including demographics, genetics, hybridization, herbivory, habitat destruction, timber harvest, trampling, climate change, and fire.

Pilar Maul is an assistant professor of biology at St. Thomas University in Miami Gardens, Florida. She earned her PhD in cell and molecular biology at the Florida Institute of Technology. Her postdoctoral work at the US Department of Agriculture and at the University of Florida focused on the identification of molecular mechanisms of chilling tolerance and the induction of somatic embryogenesis in citrus species. She has more than 15 years of experience in the use of plant tissue culture, studying various aspects of plant biology in potato, sweet potato, and other species. Currently she conducts tissue culture research in a collaborative project with Bok Tower Gardens and has collaborative research projects with the US Department of Agriculture on the genetic improvement of flower bulbs and avocado.

Leonie Monks is a research scientist with the Western Australian Department of Environment and Conservation in Perth, Western Australia. She is responsible for planning, implementing, and monitoring reintroductions of threatened plant species. Her research is focused on improving reintroduction techniques and refining reintroduction success criteria.

Jennifer Ramp Neale is director of research and conservation at Denver Botanic Gardens, Denver, Colorado. Her work focuses on conservation and restoration genetics; the incorporation of genetics and pollination in conservation and restoration planning, including the incorporation of genetic data into geographic information systems; community-level genetic examination and conservation; and demographic monitoring and pollination studies of rare and endangered flora in the Rocky Mountain region.

Valerie Pence is director of plant research at the Center for Research on Endangered Wildlife at the Cincinnati Zoo, Cincinnati, Ohio. She conducts groundbreaking research on cryopreservation and plant propagation.

Deana D. Pennington is an interdisciplinary geoscientist and informatics specialist at the University of Texas at El Paso. Her research interests revolve around cross-disciplinary study of climate change impacts on natural systems. She has conducted landscape- and regional-scale research in the areas of wildfire and harvest disturbance in forests, drought effects in semiarid grasslands, landscape change effects on biogeography and spatial–temporal patterns of biodiversity, and paleoenvironmental reconstructions. She is interested in cross-disciplinary learning, knowledge-sharing techniques, and collaborative innovation.

Cheryl Peterson has been manager of the Rare Plant Conservation Program at Bok Tower Gardens, Lake Wales, Florida since 2002. She has a bachelor's degree in microbiology and a master's degree in cell and molecular biology. Her previous research has involved marine biodiversity and invertebrate genetics, and she has studied the Florida endemic mint genus *Conradina*. She is currently working to establish protected populations, studying seed germination and monitoring populations of several rare plant species.

Jennifer Possley is a field biologist at the Center for Tropical Plant Conservation, Fairchild Tropical Botanic Garden, Miami, Florida. She maps and monitors the rare flora of Miami–Dade County, has reintroduced eight rare plant species into seventeen locations, and is currently researching the ecological effects of vegetation management on biodiversity. She has special interests in pteridophytes and invasive plants. Before joining Fairchild's research staff, Jennifer received a BA in biology from Kalamazoo College and an MS in agronomy from the University of Florida.

Andrew S. Pullin is a professor of evidence-based conservation at Bangor University, UK. He is interested in evidence-based practice in environmental management, adapting methods from the field of human health care to improve the potential of science to inform policy and management decisions. He is director of the Centre for Evidence-Based Conservation and cofounded the Collaboration for Environmental Evidence, which seeks to provide a reliable source of evidence on environmental impacts of human actions and effectiveness of environmental interventions worldwide.

Johnny Randall is the assistant director for natural areas and conservation programs at the North Carolina Botanical Garden and adjunct faculty in the Curriculum for the Environment and Ecology, both part of the University of North Carolina at Chapel Hill. His primary responsibilities are to oversee the conservation and management of natural areas and administer the garden's seedbank programs. He also does research on rare plant reintroductions and habitat restoration and rehabilitation. He is a member of the North Carolina Plant Conservation Scientific Committee and a board member and past president of the North Carolina Exotic Pest Plant Council, and he sits on several other conservation boards and committees.

Peter H. Raven is president emeritus of the Missouri Botanical Garden and George Engelmann Professor of Botany Emeritus at Washington University in St. Louis. In addition, he is a trustee of the National Geographic Society and chairman of the society's Committee for Research and Exploration. For nearly 39 years, he has headed the Missouri Botanical Garden, an institution he has nurtured to become an international, world-class center for botanical research, education, and horticulture display. He is a member of more than twenty academies of science in countries around the world and a recipient of the National Medal of Science, the highest award for scientific accomplishment in the United States. He has written numerous books and publications, both popular and scientific. He devotes much of his time to advocacy for conservation and sustainable development.

Sarah Reichard is a professor at the University of Washington in Seattle and is associate director of its Botanic Garden. Her research is focused on understanding the biology of invasive plants and using that understanding to develop risk assessment methods to prevent their introduction and spread. She coauthored a

National Academy of Sciences report, "Predicting Invasions of Non-indigenous Plants and Plant Pests." Editor of *Invasive Species in the Pacific Northwest* (University of Washington Press) and author of numerous research papers, she also served 6 years on the federal Invasive Species Advisory Committee and is on the Invasive Species Specialist Group of the International Union for the Conservation of Nature. Her newest book is *The Conscientious Gardener: Cultivating a Garden Ethic* (University of California Press). She founded and directs the Washington Rare Plant Care and Conservation Program, which works with federal, state, and local agencies to protect Washington's rare plant species.

Julissa Roncal received a BS and Licenciature from Cayetano Heredia University in Perú and a PhD from Florida International University, Miami. She recently received the Marie Curie Intraeuropean Fellowship to work at the Institut de Recherche pour le Développement in France. Her research focuses on the habitat ecology, phylogenetics, biogeography, and conservation of the palm family (Arecaceae), particularly on tribe Geonomateae. She has also conducted restoration ecology research addressing the new habitat conditions for rare and endangered species in south Florida in the face of anthropogenic change.

Gavin B. Stewart is a scientist at the Centre for Reviews and Dissemination, University of York, UK.

Weibang Sun is a professor of botany and director of Kunming Botanical Garden under Kunming Institute of Botany, Chinese Academy of Sciences. His chief research interests are conservation biology of rare and endangered Chinese plants, sustainable uses of the economically important plant resources in southwest China, and impacts of alien invasive plants on biodiversity in Yunnan province. He has published seventy articles and twelve illustrated plant books.

Pengyun Tao is the senior engineer at the Seedling Station of Wenshan Forestry Bureau, Southeast Yunnan, China. Tao evaluates and cultivates trees for reforestation.

Lauren Weisenberger is a research botanist for the Oahu Army Natural Resources Program. She specializes in research that determines the long-term storage potential of seeds of rare plant species, the reproductive biology of rare plants, and the design of plant reintroductions. She oversees the Oahu Army Seed Laboratory and Seed Bank. She is a PhD candidate in ecology, evolution, and conservation biology (botany) at the University of Hawaii, Manoa, investigating levels of inbreeding depression in small plant populations and potential for outbreeding depression in mixed-source reintroductions.

Kristie S. Wendelberger worked as a rare plant biologist at Fairchild Tropical Botanic Garden for 4 years before obtaining her MS from the University of North Carolina, Chapel Hill. She is currently a PhD student at Florida International University. Her work in Florida and North Carolina involved studying and performing experimental reintroductions on some of south Florida's rarest plant species. Her current dissertation research examines sea-level rise effects on rare coastal plant communities in Everglades National Park.

Samuel J. Wright has been a field biologist at Fairchild Tropical Botanic Garden's Center for Tropical Plant Conservation since 2001. He has a BS degree in wildlife ecology and conservation from the University of Florida. He is responsible for ecological research related to recovery and restoration projects of native endangered plant species, particularly those that occur in south Florida's coastal dune ecosystems. He works closely with natural area managers from Palm Beach, Broward, and Miami–Dade Counties in determining appropriate management techniques for native rare plants and habitats.

Huabin Yang is an engineer at the Seedling Station of Wenshan Forestry Bureau, Southeast Yunnan, China. Yang evaluates and cultivates trees for reforestation.

Yuan Zhou is an assistant professor at Kunming Botanical Garden, Kunming Institute of Botany, Chinese Academy of Sciences. Zhou's research interests include the introduction, domestication, and conservation of economically significant plants.

THE SCIENCE AND PRACTICE
OF ECOLOGICAL RESTORATION

Wildlife Restoration: Techniques for Habitat Analysis and Animal Monitoring, by Michael L. Morrison

Ecological Restoration of Southwestern Ponderosa Pine Forests, edited by Peter Friederici, Ecological Restoration Institute at Northern Arizona University

Ex Situ Plant Conservation: Supporting Species Survival in the Wild, edited by Edward O. Guerrant Jr., Kayri Havens, and Mike Maunder

Great Basin Riparian Ecosystems: Ecology, Management, and Restoration, edited by Jeanne C. Chambers and Jerry R. Miller

Assembly Rules and Restoration Ecology: Bridging the Gap Between Theory and Practice, edited by Vicky M. Temperton, Richard J. Hobbs, Tim Nuttle, and Stefan Halle

The Tallgrass Restoration Handbook: For Prairies, Savannas, and Woodlands, edited by Stephen Packard and Cornelia F. Mutel

The Historical Ecology Handbook: A Restorationist's Guide to Reference Ecosystems, edited by Dave Egan and Evelyn A. Howell

Foundations of Restoration Ecology, edited by Donald A. Falk, Margaret A. Palmer, and Joy B. Zedler

Restoring the Pacific Northwest: The Art and Science of Ecological Restoration in Cascadia, edited by Dean Apostol and Marcia Sinclair

A Guide for Desert and Dryland Restoration: New Hope for Arid Lands, by David A. Bainbridge

Restoring Natural Capital: Science, Business, and Practice, edited by James Aronson, Suzanne J. Milton, and James N. Blignaut

Old Fields: Dynamics and Restoration of Abandoned Farmland, edited by Viki A. Cramer and Richard J. Hobbs

Ecological Restoration: Principles, Values, and Structure of an Emerging Profession, by Andre F. Clewell and James Aronson

River Futures: An Integrative Scientific Approach to River Repair, edited by Gary J. Brierley and Kirstie A. Fryirs

Large-Scale Ecosystem Restoration: Five Case Studies from the United States, edited by Mary Doyle and Cynthia A. Drew

New Models for Ecosystem Dynamics and Restoration, edited by Richard J. Hobbs, and Katharine N. Suding

Cork Oak Woodlands in Transition: Ecology, Adaptive Management, and Restoration of an Ancient Mediterranean Ecosystem, edited by James Aronson, João S. Pereira, and Juli G. Pausas

Restoring Wildlife: Ecological Concepts and Practical Applications, by Michael L. Morrison

Restoring Ecological Health to Your Land, by Steven I. Apfelbaum and Alan W. Haney

Restoring Disturbed Landscapes: Putting Principles into Practice, by David J. Tongway and John A. Ludwig

Intelligent Tinkering: Bridging the Gap between Science and Practice, by Robert J. Cabin

Making Nature Whole: A History of Ecological Restoration, by William R. Jordan and George M. Lubick

The Restoring Ecological Health to Your Land Workbook, by Steven I. Apfelbaum and Alan Haney